ADVANCES IN
HEAT TRANSFER

Volume 16

Contributors to Volume 16

M. G. COOPER

BENJAMIN GEBHART

DAVID S. HILDER

W. M. KAYS

MATTHEW KELLEHER

R. J. MOFFAT

RICHARD A. W. SHOCK

JOHN R. THOME

Advances in

HEAT
TRANSFER

Edited by

James P. Hartnett

Energy Resources Center
University of Illinois
Chicago, Illinois

Thomas F. Irvine, Jr.

Department of Mechanical Engineering
State University of New York
at Stony Brook
Stony Brook, New York

Volume 16

 1984

ACADEMIC PRESS, INC.
(Harcourt Brace Jovanovich, Publishers)
Orlando • San Diego • New York • London
Toronto • Montreal • Sydney • Tokyo

ACADEMIC PRESS, INC.
Orlando, Florida 32887

United Kingdom Edition published by
ACADEMIC PRESS, INC. (LONDON) LTD.
24/28 Oval Road, London NW1 7DX

Library of Congress Cataloging in Publication Data 63-22329

ISBN 0-12-020016-3   0065-2717

PRINTED IN THE UNITED STATES OF AMERICA

84 85 86 87   9 8 7 6 5 4 3 2 1

CONTENTS

The Diffusion of Turbulent Buoyant Jets

BENJAMIN GEBHART, DAVID S. HILDER, and MATTHEW KELLEHER

Boiling of Multicomponent Liquid Mixtures

JOHN R. THOME and RICHARD A. W. SHOCK

Heat Flow Rates in Saturated Nucleate Pool Boiling—A Wide-Ranging Examination Using Reduced Properties

M. G. COOPER

A Review of Turbulent-Boundary-Layer Heat Transfer Research at Stanford, 1958–1983

R. J. MOFFAT and W. M. KAYS

CONTRIBUTORS

Numbers in parentheses indicate the pages on which the authors' contributions begin.

M. G. COOPER (157), *Department of Engineering Science, Oxford University, Oxford OX1 3PJ, England*

BENJAMIN GEBHART (1), *Department of Mechanical Engineering and Applied Mechanics, University of Pennsylvania, Philadelphia, Pennsylvania 19104*

DAVID S. HILDER (1), *Philadelphia Naval Shipyard, Philadelphia, Pennsylvania 19112*

W. M. KAYS (241), *Department of Mechanical Engineering, Thermosciences Division, Stanford University, Stanford, California 94305*

MATTHEW KELLEHER (1), *Department of Mechanical Engineering, Naval Postgraduate School, Monterey, California 93940*

R. J. MOFFAT (241), *Department of Mechanical Engineering, Thermosciences Division, Stanford University, Stanford, California 94305*

RICHARD A. W. SHOCK (59), *Heat Transfer and Fluid Flow Service, Atomic Energy Research Establishment Harwell, Didcot, Oxfordshire OX11 0RA, England*

JOHN R. THOME (59), *Department of Mechanical Engineering, Michigan State University, East Lansing, Michigan 48824*

PREFACE

The serial publication *Advances in Heat Transfer* is designed to fill the information gap between the regularly scheduled journals and university-level textbooks. The general purpose of this publication is to present review articles or monographs on special topics of current interest. Each article starts from widely understood principles and in a logical fashion brings the reader up to the forefront of the topic. The favorable response by the international scientific and engineering community to the volumes published to date is an indication of how successful our authors have been in fulfilling this purpose.

The Editors are pleased to announce the publication of Volume 16 and wish to express their appreciation to the current authors who have so effectively maintained the spirit of this serial.

The Diffusion of Turbulent Buoyant Jets

BENJAMIN GEBHART

*Department of Mechanical Engineering and Applied Mechanics,
University of Pennsylvania, Philadelphia, Pennsylvania*

DAVID S. HILDER

Philadelphia Naval Shipyard, Philadelphia, Pennsylvania

MATTHEW KELLEHER

*Department of Mechanical Engineering, Naval Postgraduate School,
Monterey, California*

I. Introduction

The cooling water discharge from a power plant into a large body of water, the thermally loaded condenser discharge from the condenser of a moving ship, and the high-temperature gas issuing from a stack or gas

1

turbine exhaust are all buoyant momentum jets. The trajectory and decay of such jets after discharge are influenced by such factors as initial jet velocity and buoyancy, ambient motion and stratification, and downstream mixing rate. Questions such as whether or not the jet will rise to a certain level, what the jet velocity and temperature will be at any point along its trajectory, or what effect ambient fluid stratification will have on behavior all require detailed and accurate analysis. There have been many contributions in the past few decades to the understanding of the mechanics of buoyant jet mixing and trajectory. The ultimate objective is to develop accurate general models that predict both trajectory and decay.

The need for such predictive models has grown. Since nuclear- and fossil-fueled power plants have thermal efficiencies on the order of 30–40%, the immense discharge of heat into either the atmosphere or a body of water has a very large effect. Sewage is often discharged as treated effluent into rivers, lakes, and oceans. The proper evaluation of the ecological impact of such discharges requires that their subsequent behavior be predictable. More stringent environmental regulations and heightened public awareness require increasing accuracy in such prediction.

The need for the prediction of jet behavior is not limited to environmental issues. Rapid advancement of the ability to detect small temperature and concentration differences, and other anomalies, may make it increasingly easy to detect many physical effects, changes, and motions in the environment. The implications for increasing knowledge of environmental, geophysical, and technological processes are enormous.

Given the wide range of applications in which jet behavior is to be analyzed, the range of possible jet and/or ambient characteristics that may be of interest is equally wide. The variables include initial jet geometry, discharge momentum, thermal and concentration loading, turbulence characteristics, as well as ambient flow conditions, turbulence, and stratification. An extremely large number of appreciably different combinations arise.

The summary and calculations here concern a single, fully turbulent, circular buoyant jet, discharged into a surrounding ambient of the same fluid. Two-dimensional trajectories are included, wherein any ambient flow is taken as parallel to the horizontal component of jet velocity. Jet encounter with an abrupt ambient discontinuity, such as a two-phase interface, is not treated here. That is, the ambient is considered infinite in extent.

Among the variables are:

(1) buoyancy effects, arising from density differences between the jet and the ambient (differences may arise from temperature and/or concentration variations);

(2) ambient density stratification, arising from vertical nonuniformity of temperature and/or concentration in the ambient;

(3) ambient flow conditions, with respect to the jet, of differing magnitude and orientation relative to the jet;

(4) initial jet discharge characteristics, including direction of momentum.

II. Characteristics of Circular Discharges

The terms *jet, momentum jet, forced plume,* and *plume* are often used to describe qualitatively the differing characteristics of a discharge penetrating an ambient medium. In general usage, *jet, momentum jet,* and *forced plume* refer to the downstream region wherein the momentum of the initial discharge is still sufficient to influence jet mixing and trajectory. A discharge in which the discharge momentum is everywhere negligible, relative to the eventual total momentum produced by buoyancy, is called a plume. In this account these will be called buoyant jets and plumes.

The jet–ambient interaction mechanisms are classified according to the following characteristics:

(1) Jet buoyancy
    (a) neutrally buoyant
    (b) buoyant (positively or negatively)
(2) Orientation of discharge
    (a) horizontal (perpendicular to the gravity field)
    (b) inclined
(3) Ambient motion
    (a) quiescent
    (b) flowing
(4) Ambient stratification
    (a) unstratified
    (b) linearly stratified
    (c) other stratifications

Independent of the jet–ambient mechanisms, each jet passes through several flow regimes along its trajectory. They are shown for an inclined submerged buoyant jet in Fig. 1. The regimes are as follows:

(1) *The zone of flow establishment.* In this region, flow characteristics are dominated by the discharge conditions. Velocity and scalar quantity profile (temperature, salinity, etc.) undergo transition from their initial discharge configurations, through a turbulent shear layer formed around

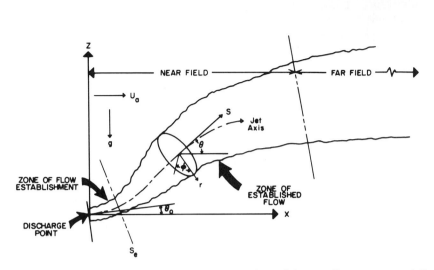

FIG. 1. Flow regimes of a buoyant momentum jet and the coordinate system and dimensions in which its trajectory and growth are described.

the jet periphery. As mixing with the ambient progresses, the turbulent shear layer grows inward and the extent of the core of undisturbed profiles becomes smaller. The zone of flow establishment ends at the point where turbulent mixing reaches the jet centerline. The jet behavior in this region is strongly influenced by initial momentum and discharge conditions and is only slightly influenced by the ambient.

(2) *The zone of established flow.* This region begins when turbulent mixing has reached the jet centerline. The motion of the jet and its physical characteristics are governed by its initial and acquired momentum, by its buoyancy, and by ambient stratification and flow conditions. Initial discharge conditions play a progressively smaller role. The flow progresses from jetlike to plumelike behavior.

(3) *The far field.* In this region the jet's initial momentum has a negligible effect, and the jet may be passively convected by ambient motions. The jet fluid may be further diffused by ambient turbulence, and the distribution of the jet as a separate entity gradually disappears.

The regions of flow establishment and established flow regimes are the near field. They are the regions of the most vigorous mechanisms. These are the concern of this account and of the calculations of trajectory and decay.

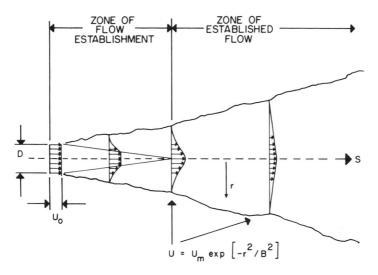

FIG. 2. Development of Gaussian velocity profiles in a momentum jet after discharge.

Experimental work, begun by Albertson et al. [1] and continued and expanded by many others, has shown that within the zone of established flow, mean velocity profiles are nearly Gaussian:

$$U = U_m \exp[-r^2/B^2] \qquad (1)$$

where U_m is the local centerline velocity, r is the radial jet coordinate, and B is a characteristic jet width. It is the radial distance at which U is equal to $1/e$ times the mean centerline value, U_m.

Profiles of jet scalar quantities, such as temperature and concentration, have also been found to be Gaussian in the zone of established flow by investigators such as Fan [2], Hoult et al. [3], and others. The profiles may be expressed as:

$$\Delta t = \Delta t_m \exp[-r^2/\lambda_t^2 B^2] \qquad (2)$$

$$\Delta c = \Delta c_m \exp[-r^2/\lambda_c^2 B^2] \qquad (3)$$

where $\Delta t = (t - t_a)$, $\Delta t_m = (t_m - t_a)$, $\Delta c = (c - c_a)$, and $\Delta c_m = (c_m - c_a)$. The values λ_t and λ_c are the relative radial spreading ratio between velocity and the scalar properties t and c. These quantities are related to the turbulent entrainment Prandtl and Schmidt number effects. Figure 2 illustrates a profile within the jet.

A coordinate system to describe the trajectory and physical dimensions of a jet system is shown in Fig. 1. The X coordinate is horizontal. The Z positive direction is vertical, opposite to the gravity vector. The streamwise coordinate S is measured along the direction of the mean centerline of the jet. The local angle between S and X, the inclination of the jet from the horizontal, is θ. The polar coordinates ϕ and r, defining the jet cross section, are normal to S. Herein any ambient medium motion is assumed to be horizontal, that is, in the direction of X.

A principal quantitative measure of relative momentum and buoyancy is the densimetric Froude number F given by

$$F = \frac{U_0}{[gD(\rho_a - \rho_0)/\rho_0]^{1/2}} \tag{4}$$

The contribution of momentum is reflected in the numerator by the discharge velocity U_0. The buoyancy effect is included in the denominator by the density difference, or units of buoyancy $(\rho_a - \rho_0)/\rho_0$. This quantity is the measure of the velocity level generated by the buoyancy force. Thus the value of the densimetric Froude number ranges from near zero for plumes to infinity for pure, nonbuoyant momentum jets. Hereafter, the term *Froude number* will be used to mean the densimetric form in Eq. (4).

III. Review

A. Modeling Schemes

Several kinds of predictive models have been developed for the motion of circular buoyant momentum jets. Although specific calculations have considered different circumstances (e.g., in origin of buoyancy, for stratified or uniform ambients and quiescent or coflowing ambients, etc.), all models are one of two kinds.

(1) *Algebraic models* are algebraic equations based on either empirical data or simplification of differential models. These most typically predict only trajectory and jet width. Some, such as the model of Shirazi *et al.* [4], also predict velocity, concentration, and temperature residuals. Data-based algebraic models tend to become unreliable when the basic conditions on which they were based, such as general temperature and salinity range of the jet and ambient, are significantly changed.

(2) *Differential models* are based on the relevant conservation equations of mass, momentum, energy, and chemical species. This modeling technique allows prediction of jet trajectory and width, as well as velocity, temperature, and concentration decay downstream in the jet. Stratification and motion of the ambient may also be accommodated. The promi-

nent differential models are entrainment, mixing length, and k–ε and eddy diffusivity turbulence modeling.

Because of their limited applicability, algebraic models are not treated here. They are sometimes useful for prediction when the jet–ambient system involved is simple and only information such as trajectory is required. However, by far the greatest effort in recent years has involved the more general and inclusive differential approach to jet modeling. In a majority of these differential models the entrainment mixing concept has been used, rather than models utilizing mixing length k–ε, or turbulent diffusion hypotheses.

Morton *et al.* [5] were the first to use the entrainment concept to develop a buoyant jet model, as previously suggested by Taylor [6]. The concept supposes that the downstream induction of quiescent ambient fluid into the moving jet is proportional to the local jet centerline velocity U_m and a characteristic jet periphery $2\pi B$. Thus

$$E \propto 2\pi B U_m$$

where E represents volumetric rate of entrainment, or ambient inflow per unit of jet length, into the jet. The definition of α is completed by

$$dQ/dS = \rho E \tag{5}$$

where Q is the total mass flow in the jet at any downstream location S. The constant of proportionality for E, α, is called the entrainment constant, or coefficient. The rate of entrainment is then written as:

$$E = 2\pi B \alpha U_m \tag{6}$$

Solutions of the governing equations for differential modeling have, in the past, been based on the following assumptions.

(1) The turbulent jet flow is steady.

(2) Since the jet flow is fully turbulent, radial molecular diffusion is neglected, compared to radial turbulent transport.

(3) Streamwise turbulent transport is a negligible downstream transport mode, compared with streamwise convective transport.

(4) The variation of fluid density throughout the flow field remains small compared to a chosen reference density. Density variation is included only in buoyancy terms. This is a Boussinesq approximation.

(5) Other fluid properties are taken constant throughout the field.

(6) Pressure is hydrostatic throughout the flow field.

(7) The jet remains axisymmetric throughout the near field. That is,

velocity, temperature, density, and salinity profiles are assumed not to develop circumferential variations.

The governing conservation equations, in the forms used in differential modeling, are presented in Table I, in the physical variables.

With the exception of Hoult *et al.* [3], all studies cited in the following discussion have assumed that the velocity, temperature, salinity, and density profiles are all Gaussian. This assumption, therefore, limits the applicability of such models to the zone of established flow.

Hoult *et al.* instead used a "top hat" velocity profile throughout, rather than a Gaussian distribution. This assumption certainly applies very near the jet origin but not downstream in the rest of the near field. The reduced form of the conservation equations for this model require the cross-stream integration, with the Gaussian assumption. Since Hoult assumed top hat profiles for both the zone of flow establishment and the zone of established flow, initial conditions are those at the jet outlet. In this method the values of ρ, t, c, and U ascribed to the jet at various points along the path are taken to be mean values for the entire jet cross section. This is a more limiting result than that with models using Gaussian profiles, where maximum values of jet properties result, and the entire cross-

TABLE I

The General Equations, in Dimensional Form, for Entrainment Modeling of
Buoyant Momentum Jets

Equation	Form
Continuity	$\dfrac{d}{dS}\left\{\displaystyle\int_0^{2\pi}\int_0^\infty Ur\,dr\,d\phi\right\} = 2\pi\alpha U_{\mathrm{m}}B = E$
Horizontal momentum	$\dfrac{d}{dS}\left\{\displaystyle\int_0^{2\pi}\int_0^\infty U^2\cos\theta\,r\,dr\,d\phi\right\} = U_{\mathrm{a}}E$
Vertical momentum	$\dfrac{d}{dS}\left\{\displaystyle\int_0^{2\pi}\int_0^\infty \rho U^2\sin\theta\,r\,dr\,d\phi\right\} = \displaystyle\int_0^{2\pi}\int_0^\infty(\rho_{\mathrm{a}}-\rho)gr\,dr\,d\phi$
Energy	$\dfrac{d}{dS}\left\{\displaystyle\int_0^{2\pi}\int_0^\infty U(t-t_{\mathrm{a}})r\,dr\,d\phi\right\} = -\dfrac{dt_{\mathrm{a}}}{dS}\displaystyle\int_0^{2\pi}\int_0^\infty Ur\,dr\,d\phi$
Concentration (or scalar species)	$\dfrac{d}{dS}\left\{\displaystyle\int_0^{2\pi}\int_0^\infty U(c-c_{\mathrm{a}})r\,dr\,d\phi\right\} = -\dfrac{dc_{\mathrm{a}}}{dS}\displaystyle\int_0^{2\pi}\int_0^\infty Ur\,dr\,d\phi$
Horizontal component of trajectory	$dX = dS\cos\theta$
Vertical component of trajectory	$dZ = dS\sin\theta$

section profile may be deduced from the appropriate Gaussian distribution.

Abraham [7] initially used the vertical and horizontal momentum equations, as well as the energy equation, to model jets discharged to a quiescent ambient. The continuity equation was not needed or used in the calculations. The solution required a prespecification of the variation of B, in (1), as a function of S. Most other models have included the continuity equation in lieu of prespecifying the B variation.

The solution of the seven equations in Table I yields values of jet centerline velocity U_m and temperature and concentration differences Δt_m and Δc_m, as well as jet width $D(S)$ and trajectory (X, Z), all as functions of S. The solution of the equations, of course, also requires that the entrainment function E be specified, that α be given. Therein lie the principal differences between entrainment models. These models fall into two general categories: those for a quiescent ambient and those for a flowing ambient.

1. Quiescent Ambient Media

Albertson et al. [1] and others have verified through measurements that for nonbuoyant momentum jets, (i.e., F = ∞), an appropriate value of α within the zone of established flow is 0.057. There seems to be little disagreement with this value, judged from numerous comparisons of differential modeling with this value and with experimental data.

Abraham [7] suggested, also on the basis of experimental evidence, that, for flows resulting largely from buoyancy, small F, the value is $\alpha = 0.085$. This is in good agreement with the suggestion of List and Imberger [8] of $\alpha = 0.082$ for pure buoyant plumes (F = 0). Fan and Brooks [9] had suggested $\alpha = 0.082$ for all flows except pure momentum jets. Fan and Brooks also recommended, on the basis of their experiments, $\alpha = 0.057$ for pure momentum jets.

In applications, however, buoyant discharges are seldom either pure jets or plumes. Typically, their flow is some stage of transition away from jet behavior toward plume behavior. Morton et al. [5] proposed to model this the whole range by:

$$\alpha = 0.057 + a_2/F_L \qquad (7)$$

where a_2 is an empirically determined constant and F_L is a local Froude number, based on the local centerline velocity U_m. The same general form was derived by Fox [10] for a vertically discharged buoyant jet.

Hirst [11] postulated that, for a buoyant discharge into a quiescent ambient, the entrainment function should depend on

(1) local mean flow conditions in the jet, i.e., U_m and B;
(2) local buoyancy within the jet, as indicated by a local Froude number;
(3) jet orientation θ_0 (see Fig. 1).

The following form was proposed:

$$\alpha = 0.057 + (0.97/F_L) \sin \theta \qquad (8)$$

This is the general form also suggested by Morton *et al.* [5], with the constants defined by fitting this function to known discharge and end-point conditions of jet flows. Another entrainment function for initially horizontal buoyant momentum jets is the jet–plume extremum fit proposed by Riester *et al.* [12]:

$$\alpha = [(0.057 \cos \theta)^2 + (0.082 \sin \theta)^2]^{1/2} \qquad (9)$$

From data on an upward buoyant jet discharged vertically downward into a quiescent ambient, Davis *et al.* [13] proposed:

$$\alpha = 0.057 + 0.083/F^{0.3} \qquad (10)$$

A summary of entrainment functions for discharges into quiescent ambient media is given in Table II.

2. Flowing Ambient Media

Hirst [14] proposed an entrainment function applicable to three-dimensional buoyant jet flow. That is, the horizontal component of initial jet velocity was not necessarily parallel to the horizontal ambient flow, but was inclined away from it, horizontally. Eliminating terms relating entirely to that entrainment mechanism, the resulting form for two-dimensional buoyant jets in an ambient at a velocity U_a becomes:

$$E = 2\pi B[a_1 + (a_2/F_L) \sin \theta][|U_m - U_a \cos \theta| + a_3 U_a \sin \theta] \qquad (11)$$

where the term $|U_m - U_a \cos \theta|$ represents the relative velocity of the jet with respect to the ambient, in the direction of initial jet flow. It is a pure "coflow" entrainment term. The last term

$$a_3 U_a \sin \theta$$

is the entrainment contribution arising from cross flow, that is, of ambient motion normal to the jet axis, as the jet turns upward because of buoyancy.

Hirst [14] specified values of a_1 and a_2 so that the flowing ambient entrainment function reduced to the quiescent ambient function if $U_a = 0$.

TABLE II

ENTRAINMENT FUNCTIONS FOR USE WITH QUIESCENT AMBIENTS (FORM: $E = 2\pi\alpha U_m B$)

Model[a]	α	$\theta = 0$	$\theta = \pi/2$	F or $F_L \to 0$	F or $F_L \to \infty$	References
1	0.057 (jets)	0.057	0.057		0.057	Albertson et al. [1]
2	0.082 (plumes)	0.082	0.082	0.082		List and Imberger [8]
3	$0.057 + (0.97/F_L) \sin \theta$	0.057	$f(F_L)$	$\infty \ (\theta \neq 0)$	0.057	Hirst [11]
4	$0.057 + 0.083/F^{0.3}$	$f(F)$	$f(F)$	∞	0.057	Davis et al. [13]
5	$[(0.057 \cos \theta)^2 + (0.082 \sin \theta)^2]^{1/2}$	0.057	0.082	$f(\theta)$	$f(\theta)$	Riester et al. [12]
6	0.085	0.085	0.085	0.085		Abraham [7]
7	$0.057 + a_2/F_L$	$f(F_L)$	$f(F_L)$		0.057	General form of Morton et al. [5]

[a] (1) Applicable to simple momentum jet; (2) applicable to simple buoyant plume; (3) applicable to buoyant jet discharged at varying angles to a quiescent ambient; (4) empirical fit for a buoyant jet discharged vertically downward into a quiescent ambient; (5) applicable to buoyant jet discharged horizontally to a quiescent ambient; (6) applicable to simple buoyant plume; (7) empirically determined coefficient a_2 vertically discharged.

A value of 9.0 was given for a_3, based on a best fit to limited data. This resulting entrainment function thus became:

$$E = 2\pi B(0.057 + (0.97/F_L) \sin \theta)[|U_m - U_a \cos \theta| + 9.0U_a \sin \theta] \quad (12)$$

Ginsberg and Ades [15] performed a least squares analysis on a large set of laboratory trajectory data. Using Hirst's entrainment model with a_1 and a_2 as specified, they found that a large variation in the value of a_3 was necessary to fit predicted results with the data. They constructed a correlation for a_3 as a function of F and the coflow ratio R:

$$a_3 = 25.810[F^{0.19464}R^{0.35155}] - 10.825 \quad (13)$$

where the coflow ratio R is defined as:

$$R = U_a/U_0 \quad (14)$$

Other entrainment functions for coflowing ambients have appeared using variations of the governing differential equations. Schatzmann [16] proposed an entrainment function similar in form to that of Hirst [14], but for use in a set of governing equations in which the Boussinesq approximation was not invoked. Fan and Brooks [9] proposed an entrainment function for coflowing ambients that included a drag term as well as the usual proportionality of entrainment to centerline velocity and jet width. It was found that the value of the drag coefficient, as well as the entrainment constant, $\alpha = 0.082$, had to be readjusted to make the prediction of the model conform with data, with each change in discharge or ambient conditions. The entrainment functions of Hirst [14], Ginsberg and Ades [15], and Schatzmann [16] are collected in Table III.

TABLE III

ENTRAINMENT FUNCTIONS FOR BUOYANT JETS ISSUING INTO FLOWING AMBIENTS (WITH
COFLOWING AMBIENTS: COMPLETE VOLUMETRIC ENTRAINMENT,
$|U^*| = |U_m - U_a \cos \theta|$)

Model[a]	Form	References
8	$E = 2\pi B[0.057 + (0.97/F_L) \sin \theta](|U^*| + 9.0U_a \sin \theta)$	Hirst [14]
9	$E = 2\pi B[0.057 + (0.97/F_L) \sin \theta](|U^*|$ $+ \{25.81[F^{0.195}(U_a/U_0)^{0.352}] - 10.83\}U_a \sin \theta)$	Ginsberg and Ades [15]
10	$E = 2\pi B[0.057 - (0.67/F_L) \sin \theta](|U^*|$ $+ 2U_a \sin \theta)/[1 + (5U_a \cos \theta)/|U_m - U_a|]$	Schatzmann [16]

[a] (8) Applicable to buoyant jet discharged at varying angles into a flowing ambient, crossflow terms omitted; (9) applicable to buoyant jet discharged at varying angles to a flowing ambient; (10) same as (9), empirically determined coefficients.

The progression of differential modeling is largely independent of the entrainment function used. First, the governing equations are integrated over the jet cross section, in ϕ and r, so that they appear in differential form, in terms of the trajectory coordinate S. They are then nondimensionalized with respect to chosen reference variables. Initial conditions are specified. The resulting equations are then numerically integrated over the desired range of the streamwise path length s. However, different downstream trajectories and decay are found, for a given initial jet, because of differences in any of the following:

(1) entrainment function used;
(2) starting length processes;
(3) initial conditions specified for the beginning of the zone of established flow;
(4) equation of state specified for the density of the fluid;
(5) computational technique.

Examples of predicted jet trajectories and physical properties are presented in Figs. 3 to 35. Comparison between models for identical or closely similar discharge and ambient conditions are presented in Figs. 9 to 13. Discussion of these predictions and comparisons will follow.

B. EXPERIMENTAL STUDIES

By far the most comprehensive and often cited set of data for buoyant water jets resulted from the decay and trajectory measurements of Fan [2]. The experiments concerned two classes of buoyant jets:

(1) Inclined jets discharged into a stagnant environment with linear density stratification.
(2) Vertical buoyant jets, $\theta_0 = 90°$, discharged into a uniform cross stream, at U_a, with no ambient stratification.

The experiments for the flows of class 1 were conducted in a 2.26 × 1.07 m tank with a depth of 0.61 m. The tank was stratified with successive 3–5 cm layers of aqueous salt solutions. Tank temperature remained constant within a 2°C range over the duration of the experiments.

Nozzle of diameters in the range 0.223–0.686 cm were used. Flow into the nozzle was provided from an unregulated head tank, which provided a discharge rate estimated to be constant within 3%. Measurements in each experimental run were limited to jet trajectory and half-width, observed photographically by use of a tracer dye premixed into the discharged fluid.

Fan [2] described one of the dilemmas of conducting experiments on jets of small physical scale. For a complete experimental check of theory, jet velocity and density must be determined. However, laboratory experiments in stratified flows are usually of small size and do not allow the time required to measure time-dependent quantities. On the other hand, jet trajectories and half-widths are determined conveniently by photography. These two measures are interrelated with other jet characteristics. Thus comparison of observed and calculated values of these quantities is expected to be indicative of the applicability of modeling calculations. The experimental discharge Froude number ranged from 10 to 60, except for three runs with a nonbuoyant momentum jet ($F = \infty$) and six runs in an unstratified ambient with Froude numbers ranging from 10 to 130.

In the second group of experiments, with vertical jets in a flowing, unstratified pure water ambient, the saline jets were actually negatively buoyant. Conductivity measurements were taken by movable probes at a number of downstream stations. The locus of stations of maximum concentration was defined as the jet centerline. The jet width was defined from concentration readings taken radially outward from the centerline.

These latter experiments were conducted in a 40 m flume, 1.1 m wide, with a water depth of 0.51 m. Flow was induced by inclining the flume. A region in the core of the flume flow, with the least shear effects from wall boundaries, was selected to introduce the jet via a nozzle. The variation of ambient velocity in this region was estimated at $+6$ to -9%. Experimental runs were made over a Froude number range of 10–80, ambient flow resulting in R = 0.0625–0.25, with discharge nozzles of diameters from 0.508 to 0.762 cm.

In an attempt to ascertain the effect of ambient turbulence and shear introduced by the restricted cross-sectional dimensions of the flume, some concentration measurements were made in which the fluid in the flume was stagnant while the jet discharge was towed through the ambient by a carriage mounted over the flume. In this case, conductivity probes were fixed to the carriage. The probes were moved to different vertical positions in successive runs.

The experimental results of Fan have often also been used by later modelers, for comparisons with calculations. The data have also been used to assess earlier hypotheses and observations regarding the Gaussian distribution of concentration in the zone of established flow.

Riester *et al.* [12] studied horizontally discharged buoyant freshwater and saltwater jets. The ambient was quiescent and unstratified. A 6.2 × 1.1 × 0.8 m tank was used, with a 0.87 cm diameter jet discharge nozzle. Jet trajectory and width were recorded photographically by a tracer dye in the jet fluid. Temperature distributions were measured by a rake of thermocouples. Jet centerline was inferred from the measured temperature distributions. The unusual characteristics of these experiments were a wide range of experimental ambient temperatures, 4.5–43.0°C, and the use of both saltwater and freshwater jets. Some possible implications of the results will be discussed in a later section.

Davis et al. [13] measured the behavior of single and multiple-port saltwater discharges directed vertically downward into a freshwater ambient. The tank water depth was 0.91 m and the tank length was 12.1 m. Concentration profiles were measured by conductivity probes. Velocity measurements were made by hot film anemometry. Several experimental runs were made with the carriage stationary and led to the proposal of the following entrainment function:

$$\alpha = 0.057 + 0.083/F^{0.3} \tag{15}$$

These same experiments verified the Gaussian nature of velocity profiles in the zone of established flow.

Shirazi et al. [4] studied buoyant jets in flowing turbulent ambients. An inclined 37 m flume was used to produce ambient flow. Turbulence was introduced by imposing a layer of varying sized rock over the flume bed. Temperature and salinity concentration were measured using conductivity probes. Turbulence was monitored by hot film anemometers.

From these data, a set of algebraic correlations were made, expressing temperature and concentration residuals, jet width, and trajectory, as functions of downstream distance, F and R. It was determined that as the level of turbulence increases, the rate of decay of centerline temperature and concentration, with respect to the streamwise coordinate, increases. Not surprisingly, it was also found that correlation of the data became more difficult as turbulence level increased.

Pryputniewicz and Bowley [17] conducted measurements of a buoyant jet discharged vertically upward into a uniform quiescent ambient. Temperature profiles were measured by a rake of thermistors in the flow field. Of interest in this study is the presentation of data showing that temperature residuals remain at the terminal rise locations, along the surface. Froude numbers of 1–50 resulted with nozzle diameters of 0.425–0.550 in. A summary of the range of parameters studied in all the experiments is presented in Table IV.

TABLE IV

SUMMARY OF THE PARAMETERS OF PAST EXPERIMENTS

Experimenter	Initial jet diameter (cm)	F	Ambient[a] 1	2	3	Measured quantities[b] 4	5	6	7	8	Max. vertical range of rise	R
Fan and Brooks [9]	0.223–0.686	10–60	X	X		X	X				0.61 m	—
Fan and Brooks [9]	0.508–0.762	10–80			X	X	X		X		0.61 m	0.0625–0.25
Riester et al. [12]	0.87	3.16–20		X		X	X			X	0.8 m	—
Davis et al. [13]	1.12–3.81	1.5–36	X			X	X	X	X		0.91 m	—
Shirazi et al. [4]	0.468–1.882	5.9–215	X		X	X	X	X	X		2 ft	0.11–3.7
Pryputniewicz and Bowley [17]	0.425–0.550	1–50	X			X	X		X		5 ft	—

[a] (1) Unstratified, (2) stratified, (3) flowing.
[b] (4) Trajectory, (5) width, (6) velocity, (7) concentration, (8) temperature.

IV. Properties and Ambient Stratification Modeling

The determination of the discharge Froude number and the buoyancy force requires the evaluation of both the temperature t and concentration c effects on density. Often the density $\rho(t, c, p)$ is a sufficiently linear function of both t and c over the range of temperature and concentration difference between the jet and the ambient. Then density differences may be estimated accurately in terms of the two volumetric coefficients of expansion,

$$\beta = -\frac{1}{\rho_r}\left(\frac{\partial\rho}{\partial t}\right)_{c,p} \quad \text{and} \quad \gamma = -\frac{1}{\rho_r}\left(\frac{\partial\rho}{\partial c}\right)_{t,p} \tag{16}$$

where ρ_r is some reference value of ρ, say $\rho_0 = \rho(t_0, c_0, p_0)$. Then the overall density differences are

$$\Delta\rho = -\beta\rho_r\,\Delta t = -\beta\rho_r(t_0 - t_a) \quad \text{and}$$

$$\Delta\rho = -\gamma\rho_r\,\Delta c = -\gamma\rho_r(c_0 - c_a) \tag{17}$$

for the independent t and c effects on density. During integration, t_a and c_a are local values. The density at some t and c, in terms of the initial jet density ρ_0, is approximated, for uniform pressure p_0, as

$$\rho(t, c, p_0) - \rho(t_0, c_0, p_0) = -\rho(t_0, c_0, p_0)[\beta(t - t_0) + \gamma(c - c_0)] \tag{18a}$$

or

$$\rho = \rho_0[1 - \beta(t - t_0) - \gamma(c - c_0)] \tag{18b}$$

In particular, the initial density difference between the local ambient and the initial jet, which appears in the Froude number, is as follows:

$$(\rho_{a0} - \rho_0)/\rho_0 = \beta(t_0 - t_a) + \gamma(c_0 - c_a) \tag{19}$$

where ρ_{a0} is the ambient density at jet discharge.

In an unstratified ambient medium, the reduced, nondimensional forms of equations will contain the following three nondimensional terms:

$$(\rho_a - \rho_m)/(\rho_a - \rho_0), \quad (t_m - t_a)/(t_0 - t_a), \quad (c_m - c_a)/(c_0 - c_a)$$

With $\lambda_t = \lambda_c$ in Eqs. (2) and (3), profiles of t and c are the same and

$$(t_m - t_a)/(t_0 - t_a) = (c_m - c_a)/(c_0 - c_a) \tag{20}$$

at all points along a trajectory at which $(t_0 - t_a) \neq 0$ and $(c_0 - c_a) \neq 0$. Thus, for constant values of β and γ in an unstratified ambient,

$$(\rho_a - \rho_m)/(\rho_a - \rho_0) = (t_m - t_a)/(t_0 - t_a) = (c_m - c_a)/(c_0 - c_a) \tag{21}$$

and defined values of β and γ are not required for integration, computation, and solution, beyond the initial calculation of actual discharge Froude number values.

A similar result arises for some special circumstances in linearly stratified ambients. If $\rho = \rho(t)$, density stratification is defined solely by a temperature stratification parameter

$$(\partial t_a / \partial z) D / (t_{a0} - t_0)$$

Having taken β constant, the downstream Froude number changes only because of changes in velocity and temperature. For such flows,

$$(\rho_a - \rho_m)/(\rho_{a0} - \rho_0) = (t_m - t_a)/(t_0 - t_{a0}) \tag{22}$$

pertains throughout. In addition, if $\rho \neq \rho(c)$, an assumption that might be made in a jet with a tracer dye only, the concentration and vertical momentum equations are not coupled. Concentration computations can proceed independently, even to the point of specifying a concentration stratification if desired.

Parallel reasoning holds true for the condition $\rho = \rho(c)$. Then a density dependence on the concentration stratification parameter would couple the concentration and vertical momentum equation. Presumably, temperature would be uniform throughout the system in such a case.

These simple, degenerate cases of the overall modeling problem are important because they represent conditions under which measurements are often taken. Modelers then specify these conditions to compare data with their analytical models. The underlying assumptions and limitations of such formulation are:

(1) With no equation of state incorporated into the computational process, only unstratified ambients can be accommodated if $\rho = \rho(t, c)$, and then only if β and γ are assumed constant.

(2) If no equation of state is included in the model, density stratification in the ambient may be specified as either

$$(\partial t_a / \partial z) D / (t_{a0} - t_0), \qquad \text{for} \quad \rho = \rho(t)$$

or

$$(\partial c_a / \partial z) D / (c_{a0} - c_0), \qquad \text{for} \quad \rho = \rho(c)$$

or

$$(\partial \rho_a / \partial z) D / (\rho_{a0} - \rho_0), \qquad \text{directly}$$

Ambients with more than one density determining property gradient cannot be accommodated, since temperature, concentration, and density residuals would be independent of each other and profiles could be dissimilar.

The more general occurrence, of an ambient medium with both temperature and concentration stratification, requires an equation of state for the solution of the governing equations. Such a temperature- and concentration-dependent density relation may be used in one of two ways:

(1) Internally in the formulation and calculational scheme, with ambient temperature and concentration gradients specified. The local density differences are computed downstream from the decaying jet temperature and concentration levels, as calculated from the energy and concentration equations.

(2) External to the actual downstream integration calculations, by using any specified ambient stratification to calculate the actual ambient density variation. The equation of density excess, or deficiency, is of the form:

$$\frac{d}{dS}\left\{\int_0^{2\pi}\int_0^\infty U(\rho_a - \rho)r\,dr\,d\phi\right\} = -\frac{d\rho_a}{dS}\left\{\int_0^{2\pi}\int_0^\infty Ur\,dr\,d\phi\right\} \quad (23)$$

The same Gaussian profile is assumed for density as for two of its constituent properties, temperature and concentration. This additional equation uncouples the temperature and concentration equations from the vertical momentum equation.

The first of these procedures was used by Hirst [14]. The second was used by Fan and Brooks [9]. Other models specify the method of use of an equation of state, only as noted in the following discussion.

Hirst specified particular constant values of β and γ throughout the downstream integration of the governing equations. The equation of state was:

$$(\rho_a - \rho)/\rho_0 = \beta(t - t_{a0}) + \gamma(c - c_{a0}) \quad (24)$$

In reality, of course, the density of water is a complex and, under some conditions, highly nonlinear, function of temperature, species concentration, and pressure—t, s, and p. Many modeling situations arise in which the ranges of t, s, and p do not permit the use of constant values of either β or γ, or the omission of the pressure dependence of density. Table V shows some of the variation of β, at $p = 1$ bar, encountered over a very small but commonly occurring temperature range, for both saline water at 35 parts per thousand (typical of seawater) and fresh water.

The model of Riester et al. [12] attempted to account for the effect of variable β by expressing it as a polynomial.

TABLE V

Variation of β over Limited Temperature Range, for Pure and Saline Water[a]

Salinity (parts per thousand)	Temperature (°C)									
	−1	0	1	2	3	4	5	6	7	8
35	3.70	5.01	6.27	7.49	8.67	9.83	10.97	12.09	13.19	14.29
0	—	—	—	−3.31	−1.77	0.05	1.68	3.22	4.66	6.05

[a] $\left(-\dfrac{1}{\rho}\left(\dfrac{\partial\rho}{\partial t}\right)_p \times 10^5 \,°C^{-1} \text{ for pressure} = 1 \text{ atm}\right).$

$$\beta = a_1 + a_2 t + a_3 t^2 \tag{25}$$

where the coefficients a_1, a_2, and a_3 were evaluated at a temperature pertinent to a given circumstance. The discharge Froude number was redefined as

$$F_a = U_0/[gD\beta(t_0 - t_a)]^{1/2} \tag{26}$$

where β is the value appropriate to any particular jet in water. The need for an accurate equation of state was stressed, if the initial temperature difference between the jet and the ambient is greater than 3°C. This is a very significant matter when laboratory experiments are made at high temperature differences, to simulate flow of low temperature difference. Also, saltwater jets cannot in general simulate freshwater jets operating over large temperature differences, since the saline density effect does not generally duplicate the temperature–density characteristics of a freshwater system. This last observation indicated that although the proposed model attempted to compensate for the variable temperature effects on density, no dependence on salinity was included except for that which may have been implicit in the choice of the three coefficients in Eq. (25).

A convenient equation of state for buoyant jet modeling would correlate density comprehensively as a function of temperature, species concentration, and pressure. It should be applicable to all temperature, concentration, and pressure ranges of interest. Also, for most modeling of fresh- and saltwater systems, the concentration variable should be salinity. A choice is made in the following section.

V. General Formulation

The foregoing modeling system assumptions regarding the flow field were used in the subsequent calculations outlined in the following pages. Generally, they assume a steady, axisymmetric jet with negligible molecular transport, negligible steamwise turbulent transport, and small curvature effects, diffusing in a hydrostatic pressure field. The governing equations in the zone of established flow follow below and are summarized in Table I.

Continuity of mass equates the downstream rate of change in total mass of the jet to the rate of fluid entrainment. The entrainment concept results in the following:

$$\frac{d}{dS}\left\{\int_0^{2\pi}\int_0^\infty \rho U r\, dr\, d\phi\right\} = \rho E = \rho(2\pi\alpha U_m B) \tag{27a}$$

The first Boussinesq approximation, of constant density, except in the buoyancy force, yields:

$$\frac{d}{dS}\left\{\int_0^{2\pi}\int_0^\infty U r\, dr\, d\phi\right\} = 2\pi\alpha U_m B \tag{27b}$$

Horizontal momentum is conserved in a hydrostatic pressure field. Thus the change in horizontal momentum within the jet is equal to the horizontal momentum of the fluid entrained:

$$\frac{d}{dS}\left\{\int_0^{2\pi}\int_0^\infty \rho U^2 \cos\theta\, r\, dr\, d\phi\right\} = \rho U_a E \tag{28a}$$

Again, invoking the Boussinesq approximation concerning density level,

$$\frac{d}{dS}\left\{\int_0^{2\pi}\int_0^\infty \rho U^2 \cos\theta\, r\, dr\, d\phi\right\} = U_a E \tag{28b}$$

The change in the vertical momentum of the jet is the result of the action of the buoyancy force over the horizontal cross section of the jet:

$$\frac{d}{dS}\left\{\int_0^{2\pi}\int_0^\infty \rho U^2 \sin\theta\, r\, dr\, d\phi\right\} = \int_0^{2\pi}\int_0^\infty (\rho_a - \rho)gr\, dr\, d\phi \tag{29}$$

In a stratified ambient, the ambient temperature, as well as the temperature of the jet, may be variable along the trajectory. The energy equation relates the change in energy of the system, as expressed by a temperature excess or deficiency relative to the ambient, to the rate at which the ambient temperature is changing.

$$\frac{d}{dS}\left\{\int_0^{2\pi}\int_0^\infty C_p\rho U(t - t_a)r\, dr\, d\phi\right\} = -\frac{dt_a}{dS}\int_0^{2\pi}\int_0^\infty \rho C_p U r\, dr\, d\phi \tag{30a}$$

Assuming constant specific heat and the uniform density,

$$\frac{d}{dS}\left\{\int_0^{2\pi}\int_0^\infty U(t - t_a)r\, dr\, d\phi\right\} = -\frac{dt_a}{dS}\int_0^{2\pi}\int_0^\infty U r\, dr\, d\phi \tag{30b}$$

Similarly, the change in concentration excess relative to the ambient is related to the rate at which the ambient reference concentration is changing:

$$\frac{d}{dS}\left\{\int_0^{2\pi}\int_0^\infty \rho U(c - c_a)r \, dr \, d\phi\right\} = -\frac{dc_a}{dS}\int_0^{2\pi}\int_0^\infty \rho Ur \, dr \, d\phi \quad (31a)$$

Again, with uniform density:

$$\frac{d}{dS}\left\{\int_0^{2\pi}\int_0^\infty U(c - c_a)r \, dr \, d\phi\right\} = -\frac{dc_a}{dS}\int_0^{2\pi}\int_0^\infty Ur \, dr \, d\phi \quad (31b)$$

Geometrically, the incremental trajectory of the jet may be described by:

$$dX = dS \cos\theta \quad (32a)$$

$$dZ = dS \sin\theta \quad (32b)$$

It will be assumed that the Gaussian velocity, temperature, and concentration profiles cited previously are valid for the zone of established flow. In Appendix A the nondimensionalization scheme is shown. The di-

TABLE VI

Nondimensional Differential Form of the Governing Equations for an Unstratified Ambient

Equation	Quiescent ambient	Flowing ambient
Continuity	$\dfrac{d}{ds}(u_m b^2) = 2\alpha u_m b$	$\dfrac{d}{ds}(u_m b^2) = 2\alpha b[|u_m - R\cos\theta| + a_3 R\sin\theta]$
Horizontal momentum	$\dfrac{d}{ds}(u_m^2 b^2 \cos\theta) = 0$	$\dfrac{d}{ds}(u_m^2 b^2 \cos\theta)$ $= 4R\alpha b[|u_m - R\cos\theta| + a_3 R\sin\theta]$
Vertical momentum	$\dfrac{d}{ds}(u_m^2 b^2 \sin\theta)$ $= \left(\dfrac{\rho_a - \rho_m}{\rho_{a0} - \rho_0}\right)\dfrac{2\lambda^2 b^2}{F^2}$	$\dfrac{d}{ds}(u_m^2 b^2 \sin\theta)$ $= \left(\dfrac{\rho_a - \rho_m}{\rho_{a0} - \rho_0}\right)\dfrac{2\lambda^2 b^2}{F^2}$
Energy	$\dfrac{d}{ds}\left(u_m \dfrac{\Delta t_m}{\Delta t_0}\dfrac{\lambda^2 b^2}{(\lambda^2 + 1)}\right) = 0$	$\dfrac{d}{ds}\left(u_m \dfrac{\Delta t_m}{\Delta t_0}\dfrac{\lambda^2 b^2}{(\lambda^2 + 1)}\right) = 0$
Concentration	$\dfrac{d}{ds}\left(u_m \dfrac{\Delta c_m}{\Delta c_0}\dfrac{\lambda^2 b^2}{(\lambda^2 + 1)}\right) = 0$	$\dfrac{d}{ds}\left(u_m \dfrac{\Delta c_m}{\Delta c_0}\dfrac{\lambda^2 b^2}{(\lambda^2 + 1)}\right) = 0$
Horizontal trajectory	$\dfrac{d}{ds}(x) = \cos\theta$	$\dfrac{d}{ds}(x) = \cos\theta$
Vertical trajectory	$\dfrac{d}{ds}(z) = \sin\theta$	$\dfrac{d}{ds}(z) = \sin\theta$

mensionless equations for unstratified and stratified ambients are collected in Tables VI and VII, respectively, for both quiescent and flowing ambients. These equations, combined with the density information (e.g., the relation for saline water in Appendix B and the entrainment functions in Tables II and III), are the basis of all subsequent calculations.

Flow Establishment

The preceding equations apply only in the zone of established flow, which begins at S_e (see Fig. 1). Therefore the end of the region of flow establishment must be specified from other observations. The procedures followed here in calculations are based on the experimental work of Abraham [7], for λ_t and λ_c, and on the analytical development by Hirst [18], for B_e. They are also consistent with assumptions made in Hirst [11, 14] and

TABLE VII

NONDIMENSIONAL DIFFERENTIAL FORM OF THE GOVERNING EQUATIONS FOR A
STRATIFIED AMBIENT

Equation	Quiescent ambient	Flowing ambient
Continuity	$\dfrac{d}{ds}(u_m b^2) = 2\alpha u_m b$	$\dfrac{d}{ds}(u_m b^2) = 2\alpha b[\|u_m - R\cos\theta\| + a_3 R\sin\theta]$
Horizontal momentum	$\dfrac{d}{ds}(u_m^2 b^2 \cos\theta) = 0$	$\dfrac{d}{ds}(u_m^2 b^2 \cos\theta)$ $= 4R\alpha b[\|u_m - R\cos\theta\| + a_3 R\sin\theta]$
Vertical momentum	$\dfrac{d}{ds}(u_m^2 b^2 \sin\theta)$ $= \left(\dfrac{\rho_a - \rho_m}{\rho_{a0} - \rho_0}\right)\dfrac{2\lambda^2 b^2}{F^2}$	$\dfrac{d}{ds}(u_m^2 b^2 \sin\theta)$ $= \left(\dfrac{\rho_a - \rho_m}{\rho_{a0} - \rho_0}\right)\dfrac{2\lambda^2 b^2}{F^2}$
Energy	$\dfrac{d}{ds}\left(u_m \dfrac{\Delta t_m}{\Delta t_0}\dfrac{\lambda^2 b^2}{(\lambda^2 + 1)}\right)$ $= -\dfrac{dt_a}{\Delta t_0\, ds}\{u_m b^2\}$	$\dfrac{d}{ds}\left(u_m \dfrac{\Delta t_m}{\Delta t_0}\dfrac{\lambda^2 b^2}{(\lambda^2 + 1)}\right) = -\dfrac{dt_a}{\Delta t_0\, ds}\{u_m b^2\}$
Concentration	$\dfrac{d}{ds}\left(u_m \dfrac{\Delta c_m}{\Delta c_0}\dfrac{\lambda^2 b^2}{(\lambda^2 + 1)}\right)$ $= -\dfrac{dc_a}{\Delta c_0\, ds}\{u_m b^2\}$	$\dfrac{d}{ds}\left(u_m \dfrac{\Delta c_m}{\Delta c_0}\dfrac{\lambda^2 b^2}{(\lambda^2 + 1)}\right) = -\dfrac{dc_a}{\Delta c_0\, ds}\{u_m b^2\}$
Horizontal trajectory	$\dfrac{d}{ds}(x) = \cos\theta$	$\dfrac{d}{ds}(x) = \cos\theta$
Vertical trajectory	$\dfrac{d}{ds}(z) = \sin\theta$	$\dfrac{d}{ds}(z) = \sin\theta$

Ginsberg and Ades [15]. The initial conditions, at s_e, are then taken to be:

$$U_{m,e} = U_0 \tag{33a}$$

$$\lambda_t = \lambda_c = 1.16 \tag{33b}$$

$$B_e = D/\sqrt{2} \tag{33c}$$

$$\frac{(t_m - t_a)_e}{(t_m - t_a)_0} = \frac{(\lambda_t^2 + 1)}{2\lambda_t^2} = \frac{(c_m - c_a)_e}{(c_m - c_a)_0} \tag{33d}$$

The values of S_e, beyond the point of discharge are taken from Abraham [7], to be:

$$S_e/D = 6.2, \qquad\qquad F^2 \geq 40$$

$$S_e/D = 3.9 + 0.057F^2, \qquad 5 \leq F^2 < 40$$

$$S_e/D = 2.075 + 0.425F^2, \qquad 1 \leq F^2 < 5$$

$$S_e/D = 0, \qquad\qquad 0 \leq F^2 < 1$$

All that now remains to be specified to permit integration of the equations downstream along the S coordinate are the specific entrainment functions to be used and the specification of any ambient stratification condition. The density equation of state in water, when needed, is that given in Appendix B.

The calculations were done by IBM 360 and IBM 3033 computing systems. The method of integration is a trapezoidal rule expression of the governing equations with adjustable step size. Two versions of the general method were used here. The first evaluated flows in unstratified ambients, or in ambients characterized by linear gradients in temperature, concentration, or density itself. No explicit equation of state is then necessary. However, F is related to other given conditions by β and/or γ. The second version, with the same computational technique, uses an accurate density equation.

Since one objective is to develop a predictive model applicable to large-scale discharges, in open water generally, the density equation used is the simpler one of Gebhart and Mollendorf [19]. It correlates the temperature dependence of density as an expansion around the density extremum temperature, at any given level of pressure and salinity. The resulting expression is fitted to a comprehensive set of experimental data over temperature, salinity, and pressure levels to $t = 20°C$, $s = 40‰$, and $p = 1000$ bars absolute. Density is in kilograms per cubic meter. This equation is given in Appendix B.

For each of these latter calculations, the density equation was first used to establish the density field in the ambient medium, stratified or not. At least one reference temperature and salinity level and appropriate gradients, or the complete temperature and salinity fields, must be specified. The water surface pressure is assumed to be 1 atm. The relation is first used at the surface, and the density and pressure are calculated progressively downward, integrating the hydrostatic equation, to determine $p_a(z)$.

Once the ambient density distribution with depth $\rho_a(z)$ is established, the density relation is used during trajectory calculations to determine the local density from local jet temperature and concentration values, at the local ambient pressure. The ambient density and variation are assumed to be linear between the points calculated for the ambient along the trajectory.

The density relation is also used to calculate the initial density difference between the jet and the ambient at the point of discharge, $\rho_0 - \rho_{a0}$, if the jet and salinity level are specified, or to calculate the initial temperature of the jet, given Froude number and density. The results of using such an accurate density relation are:

(1) the full density dependence on temperature, salinity, and pressure is accurately taken into account in all calculations;

(2) the actual ambient density field of either a hypothetical or real circumstance is accurately established, utilizing appropriate temperature and salinity inputs for the environment under consideration;

(3) the relation is readily applicable to computerized analysis and is computationally compact.

VI. Comparative Calculations, Jets in Unstratified Quiescent Ambients

Comparison of recent and earlier jet trajectory and decay calculations, using the same entrainment functions, has shown close agreement, even though different numerical techniques were used in different studies. However, such comparisons are not possible for the new results herein, which incorporate additional effects such as the use of an accurate density equation, with and without stratification. The jet and quiescent ambient system represents by far the most frequently studied circumstance, both in experiments and in modeling.

This section compares the predictions of entrainment models 1, 2, 3, and 5 with each other, over a Froude number range F = 1–200, in terms of trajectories, growth, and decay. There is comparison with the principal data available. The following section compares flowing ambient models 8 and 9 with each other and with some data. The next section then investigates the effects of the most simple and the more realistic conditions of

stratified quiescent ambients. The major purposes of all these comparisons is to assess the models, relative to each other and to the very sparse
data, and to show the effects of more realistic stratification modeling.

A. THE SIMPLEST ENTRAINMENT MODELS

The earlier models, entries 1 and 2 in Table II, nominally apply for a
momentum jet and a buoyant plume, respectively. However, both have
been applied to different circumstances. In this section, newly calculated
trajectories are compared with past ones, over a range of F, and with
some of the data.

For horizontally discharged buoyant jets, Fig. 3 compares the newly
calculated trajectories, for $F = 10$ and 20, with the early results of Abraham [7], both using $\alpha = 0.085$. The disagreement between the two is
relatively small and consistent. This suggests a cumulative effect of small
differences in the calculational procedure. Present calculations predict
slightly less horizontal penetration into the ambient at the Froude numbers compared. Figure 4 shows the uniformly close correspondence between the present calculations and those of Hirst [11, 14], both using

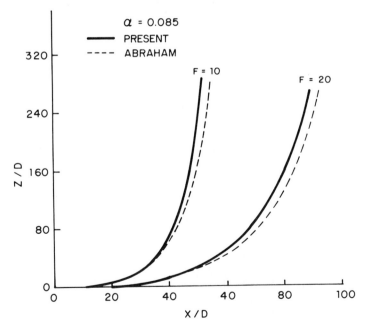

FIG. 3. Present trajectory results, compared with calculations of Abraham [7], both with
$\alpha = 0.085$.

Hirst's entrainment function, entry 3 in Table II, for F = 4, 6, 8, and 10. Agreement is generally good, although the pattern of the difference changes at large F. Figure 5 again compares the present calculations with those of Hirst, but with the two entrainment functions α = 0.057 and 0.082 at F = 33. Agreement is again close during early penetration, with increasing difference farther downstream, as for F = 10.

A comparison of the new and the Hirst trajectory calculations, with the data of Fan [2] at F = 14.42, in a heated horizontal jet, is given in Fig. 6. Again, the calculations are reasonably close to each other. However, the data lie between the two calculations, suggesting the inapplicability of either particular entrainment value for this buoyant circumstance.

Present calculations are compared with those of Fan and Brooks [9], using α = 0.082, in Fig. 7, at large values of F. Again, previous calculations predict less horizontal jet penetration.

Next, two kinds of comparisons are made, in connection with both the data and the resulting entrainment model of Riester *et al.* [12]. The data is at F = 8 with heated jets into ambients at t_a = 4, 10, and 21°C. The model is entry 5 in Table II. In Fig. 8 the calculated curves are seen, along with the data points. They show reasonably good agreement for high-tempera-

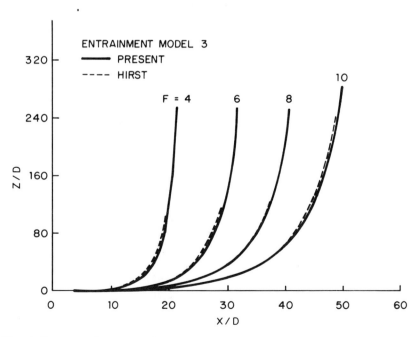

FIG. 4. Present trajectory results, compared with those of Hirst [11, 14], with the same entrainment function, entry 3 in Table II.

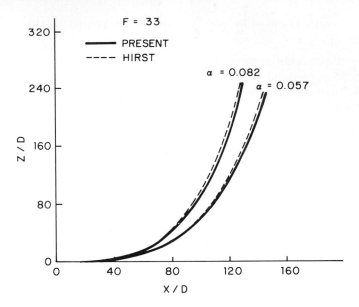

Fig. 5. Present results compared with those of Hirst [11, 14], using $\alpha = 0.057$ and 0.082, jet and plume values.

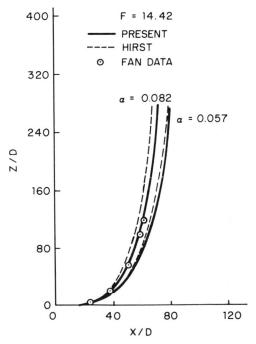

Fig. 6. Comparison of present and Hirst calculations, for $\alpha = 0.057$ and 0.082, with the data of Fan [2].

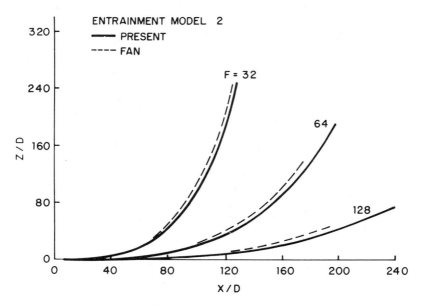

FIG. 7. Comparison of present results with those of Fan and Brooks [9], for $\alpha = 0.082$.

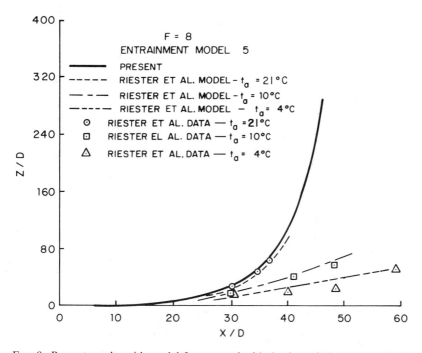

FIG. 8. Present results with model 5 compared with the data of Riester *et al.* [12].

ture ambients, where conventional values of β resulted in little adjustment of the Froude number. Progressively poorer comparison was evident with their prediction for lower-temperature ambients, where the Froude number adjustment became larger.

The foregoing comparisons of models, calculations, and data indicate several things. Calculated trajectories are not very sensitive to the differences in methods used, from the earliest to the present. The small variations in predicted trajectories apparently result at least partially from calculational procedures (in Figs. 3 through 7, for instance). However, the method actually used by earlier modelers is unknown. It is significant that large differences in trajectories do not become apparent until far downstream. Then the cumulative effect of small differences in predicted entrainment is manifested by conserved momentum in a very low-velocity flow.

B. Later Entrainment Models, Detailed Comparisons

The first five entrainment coefficients in Table II are next compared, for the values F = 1, 2, 4, 6, 8, 10, 50, 100, 150, and 200. This range of F covers an immense range of applications. The ambient is assumed un-

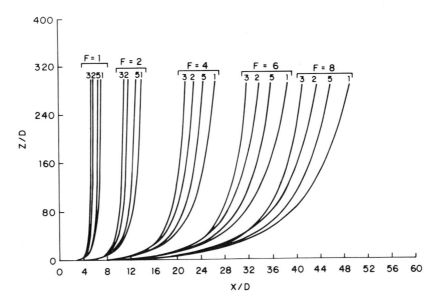

Fig. 9. Comparison of four entrainment models in a quiescent ambient medium at low F values. The numbered labels on the curves indicate the entrainment model in Table II.

stratified and, still, no particular fluid properties are incorporated into the calculations. That is, β and/or γ merely appear in F.

Downstream trajectories, or penetration, are seen in Figs. 9 and 10 for F = 1–8 and F = 10–200, respectively. Note that model 4 is only used for F = 50–200. Then the entrainment coefficient falls into the range of 0.057 $\leq \alpha \leq$ 0.082. These results show the tremendous range of behavior; from F = 1 with approximately equal initial momentum and buoyancy, to F = 200, for which buoyancy is relatively very small. Note the different horizontal, X/D, scales on Figs. 9 and 10. For F = 1, entrainment quickly dilutes the horizontal penetrative power while constant buoyancy continues to cause essentially upward motion. For F = 200, penetration is little influenced by buoyancy, as the small buoyancy effect is quickly mixed into a large mass by entrainment.

The entrainment functions that specifically relate to the nonbuoyant jet and plume, 1 and 2, often show wide downstream divergence from the mean of the other models, 3, 4, and 5. This is particularly true for the buoyant plume entrainment function, $\alpha = 0.082$, at large F. It consistently predicts higher entrainment and resulting higher trajectory at all but the lowest Froude numbers. The Albertson *et al.* [1] momentum jet value, $\alpha = 0.057$, consistently predicts the highest entrainment and, therefore, the

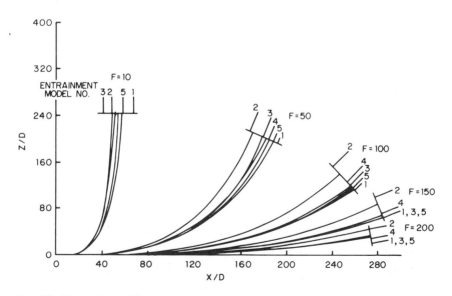

FIG. 10. Comparison of five entrainment models in a quiescent ambient medium at high F values. The numbered labels on the curves indicate the entrainment model in Table II, except model 4 is omitted for F = 10.

flattest trajectory, at all values of F. The "end-point" correlation of Riester *et al.* [12], model 5, is also seen to predict lower entrainment than the
average, for all values of F.

The generally accepted values of entrainment functions for the extrema
of Froude number are $\alpha = 0.082$ for $F = 0$ and $\alpha = 0.057$ for $F = \infty$. The
model that most closely reflects the Froude number–entrainment relation
over the range of F considered here is the function of Hirst [14], model 3.
Hirst's function actually predicts the most buoyant, most entraining behavior for the lowest Froude numbers (≤ 10).

On the other hand, all models predict strikingly similar behavior in the
decays of centerline velocity and temperature level, as seen in Figs. 11
and 12. The decay of these quantities, when plotted in terms of path
length S, are not strongly dependent on either F or the entrainment function used. In Fig. 11 are plotted the centerline velocity decay, U_m/U_0, for
most of the jets depicted in Fig. 10, that is, for $F = 10, 50, 200$. There is,
for all conditions and models, an extremely rapid initial velocity decrease
immediately after discharge. For all jets, $U_m/U_0 < 0.4$ by about 10 diameters downstream. Then $U_m/U_0 < 0.2$ by 30 diameters along the trajectory

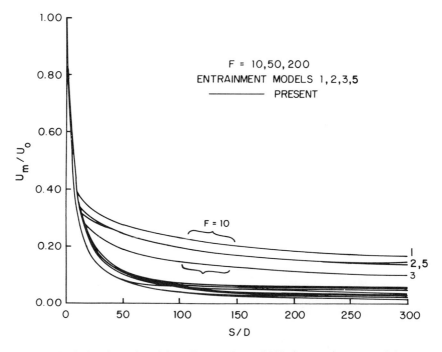

FIG. 11. Velocity decay for 12 jets, F = 10, 50, and 200, for entrainment models 1, 2, 3, and 5.

for $F = 50$ and 200. Still further downstream, the residual velocities decrease much less rapidly. Figure 11 also shows that higher velocities persist downstream at smaller values of F (i.e., for relatively more buoyant or less vigorous jets).

Close inspection of Fig. 11 reveals an apparent anomaly. For $F = 10$, model 3 predicts the lowest residual velocity along most of the trajectory. The highest velocity persists for the buoyant plume entrainment function $\alpha = 0.082$. Yet Fig. 10, for $F = 10$, indicates that these two entrainment functions predict the highest, most buoyant trajectory, with model 3 being the highest. The behavior of model 3, on the one hand displaying buoyant characteristics indicative of low levels of entrainment and on the other hand showing the rapid velocity decay characteristic of high entrainment, is very different from the apparently more consistent prediction of the buoyant plume function.

The reason for this behavior is apparent from close examination of the temperature decay in Fig. 13, an enlargement of the $F = 10$ results on Fig. 10. In the first 40 diameters of trajectory, model 3 is seen to be lower, because of more rapid entrainment, in comparison to the buoyant plume

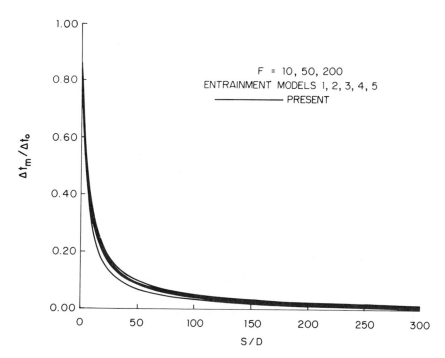

FIG. 12. Temperature difference decay for a group of 15 jets, $F = 10$, 50, and 200, for entrainment models 1 through 5, except model 4 is omitted for $F = 10$.

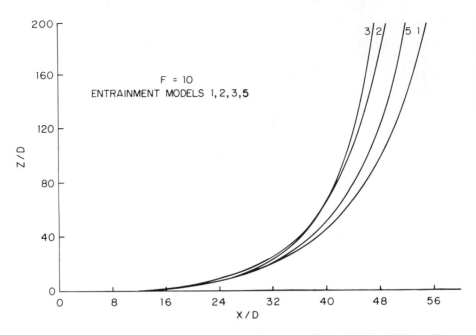

Fig. 13. Enlargement of entrainment model trajectory predictions for F = 10, in a quiescent ambient medium.

model, model 2. It then curves rapidly upward, displaying the most buoyant trajectory of all models, farther downstream. The velocity level apparently decreased early because of rapid entrainment, resulting in lower values of F_L. The second term in α (see Table II) eventually became larger than 0.082. The jet still retained enough buoyancy to cause the sharp upturn characteristic of the later trajectory, in model 3. These results are also indicative of the importance of the early level of entrainment in determining properties much further downstream.

The most uniform behavior among the group of model and Froude number calculations arose in downstream temperature decay, as seen in Fig. 12. For F = 10, 50, and 200, all temperature residuals are less than 0.2 at 25 diameters downstream and less than 0.1 at 50 diameters. The maximum spread of residuals at any location is only about 0.04 at 30 diameters. These decay curves are in close agreement with existing data. We note that, in this formulation, concentration residuals must decay in exactly the same way.

This behavior is further interpreted in Fig. 13, an enlargement of the results in Fig. 10, for F = 10. In the first 25 diameters model 3 results are lower, because of more rapid entrainment. It then becomes higher, more buoyant, downstream. Early rapid entrainment decreased veloc-

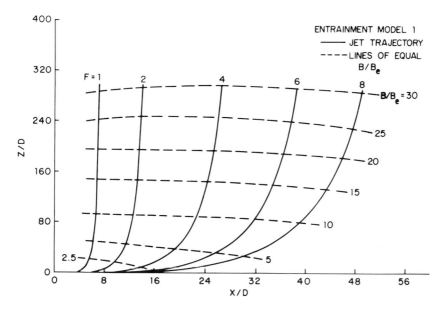

FIG. 14. Downstream radial growth of buoyant momentum jets for F = 1, 2, 4, 6, and 8 in a quiescent ambient medium.

ity and buoyancy then operated to raise the slowed jet. This result indicates the complicated interactions of early entrainment and later buoyant rise.

Downstream radial or diameter growth of jets, using entrainment model 1, are shown in Fig. 14 for F = 1, 2, 4, 6, and 8 and in Fig. 15 for F = 10, 50, 100, 150, and 200. These results illustrate that the physical extent of a jet generally increases more rapidly downstream with increasing F, that is, with increasing initial velocity. Very different characteristics are apparent in the low and high F ranges. In the low F range the size, when $Z/D \cong 80$, becomes very insensitive to F. The entrainment is largely driven by initial buoyancy effects. In the high F range, say 100–200, the size again remains insensitive to F, throughout the trajectory. Now the entrainment is driven almost entirely by initial momentum effects. This is a very considerable difference in downstream behavior.

In summary, the various past entrainment models for jets in a quiescent ambient medium predict reasonably similar behavior. The most notable exceptions are the results using the buoyant plume entrainment function, model 2, at higher F, that is, for vigorous jets. Such conditions are well outside the range of intended use. Further, all the models are in reasonably close agreement with meager existing data. All such data were measured with very small-diameter jets. Thirty years of entrainment function

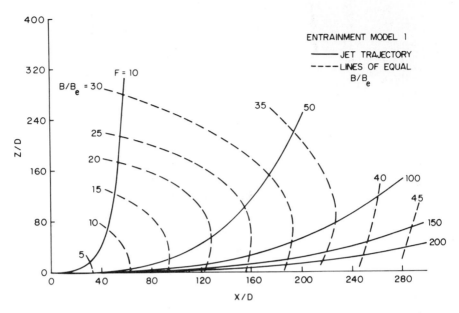

FIG. 15. Downstream radial growth of buoyant momentum jets for F = 10, 50, 100, 150, and 200 in a quiescent ambient medium.

modeling have not resulted in major changes in models of either downstream decay rates or trajectory.

VII. Comparative Calculations, Jets in Unstratified Flowing Ambients

The entrainment functions proposed for a buoyant momentum jet in a flowing ambient are models 8, 9, and 10 in Table III. The following comparisons do not show the good agreement, between different models, characteristic of the quiescent ambient medium calculations.

Figure 16 shows the trajectory predicted by the two entrainment functions 8 and 9 for an initially vertical jet, $\theta_0 = 90°$, in a horizontal cross flow. The conditions are $R = U_a/U_0 = 0.25$ and 0.125, for F = 2.83, in an unstratified ambient. The set of values of R = 0.25 and F = 2.83 is especially significant in comparing 8 and 9 because it is one of the few conditions in which the models are known to be in close agreement. For all F less than 2.83, model 9 predicts a higher, more buoyant trajectory than model 8. Conversely, at F > 2.83, model 9 generally predicts lower trajectory and somewhat more rapid decay of velocity, temperature, and concentration. For F = 2.83, model 9 predicts lower trajectory for all R > 0.25, and vice versa.

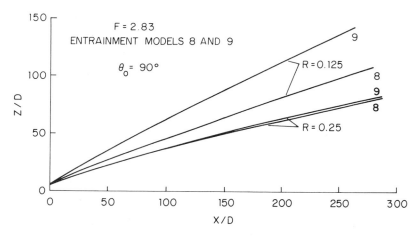

FIG. 16. Comparison of entrainment models 8 and 9 for vertical jets, $\theta_0 = 90°$, in flowing ambients, for F = 2.83 and R = U_a/U_0 = 0.125 and 0.25.

Figure 16 also shows the effect of lessening the ambient flow rate, by a factor of 2, R = 0.125, also for a jet of F = 2.83. The difference between the models in predicted jet rise has now become very large. This is a much bigger difference than those found between the predictions of different entrainment models in quiescent ambients. We will see that this large disagreement is only the first example of a systematic inadequacy in cross-flow entrainment modeling to date.

Figure 17 illustrates that large disparities between the trajectory predictions of the flowing ambient entrainment models are not limited to discharge normal to a cross flow. These trajectories are for a discharge angle $\theta_0 = 45°$, again for F = 2.83 and R = 0.125, 0.25, and 0.375. The same behavior is evident. There is reasonably good agreement for F = 2.83 and R = 0.25, as in Fig. 16. However, model 8 predicts a higher trajectory for R = 0.375 and again a lower trajectory for R = 0.125.

As F and R are increased, the upward penetration of the jet decreases. This is due to more rapid entrainment and lower initial buoyancy of the high Froude number jets, as well as the stronger horizontal "sweeping" effect of the more rapidly flowing ambients, at higher R. For most flow conditions, however, an increase of these effects does not lessen the differences between the trajectories predicted by the two entrainment models.

Figure 18 depicts another condition, F = 10, R = 0.125, and $\theta_0 = 90°$, of relatively good agreement between the two models. Also shown are data of Fan at these same values of the parameters. However, the other condition in Fig. 18, for R = 0.25, shows that model 8 diverges from both

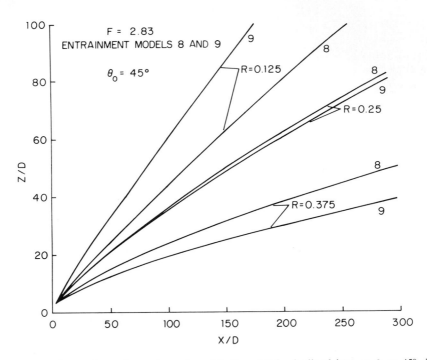

FIG. 17. Comparison of entrainment models 8 and 9 for inclined jets, at $\theta_0 = 45°$, in horizontally flowing ambients. $R = U_a/U_0 = 0.125$, 0.25, and 0.375, for $F = 2.83$.

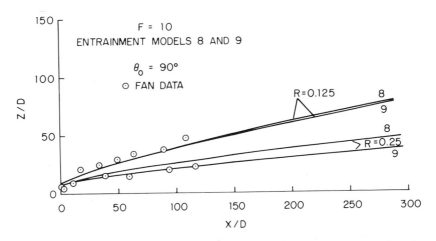

FIG. 18. Comparison of models 8 and 9 with the data of Fan, for $F = 10$, $\theta_0 = 90°$, and $R = 0.125$ and 0.25.

model 9 results and data, at these same values of F and θ_0. Figure 19 shows the disagreement between the models at $\theta_0 = 45°$. For higher Froude numbers, large disparities persist, over the range of R, as shown in Figs. 20 and 21 for F = 20 and 40.

Next, the results are given for initial pure "coflow," at $\theta_0 = 0$, that is, for the jet discharged horizontally parallel to the ambient flow. Figures 22 and 23 show the trajectories at F = 20 and 40 for R = 0.125 and 0.25. The same characteristic disagreement between the models is seen again as was seen for $\theta_0 > 0$. Here it is 15–40%, in rise. For larger values of R, 0.4 and 0.8, the results are plotted in Figs. 24 and 25. Disagreement approaches 100%. The two models are inconsistent throughout. The higher R values above are beyond the range of the data used by Ginsberg and Ades in constructing model 9.

In summary, the disagreement between the jet trajectory predictions of the two flowing ambient entrainment models increases in flows with increasing vertical flow tendencies. This results either from initial vertical momentum or from buoyancy. Clearly, the models are very sensitive to how the cross-flow entrainment is accounted for. This is where they differ

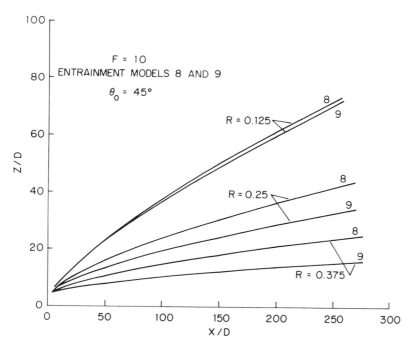

FIG. 19. Comparison of models 8 and 9, for F = 10 and $\theta_0 = 45°$, for three values of R.

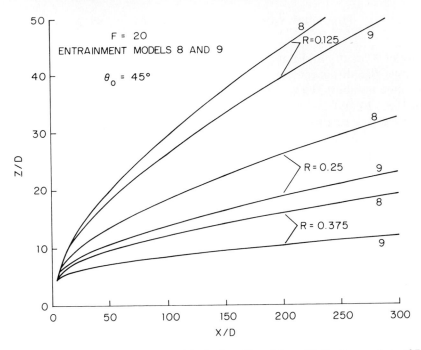

FIG. 20. Comparison of models 8 and 9, for $F = 20$ and $\theta_0 = 45°$, for three values of R.

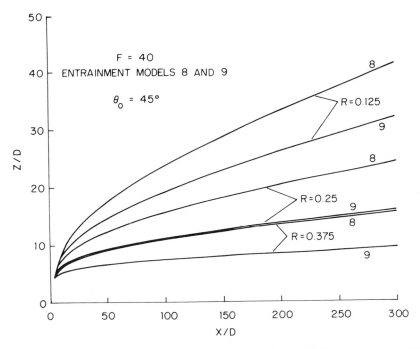

FIG. 21. Comparison of models 8 and 9, for $F = 40$ and $\theta_0 = 45°$, for three values of R.

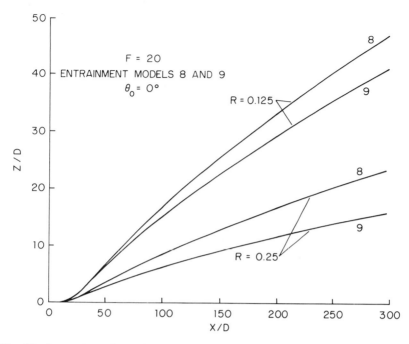

FIG. 22. Comparison of models 8 and 9 for pure initial coflow, $\theta_0 = 0°$, for $F = 20$, for two values of R.

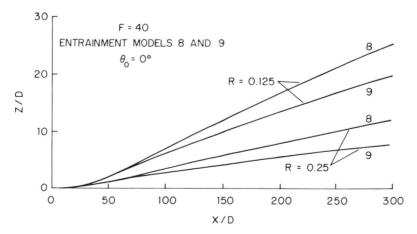

FIG. 23. Comparison of models 8 and 9 for pure initial coflow, $\theta_0 = 0°$, for $F = 40$, for two values of R.

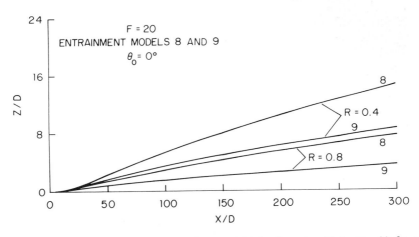

FIG. 24. Comparison of models 8 and 9 for pure initial coflow, $\theta_0 = 0°$, for F = 20, for two values of R.

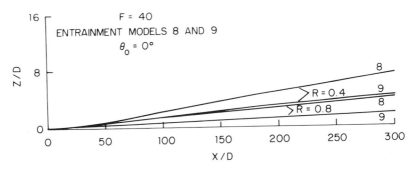

FIG. 25. Comparison of models 8 and 9 for pure initial coflow, $\theta_0 = 0°$, for F = 40, for two values of R.

most, and the difference is very large. For jets having low Froude number and R, the difference in vertical rise found is usually many diameters in magnitude. This disagreement is found at all discharge angles. At higher values of both F and R, the difference in predicted vertical rise becomes small, but relative to the total rise of the jet, the difference may be many times greater than that for lower values of F and R.

VIII. Effects of Ambient Stratifications

A stably stratified quiescent ambient, in which density increases with increasing depth, has the general effect of restricting the vertical motion

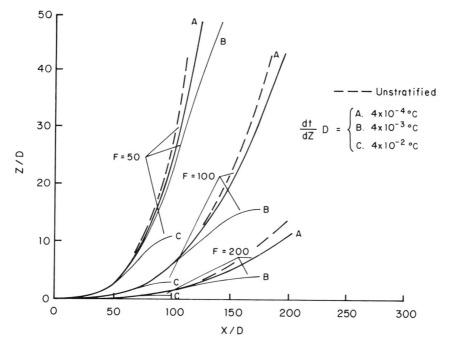

FIG. 26. Buoyant jet behavior in a quiescent stratified ambient, horizontal discharges at F = 50, 100, and 200, at three temperature stratification levels, model 3. The dashed curve at each value of F is the unstratified trajectory.

of an initially buoyant jet. This results from the combined effects of stratification and of jet buoyancy decrease, per unit mass, resulting from entrainment of ambient fluid. This change of trajectory arises regardless of the direction of buoyancy, the ambient stratification, or the magnitude of the stratification.

Figure 26 illustrates several of the effects of stratification and of its strength on trajectory for a buoyant horizontal discharge at F = 50, 100, and 200, using entrainment model 3 in Table II. Stratification is assumed to be entirely in the form of a vertical temperature gradient, as expressed in the following parameter, for curves *A, B,* and *C,* respectively:

$$D(dt_a/dZ) = 4 \times 10^{-4}, \qquad 4 \times 10^{-3}, \qquad \text{and} \qquad 4 \times 10^{-2} \text{ °C}$$

In the simple uniform properties model, used in this first example, the variation of t_a affects the buoyancy only through the temperature differences. The value of β is a constant and the hydrostatic pressure effect on density is still ignored.

These three conditions, *A, B,* and *C,* for a jet of 1 m initial diameter, amount to ambient stratifications of 4×10^{-4}, 4×10^{-3}, and 4×10^{-2} °C/m, respectively. The effects of even small stratification are large, because of the rapid dilution arising from entrainment. For the least stratification, A, the jets follow a trajectory similar to that in the unstratified ambient, with only a slight vertical restriction of the trajectory, for all three values of F. For B, at F = 50, a noticeable change in trend has arisen near the end of the trajectory. A rebending of the path has arisen. The point of inflection in this curve is very significant. In the calculations this was the vertical level at which the jet had become neutrally buoyant, because of entrainment, on the one hand, and the decreasing density of the surrounding fluid, on the other. Further upward rise beyond this point is due solely to the momentum then existing in the jet. This momentum is gradually decreased by the downward force of negative buoyancy.

Stratification C, at F = 50, shows a further effect. The jet has become vertically "trapped," in the stronger stratification. The trajectory has undergone complete rebending, from horizontal flow with entrainment to no buoyancy. Usually, in entrainment modeling, computations are stopped at the point of maximum rise. It is assumed that all but the most vigorous jets will have acquired "far field" characteristics by this time.

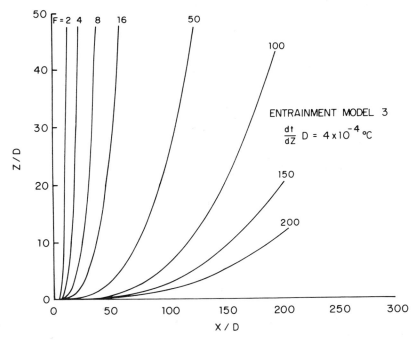

Fig. 27. Trajectories at a stratification of $(dt_a/dZ)D = 4 \times 10^{-4}$ °C, model 3.

The assumption of an axisymmetric jet shape, already uncertain in a strongly stratified ambient, is even more questionable beyond this point. Also, the initial momentum and buoyancy have likely become largely irrelevant to further diffusion of the effects, in most actual environments.

Trajectories are also shown in Fig. 26 for the same stratifications at higher discharge momentum, F = 100 and 200. The characteristics are similar to those at F = 50, except that the vertical range of rise is drastically reduced. This is because the increasing momentum, at increasing F, results in much higher entrainment and an attendant loss of positive buoyancy, per unit of jet mass, much sooner along the trajectory.

The intuitive reasoning that the trajectories of initially more buoyant, or low momentum, jets are less affected by stratification is supported by these calculations. Figures 27, 28, and 29 compare the trajectories of Froude number discharges in the range F = 2–200, in an ambient having the same three different stratification levels, A, B, and C. The jets at lower values of F are much less affected over the same three different stratification levels. Those at higher values of F show varying degrees of trajectory flattening, vertical trapping, and terminal rise. Given strong enough stratification in the ambient, of course, any initially buoyant jet eventually will

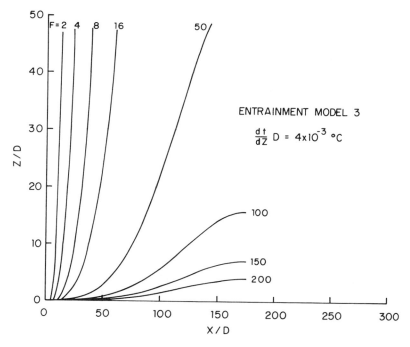

FIG. 28. Trajectories at a stratification of $(dt_a/dZ)D = 4 \times 10^{-3}$ °C, model 3.

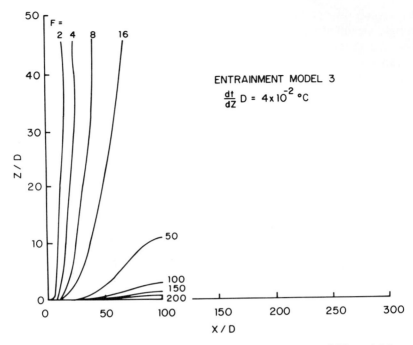

FIG. 29. Trajectories at a stratification of $(dt_a/dZ)D = 4 \times 10^{-2}$ °C, model 3.

experience negative buoyancy and terminate its rise at some elevation in a sufficiently extensive ambient. No vertical overshoot was found in any of the calculations.

The preceding method and results are typical of past jet modeling in a stratified ambient. Stratification, as indicated in the earlier discussion of equations of state, is usually assumed to be caused by a single gradient, either of temperature, salinity, or simple density. Rarely, as in Hirst [14], are temperature and salinity gradients treated together.

The capability of a comprehensive equation of state to describe actual water ambient conditions more realistically and the behavior of a jet within it is illustrated in the assembly of information in Figs. 30–35. For all these calculations the initial jet diameter and velocity are 4 m and 10 m/sec, respectively. Figure 30, as an example, characterizes the temperature, salinity, and density stratification of an area of the northern Pacific ocean, for the first 500 m of the water column. In June, temperature and salinity data for a point, 53° 04′ N, 175° 35′ W, were taken from Barstow et al. [20], as and approximated by a sequence of five temperature and

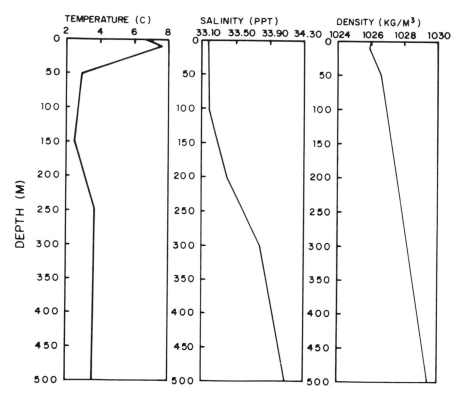

FIG. 30. Temperature, salinity, and density for the northern Pacific ocean, 53° 04' N, 175° 35' W, June, 1966.

four salinity gradients, respectively. The Gebhart–Mollendorf [19] relation was then used to establish the corresponding density field shown, which also includes the effect of hydrostatic pressure on density.

In Fig. 31 this stratified density variation is repeated. Also shown are the trajectories of five horizontal buoyant momentum discharges into that quiescent ambient, at 200 m depth. The jets are assumed to have an initial salinity equal to that of the ocean at 200 m depth, $s_{0a} = 33.42‰ = s_0$ but a higher temperature than $t_{0a} = 3.08°C$. Five values of t_0 were chosen to result in the values F = 10, 20, 30, 40, and 50. However, as the jet rises, the net buoyancy is from a combination of temperature and salinity differences. The buoyancy force along the trajectory is calculated from the equation of state in Gebhart and Mollendorf [19], based on the local values of jet and ambient temperature, salinity, and pressure.

All five of these jets experience the progression from positive to nega-

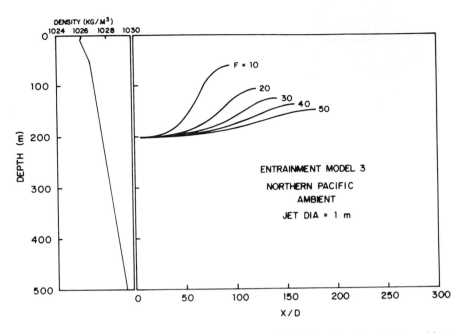

FIG. 31. Trajectory of jets at F = 10, 20, 30, 40, and 50, of 1-m diameter, discharged into the stratified ambient pictured in Fig. 30, model 3.

tive buoyancy, as well as a level of maximum upward penetration. All have a rather smooth curvature and recurvature, because of the relatively constant density gradient encountered in the vertical range of penetration. Again, no overshoot occurs.

Figure 32 illustrates another characteristic oceanic temperature–salinity–density field, from the tropical Atlantic at a point 02° 03′ S, 39° 20′ W. The data for these distributions were taken February 1972 and are from Neumann and Pierson [21]. Again, five jets—F = 1, 20, 30, 40, and 50—are assumed to be discharged into this density field at a depth of 200 m, where $s_{0a} = 35.10‰$ and $t_{0a} = 15.05°C$. Their trajectories are shown in Fig. 33. The vertical penetration of these jets is significantly less than that calculated for the Northern Pacific density profile, because of the much stronger stable density gradient near the surface in the tropical ocean.

An October 1979 temperature–salinity–density field for the Arctic Ocean at 81° 28′ N, 8° 05′ E is shown in Fig. 34. Data are from Johannessen et al. [22]. The strong salinity decrease near the surface is due to melt from the floating Arctic ice pack. The behavior of jet discharges into this ambient, where $s_{0a} = 34.75‰$ and $t_{0a} = 1.91°C$, as shown in Fig. 35, is

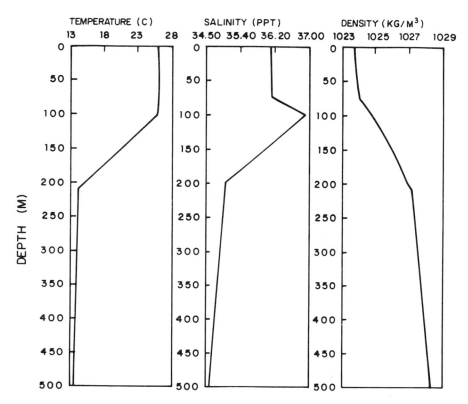

FIG. 32. Temperature, salinity, and density profiles for the tropical Atlantic ocean, 02°03'S, 39°20'W, February, 1927.

especially interesting, compared to Figs. 31 and 33. The recurvature rates encountered vary considerably. For F = 30, the jets follow a smooth curvature and recurvature. The F = 10 and F = 20 jets, however, show a much sharper recurvature. This arises because they encounter the steeper density gradient above the 100 m depth.

These examples illustrate the advantage of using a full equation of state and actual stratifications in environmental entrainment modeling. The dependence of the model on assumed values of β and γ is eliminated, as is the requirement to express ambient stratification in terms of constant temperature, salinity, or density gradients. The accuracy of the ambient density characterization is limited only by the availability and spacing of the field data. Ultimately, the use of a full equation of state can reduce one of the primary uncertainties in entrainment modeling—that of evaluating the temperature–pressure–density relationship.

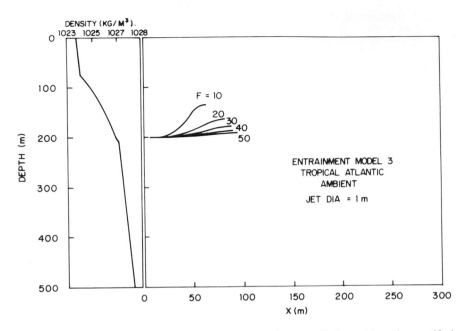

Fig. 33. Trajectory of jets at $F = 10$–50, of 1-m diameter, discharged into the stratified ambient pictured in Fig. 32, model 3.

IX. Summary and Conclusions

In reviewing the progress to date in entrainment modeling of submerged buoyant momentum jets, some rather striking matters have become apparent. The first of these is the very small physical scale and very small number of experimental measurements of jet trajectory and decay that have been made. This small and restricted data base is all that underlies the development of all jet modeling schemes. The discharge diameters were mostly about 1 cm, the largest being 3.9 cm. The entire measured jet trajectories in quiescent ambients were, at most, only a few meters in length. Longer lengths were used in some of the experiments done in flowing ambients. However, even these studies used similarly small-diameter discharges. These limitations, of course, arose, at least in part, through the common limitation in size that applies to experimental work with small resources.

This data base and the resulting models are often used to approximate the behavior of real jets, such as power plant discharges and sewage outfalls. Such flows are commonly hundreds to thousands of times larger. Such upscaling in size, from the information base to the applications,

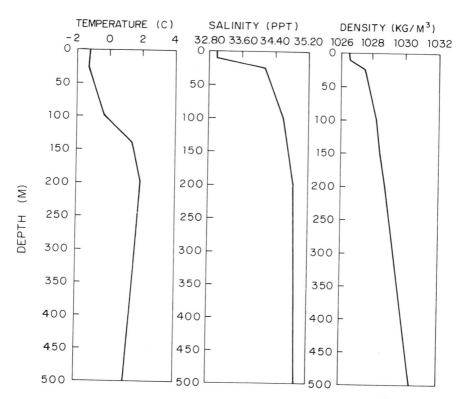

FIG. 34. Temperature, salinity, and density profiles for the Arctic ocean, 81° 28′ N, 8° 05′ E, October, 1979.

implies a faith in the similarity properties of the densimetric Froude number. Whether or not this is justified is not known, because of a lack of adequate physical data. Further, such a large-scale difference between the data and the modeling bases and the real system projection, in the case of an underwater momentum jet, does not necessarily assure similar scaling in turbulence within the jet and in the immediately surrounding ambient. Fan [2] found that the effect of ambient turbulence on jet behavior was profound. It is not unreasonable to expect that the scale of turbulence within the jet, or the relative scale between ambient turbulence and jet dimensions, may have equally important effects.

The second striking aspect of experimental and analytical work is the relatively low densimetric Froude number range used. Admittedly, more buoyant, lower Froude number jets are more "interesting" in terms of possible trajectories. They may also be of more importance in numerous environmental modeling situations. Froude numbers in the range of 10–40

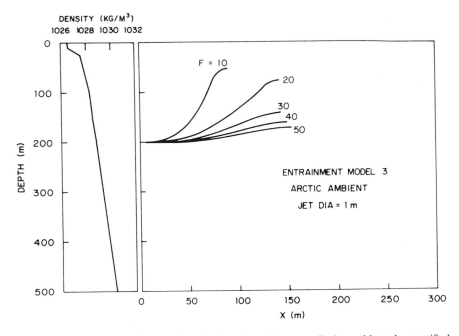

FIG. 35. Trajectories of jets at F = 10–50, of 1-m diameter, discharged into the stratified ambient pictured in Fig. 34, model 3.

are common in the literature. Values as high as 100 are rare. There are many realistic circumstances, however, in which entrainment modeling might be used, where high Froude numbers apply. Many thermal and concentration wake and conduit discharges result in very high values of the Froude number.

A third limitation in the present state of entrainment modeling is the inconsistent predictive quality of the collection of available flowing ambient entrainment functions in their suggested ranges of applicability. A fourth shortcoming is the range of R in which data are available to guide modeling. Very few experimental data exist for R greater than 0.25. Almost no data exist for an initially coflowing ambient. The data are not sufficient to select definitively either a general entrainment function or even one valid over a restricted range of F and R.

Given the limitations of the current background data to support modeling, a high level of confidence in predicting large-scale jet trajectories, involving conditions of ambient flow, turbulence, or high Froude number discharges, is unwarranted. There are simply too many unknowns to proceed confidently, because of current limits in both methods and background information.

X. Review of Other Studies

Ero [23] employed an entrainment function that varies with the local densimetric Froude number to calculate the characteristics of buoyant plumes in an atmospheric inversion. Morton [24] extended his earlier work (Morton *et al.* [5]) to investigate the general behavior of forced plumes in both uniform and stably stratified environments. In a series of papers Madni and Pletcher [25–27] and Hwang and Pletcher [28] used a finite difference method to analyze the behavior of turbulent buoyant jets in a variety of ambients, including quiescent, coflowing, and cross flowing, both with and without stratification. Savage and Chan [29] studied the behavior of a hot, vertical, laminar jet flowing into a quiescent cold ambient, by means of a perturbation series expansion as well as integral methods using hyperbolic secant distributions for velocity and temperature. Yih and Wu [30] analyzed the behavior of round buoyant plumes. For laminar flow they employed a perturbation expansion about the exact solutions for $Pr = 1$ and $Pr = 2$. An eddy viscosity formulation was used for the turbulent case to obtain exact solutions for turbulent Prandtl numbers of 1.1 and 2, and approximate solutions were obtained based on these exact solutions. Sforza and Mons [31] used Reichardt's turbulence hypothesis to analyze the mixing of various gas jets with quiescent ambient air. They also obtained measurements of the jet properties. Experiments with a downward-directed saltwater jet issuing into fresh water were performed by Wright [32]. In these experiments the jet was also towed to investigate the effects of a cross flow. Chu and Goldberg [33] used a simple one-constant entrainment model to investigate the behavior of buoyant jets and plumes in a cross-flowing ambient. They also performed a series of experiments using very small jets (of the order of 1.8 mm diameter) injected vertically downward. Chen and Nikitopoulos [34] used a k–ε turbulence model to analyze the behavior of buoyant jets in a quiescent ambient.

XI. Appendix A

Dimensionless Variables

The following dimensionless quantities are used in the nondimensionalization of the governing equations:

$$u = U/U_0 \qquad \Delta t_a / \Delta t_0 = (t_{a2} - t_{a1})/(t_0 - t_{a0})$$

$$u_m = U_m/U_0 \qquad \Delta c_a / \Delta c_0 = (c_{a2} - c_{a1})/(c_0 - c_{a0})$$

where the subscripts 1 and 2 represent adjacent points of integration downstream in the jet. Also,

$$b = B/D \qquad \Delta t_m/\Delta t_0 = (t_m - t_a)/(t_0 - t_{a0})$$

$$s = S/D \qquad \Delta c_m/\Delta c_0 = (c_m - c_a)/(c_0 - c_{a0})$$

$$x = X/D \qquad R = U_a/U_0$$

$$z = Z/D$$

XII. Appendix B

The Gebhart–Mollendorf [19] relation for the density of saline water is of the form

$$\rho(t, s, p) = \rho_m(s, p)[1 - \alpha(s, p)|t - t_m(s, p)|^{q(s,p)}]$$

where t is temperature (°C), s is salinity (‰), p is pressure (bars absolute), $\rho_m(s, p)$ is the density extremum at the given values of s and p, $t_m(s, p)$ is the temperature corresponding to the density extremum for the same s and p values, $\alpha(s, p)$ is a temperature term coefficient, and $q(s, p)$ is a temperature term exponent. These values are in turn given by

$$\rho_m(s, p) = \rho_m(0, 1)[1 + f_1(p) + sg_1(p) + s^2h_1(p)]$$

$$\alpha(s, p) = \alpha(0, 1)[1 + f_2(p) + sg_2(p) + s^2h_2(p)]$$

$$t_m(s, p) = t_m(0, 1)[1 + f_3(p) + sg_3(p) + s^2h_3(p)]$$

$$q(s, p) = q(0, 1)[1 + f_4(p) + sg_4(p) + s^2h_4(p)]$$

where

$$f_i(p) = \sum_{j=1}^{3} f_{ij}(p - 1)^j, \qquad g_i(p) = \sum_{j=0}^{3} g_{ij}(p - 1)^j$$

$$h_i(p) = \sum_{j=0}^{3} h_{ij}(p - 1)^j$$

$$\rho_m(0, 1) = 999.972 \text{ kg/m}^3$$

$$t_m(0, 1) = 4.029325°C$$

$$\alpha(0, 1) = 9.297173 \times 10^{-6} \text{ °C}^{-q}$$

$$q(0, 1) = 1.894816$$

and the values of f_{ij}, g_{ij}, and h_{ij} are given in tabular form in Table B.I.

TABLE B.I

VALUES OF f_{ij}, g_{ij}, AND h_{ij} IN THE GEBHART–MOLLENDORF RELATION FOR DENSITY OF WATER

Pressure function	j			
	0	1	2	3
f_{1j}	—	4.960998E-05	-2.601973E-09	7.842619E-13
f_{2j}	—	1.377584E-04	1.497648E-06	2.903240E-10
f_{3j}	—	-5.430000E-03	7.720181E-07	-7.038846E-10
f_{4j}	—	-1.118758E-04	-1.238393E-07	5.857253E-11
g_{1j}	7.992252E-04	-5.194896E-08	1.031185E-10	-2.979653E-14
g_{2j}	1.623355E-02	1.129961E-05	-8.053248E-08	6.966452E-12
g_{3j}	-5.265509E-02	7.496781E-05	-2.792053E-07	1.411138E-10
g_{4j}	-3.136530E-03	2.983937E-06	4.453557E-09	-2.937601E-12
h_{1j}	1.918334E-07	1.347190E-09	-2.203133E-12	1.112440E-15
h_{2j}	-4.565866E-04	-4.352912E-07	1.978675E-09	-9.079379E-13
h_{3j}	0.000000	-3.683650E-06	7.694077E-09	-4.561113E-12
h_{4j}	7.599378E-05	-8.718915E-08	-4.166970E-11	5.870105E-14

The relation is fitted in the range of temperature to 20°C, pressure to 1000 bars absolute, and salinity to 40‰ (i.e., 40 parts per thousand).

Comparison of the density predicted by this relation with the data of Chen and Millero [35] shows that the overall rms difference between the relation and this data, for pure and saline water at all temperatures and pressures considered, is reported to be 9.0 ppm.

NOMENCLATURE

B characteristic jet width

B_e value of B at S_e, $B_c = D/\sqrt{2}$

b dimensionless jet width, B/D

c concentration

D jet discharge diameter

$D(S)$ local jet diameter, at S/D

E local volumetric entrainment rate

F densimetric Froude number, $U_0/\sqrt{gD(\rho_a - \rho_0)/\rho_0}$

F_L local densimetric Froude number, $U_m^2/gB(\rho_a - \rho_m)/\rho_0$

g gravitational acceleration

Pr Prandtl number

p pressure

Q jet mass flow rate

R ambient flow ratio, U_a/U_0

r radial jet coordinate

S, s streamwise coordinate of jet velocity, $s = S/D$

s salinity, ‰ (parts per thousand)

t temperature

U streamwise jet velocity

U_0 jet discharge velocity

U^* relative local velocity, $U_m - U_a \cos \theta$

u dimensionless streamwise jet velocity, U/U_0

X, x horizontal Cartesian coordinate, $x = X/D$

Z, z vertical Cartesian coordinate, $z = Z/D$

Greek Symbols

α	entrainment coefficient	θ	local angle of inclination from hori-
β	volumetric coefficient of thermal		zontal
	expansion	λ	relative spreading ratio
γ	volumetric coefficient of concentra-	ρ	density
	tion expansion	ϕ	azimuthal jet angle
$\Delta(\ )$	$(\ )_i - (\ )_j$, along the trajectory		

Subscripts

a	ambient	0	jet discharge
c	concentration	p	pressure
e	beginning of zone of established	r	reference value
	flow	t	temperature
m	jet centerline		

Acknowledgments

The authors wish to recognize here the support of their institutions in this effort, the support of the Naval Sea Systems Command in the initial study, the support of NSF Grant MEA 8200613 (for the first author), and Ernest Chiu for his diligent and very helpful work in the final stages of manuscript completion.

References

1. M. L. Albertson, Y. B. Dai, R. A. Jensen, and H. Rouse, Diffusion of submerged jets. *Trans. Am. Soc. Civ. Eng.* **115,** 639–697 (1950).
2. L. H. Fan, "Turbulent Buoyant Jets into Stratified and Flowing Ambient Fluids," Rep. No. KH-R-15. W. M. Keck Lab., Calif. Inst. Technol., Pasadena, 1967.
3. D. P. Hoult, J. A. Fay, and L. J. Forney, A theory of plume rise compared with field observations. *J. Air Pollut. Control Assoc.* **19,** 585–590 (1969).
4. M. A. Shirazi, R. S. McQuivey, and T. N. Keefer, Heated water jet in coflowing turbulent stream. *J. Hydraul. Div., Am. Soc. Civ. Eng.* **100,** 919–934 (1974).
5. B. R. Morton, G. I. Taylor, and J. S. Turner, Turbulent gravitational convection from maintained and instantaneous sources. *Proc. R. Soc. London, Ser. A* **234,** 1–23 (1956).
6. G. I. Taylor, "Dynamics of a Mass of Hot Gas Rising in Air," MDDC 919, LADC 276, U.S. At. Energy Comm., Oak Ridge, Tennessee, 1945.
7. G. Abraham, Horizontal jets in stagnant fluid of other density. *J. Hydraul. Div., Am. Soc. Civ. Eng.* **9,** 139–154 (1965).
8. E. J. List and J. Imberger, Turbulent entrainment in buoyant jets and plumes. *J. Hydraul. Div., Am. Soc. Civ. Eng.* **99,** 1461–1474 (1973).
9. L. H. Fan and N. H. Brooks, "Numerical Solutions to Turbulent Buoyant Jet Problems," Rep. No. KH-R-18. W. M. Keck, Lab., Calif. Inst. Technol., Pasadena, 1969.
10. D. G. Fox, Forced plume in a stratified fluid. *J. Geophys. Res.* **75,** 6818–6835 (1970).
11. E. A. Hirst, Buoyant jets discharged to quiescent stratified ambients. *J. Geophys. Res.* **76,** 7375–7384 (1971).

12. J. B. Riester, R. A. Bajura, and S. H. Schwartz, Effects of water temperature and salt concentration on the characteristics of horizontal buoyant submerged jets. *J. Heat Transfer* **102,** 557–562 (1980).

13. L. R. Davis, M. A. Shirazi, and D. L. Slegel, Measurement of buoyant jet entrainment from single and multiple sources. *J. Heat Transfer* **100,** 442–447 (1978).

14. E. A. Hirst, Analysis of round, turbulent, buoyant jets discharged to flowing stratified ambients. *Oak Ridge Nat. Lab. [Rep.] ORNL (U.S.)* **ORNL-4685** (1971).

15. T. Ginsberg and M. Ades, A correlation for the entrainment function for near-field thermal plume analysis. *Trans. Am. Nucl. Soc.* **21,** 87–88 (1975).

16. M. Schatzmann, An integral model of plume rise. *Atmos. Environ.* **13,** 721–731 (1979).

17. R. J. Pryputniewicz and W. W. Bowley, An experimental study of vertical buoyant jets discharged into water of finite depth. *J. Heat Transfer* **97,** 274–281 (1975).

18. E. A. Hirst, Analysis of buoyant jets within the zone of flow establishment. *Oak Ridge Nat. Lab. [Rep.] ORNL-TM (U.S.)* **ORNL-TM-3470** (1971).

19. B. Gebhart and J. C. Mollendorf, A new density relation for pure and saline water. *Deep-Sea Res.* **24,** 831–848 (1977).

20. D. Barstow, W. Gilbert, K. Park, R. Still, and X. Wyatt, "Hydrographic Data from Oregon Waters 1966." Dep. Oceanogr., Oregon State Univ., Corvallis, 1966.

21. G. Neumann and W. J. Pierson, "Principles of Physical Oceanography." Prentice-Hall, Englewood Cliffs, New Jersey, 1966.

22. J. A. Johannessen *et al.* "A CTD-Data Report from the Norsex Marginal Ice Zone Program North of Svalbard in September–October 1979. Univ. Bergen, R. Norw. Counc. Sci. Ind. Res., Bergen, Norway, 1980.

23. M. I. O. Ero, Entrainment characteristics of buoyant axisymmetric plumes in atmospheric inversions. *J. Heat Transfer* **99,** 335–338 (1977).

24. B. R. Morton, Forced plumes. *J. Fluid Mech.* **5,** 151–163 (1959).

25. I. K. Madni and R. H. Pletcher, Prediction of jets in coflowing and quiescent ambients. *J. Fluids Eng.* **97,** 558–567 (1975).

26. I. K. Madni and R. H. Pletcher, Prediction of turbulent forced plumes issuing vertically into stratified or uniform ambients. *J. Heat Transfer* **99,** 99–104 (1977).

27. I. K. Madni and R. H. Pletcher, Buoyant jets discharging nonvertically into a uniform, quiescent ambient—A finite-difference analysis and turbulence modeling. *J. Heat Transfer* **99,** 641–647 (1977).

28. S. S. Hwang and R. H. Pletcher, Prediction of buoyant turbulent jets and plumes in a cross flow. *Heat Transfer, Int. Heat Transfer Conf., 6th, Toronto* **1,** 109–114 (1978).

29. S. B. Savage and G. K. C. Chan, The buoyant two-dimensional laminar vertical jet. *Q. J. Mech. Appl. Math.* **23,** 413–430 (1970).

30. C. S. Yih and F. Wu, Round buoyant laminar and turbulent plumes. *Phys. Fluids* **24,** 794–801 (1981).

31. P. M. Sforza and R. F. Mons, Mass, momentum, and energy transport in turbulent free jets. *Int. J. Heat Mass Transfer* **21,** 371–384 (1978).

32. S. J. Wright, Mean behavior of buoyant jets in a crossflow. *J. Hydraul. Div., Am. Soc. Civ. Eng.* **103,** 499–513 (1977).

33. V. H. Chu and M. B. Goldberg, Buoyant forced plumes in cross flow. *J. Hydraul. Div., Am. Soc. Civ. Eng.* **100,** 1203–1214 (1974).

34. C. J. Chen and C. P. Nikitopoulos, On the near field characteristics of axisymmetric turbulent buoyant jets in a uniform environment. *Int. J. Heat Mass Transfer* **22,** 245–254 (1979).

35. C. T. Chen and F. J. Millero, The specific volume of seawater at high pressures. *Deep-Sea Res.* **23,** 595–612 (1976).

Boiling of Multicomponent Liquid Mixtures

JOHN R. THOME

*Department of Mechanical Engineering, Michigan State University,
East Lansing, Michigan*

RICHARD A. W. SHOCK

*Heat Transfer and Fluid Flow Service, Atomic Energy Research Establishment Harwell,
Didcot, Oxfordshire, England*

I. Introduction

Boiling is a physical process of great practical significance and has been the subject of intensive research for the past several decades. Nuclear power vapor-generator design and safety and the sharp rise in energy costs have spurred many of the research initiatives. Most of the research efforts have been expended on the boiling characteristics of single-component liquids (e.g., water), but mixture boiling research has engineering relevance to the design of two-phase heat exchange equipment in the chemical and petrochemical processing industries, the refrigeration industry, the air separation industry, and the liquid natural gas (LNG) industry. In addition, it has applications to the pharmaceutical industry and to high heat flux cooling in the electronics industry.

Boiling of binary and multicomponent liquid mixtures, the subject of the state-of-the-art review, is markedly different from single-component boiling. (In this review we assume understanding of single-component boiling, which has been reviewed by Rohsenow [1]). Since the thermodynamics of vapor–liquid phase equilibrium of mixtures allows the vapor and liquid phases to be of differing compositions, the boiling of a liquid mixture is distinct from single-component boiling in that the driving force for heat transfer is in turn governed by mass transfer. Thus the evaporation rate can be severely retarded in the mixtures because the rate of mass diffusion is usually much slower than that of heat diffusion in the liquid phase.

The purpose of this article is to review the significant advances in the understanding of the boiling process germane to mixtures. The scope of this review is limited to miscible systems. The effects of impurities (such as lubricating oils in refrigerants), surfactants, soluble salts, and dissolved gases will be excluded from consideration. Both pool and convective boiling will be addressed.

The ultimate goals of the study of boiling of liquid mixtures are (1) to predict the superheat required for the inception of boiling, (2) to predict their heat transfer coefficients with a reasonable degree of accuracy, and (3) to predict the variation in the peak nucleate and dry out heat fluxes with composition. These goals are closer to being met for pool boiling than they are for convective boiling.

The topics discussed in this survey are presented in much the same order as when an increasing heat flux is applied to a surface to cause boiling. Thus bubble nucleation or boiling incipience is considered first, since this defines the criteria required for boiling to commence. Then the growth of the vapor bubbles from vapor nuclei up to and including their departure from the surface is discussed. This is followed by a survey of the heat transfer mechanisms, resulting from the vaporization process, which affect the nucleate pool boiling curve, and this leads to a discussion of equations for predicting nucleate pool boiling heat transfer coefficients. The peak nucleate heat flux in pool boiling is examined prior to a description of film boiling. Several aspects of convective boiling are then presented.

Most of the research done on mixture boiling has dealt with binary mixtures, since these are the easiest to study. Therefore the approach to the subject is to understand first binary mixture boiling and then to extend this knowledge to multicomponent boiling, noting that multicomponent mixtures are the rule rather than the exception in industrial processes.

II. Fundamentals of Vapor–Liquid Phase Equilibria

A working knowledge of the elementary principles of vapor–liquid phase equilibria is required for the understanding of mixture boiling. Only a brief summary of binary mixture phase equilibria is given here. Vapor–liquid phase equilibrium data for binary mixtures and several multicomponent mixtures have been compiled in several texts [2–4]. Their prediction is beyond the scope of the present review but is discussed by, for example, Prausnitz [5].

Phase equilibrium thermodynamics is the study of the relationship between the various pertinent physical properties (e.g., pressure, temperature, and composition) that exists when two or more phases (vapor, liquid, solid) in contact with one another are in a state of thermodynamic equilibrium. It is necessary to be able to predict the various properties in each phase knowing only a limited number of independent properties of the system. For example, given the saturation pressure and composition in the liquid phase for a vapor–liquid mixture system, the saturation temperature and the composition of the vapor phase are uniquely defined and determinable (at least in principle).

Phase equilibrium diagrams are used to visualize the relationship between temperature, pressure, and the compositions in the two phases. Figure 1 depicts the phase equilibrium diagram at constant pressure for an ideal binary mixture system. Saturation temperature is plotted on the vertical axis. Mole fractions of the more volatile component in the liquid and vapor phases are plotted on the horizontal axis. The more volatile component, sometimes also called the lighter component, is that with the lower boiling point at the pressure of interest. The dew point line denotes the variation in equilibrium vapor mole fraction with saturation temperature. The bubble point line depicts the functional dependency of the liquid mole fraction on the saturation temperature. It is evident from the diagram that $\tilde{y}_i \geq \tilde{x}_i$ for the more volatile component and $\tilde{y}_j \leq \tilde{x}_j$ for the less volatile component. This is expected intuitively since the more volatile component is above its normal boiling point while the reverse is true for the less volatile component.

Figure 2 illustrates a temperature–composition phase diagram for a binary mixture system forming an azeotrope. Such behavior is caused by significant differences in the intermolecular attractions between molecules of component i for one another, molecules of component j for one another, and molecules of i and j for each other. At the azeotrope the compositions of the liquid and vapor phases are identical. To the left side

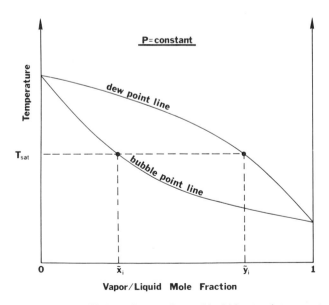

FIG. 1. Phase equilibrium diagram for an ideal binary mixture system.

of the azeotrope $\bar{y}_i \geq \bar{x}_i$ and to the right $\bar{y}_i \leq \bar{x}_i$. The slope of the bubble point line changes from negative to positive as the azeotrope is passed from left to right. However, the product $(\bar{y}_i - \bar{x}_i)(dT/d\bar{x}_i)$ is always negative as a consequence. The azeotrope behaves like a single-component liquid, since the compositions in both phases are the same.

Ideal systems are those that follow Raoult's law. This law states that the vapor mole fraction of any constitutive component i, $\bar{y}_i$ is directly proportional to its liquid mole fraction $\bar{x}_i$ via the ratio of its saturation vapor pressure at the same temperature p_i to the total system pressure p. In equation form this is written as

$$\bar{y}_i = \bar{x}_i(p_i/p) \tag{1}$$

Equation (1) is of practical use for predicting equilibrium in mixtures such as those in Fig. 1 but is of limited general use because of the assumptions from which it is derived (i.e., that the vapor and liquid phase solutions are assumed to be ideal solutions and the vapors are assumed to behave as ideal gases).

Nonideal mixtures are those that do not conform closely to Raoult's law. The methods for predicting equilibrium states in these types of mixtures become much more complex and the reader is referred to Prausnitz [5].

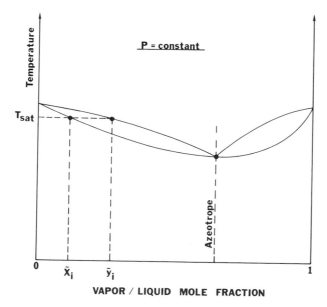

FIG. 2. Phase equilibrium diagram for an azeotropic mixture system.

III. Inception of Boiling

The inception of boiling on a heated surface is a fundamental aspect of the overall boiling process. Studies have shown [6–8] that vapor bubbles growing on a heated wall originate from small vapor nuclei trapped in pits and cracks on the order of 0.1–10 μm in diameter in the surface of the heated wall as shown in Fig. 3. The criteria for activation of these vapor nuclei are useful for the prediction of the wall superheat or heat flux required to initiate boiling. This, in turn, affects the number of boiling sites active in nucleate boiling.

A. EQUILIBRIUM SUPERHEAT

Figure 4 depicts the bubble nucleation model for a vapor nucleus at the mouth of a conical cavity in the surface of a heated wall. For mechanical equilibrium to exist, the pressure difference between the inside and out-side of the bubble must be balanced by the surface tension force that causes the bubble to adhere to the wall. A force balance gives the Laplace equation:

$$(p_g - p_l)\pi R^2 = 2\pi R\sigma \tag{2}$$

For thermal equilibrium to be satisfied, the saturation temperature of the vapor must equal that of the surrounding liquid. Since the vapor pressure inside the bubble is greater than that of the bulk, as given by Eq. (2), its saturation temperature must also be higher. Hence the bulk liquid must be uniformly superheated above its own saturation temperature to equal the temperature of the vapor for thermal equilibrium to be satisfied. Using the Clausius–Clapeyron equation,

$$\left(\frac{dp}{dT}\right)_{sat} = \frac{\Delta \bar{h}}{T_{sat}(\bar{v}_g - \bar{v}_l)} \tag{3a}$$

the principles of thermal and mechanical equilibrium, and the Kelvin equation to relate the vapor pressure inside a curved interface p_g to that

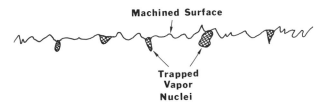

Machined Surface

Trapped Vapor Nuclei

FIG. 3. Vapor nuclei trapped in pits and cracks in a surface.

for the normal flat planar interface $p_{g\infty}$ (where $R = \infty$)

$$p_g/p_{g\infty} = \exp(2\sigma\tilde{v}_g/R\tilde{R}T_{sat})$$ (3b)

one can arrive at the equilibrium superheat equation

$$\Delta T_{sat} = \frac{2\sigma}{R(dp/dT)_{sat}}$$ (4)

The equation states that the superheat of the liquid required to maintain equilibrium is directly dependent on the surface tension and is inversely proportional to the cavity radius and the slope of the vapor pressure curve.

Haase [9] has shown that applying the Gibbs–Duhem equation to the liquid and vapor phases of a binary mixture system gives the slope of the vapor pressure curve as

$$\left(\frac{dp}{dT}\right)_{sat} = \frac{\Delta\tilde{s}}{\Delta\tilde{v}} + \left(\frac{\tilde{y} - \tilde{x}}{\Delta\tilde{v}}\right)\left(\frac{\partial^2\tilde{g}}{\partial\tilde{x}^2}\right)\frac{d\tilde{x}}{dT_{sat}}$$ (5a)

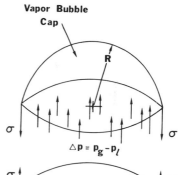

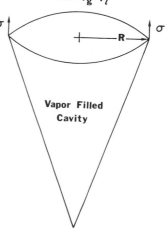

FIG. 4. Bubble nucleation model for a conical cavity.

where for components 1 and 2

$$\Delta \bar{s} = \bar{y}_1(\bar{s}_{g1} - \bar{s}_{l1}) + (1 - \bar{y}_1)(\bar{s}_{g2} - \bar{s}_{l2}) \tag{5b}$$

and

$$\Delta \bar{v} = \bar{y}_1(\bar{v}_{g1} - \bar{v}_{l1}) + (1 - \bar{y}_1)(\bar{v}_{g2} - \bar{v}_{l2}) \tag{5c}$$

This equation is equivalent to the extended Clapeyron equation derived by Stein [10] and Malesinsky [11] and given as

$$\left(\frac{dp}{dT}\right)_{sat} = \left(\frac{\partial p}{\partial T}\right)_{\bar{x}} + \frac{p(\bar{y} - \bar{x})}{\bar{R}T_{sat}} \left(\frac{\partial \bar{x}}{\partial T}\right)_p \left(\frac{\partial^2 \bar{g}}{\partial \bar{x}^2}\right)_{T,p} \tag{6}$$

which assumes ideal gas behavior of the vapor and $\bar{v}_g \gg \bar{v}_l$. However, when we consider the vapor–liquid equilibrium situation we are taking the situation before bubble growth and the change in $\bar{x}$ commence. Hence

$$\Delta \bar{x} = 0 \tag{7}$$

and Eq. (6) reduces to

$$(dp/dT)_{sat} = (\partial p/\partial T)_{\bar{x}} \tag{8}$$

Note here that there is an error in the work of Shock [12] in this respect. Shock considered the effect of small changes in composition on $(\partial p/\partial T)_{\bar{x}}$ in binary mixtures. He showed, for example, that addition of small amounts of ethanol to benzene and to water causes an increase in $(\partial p/\partial T)_{\bar{x}}$, which could be expected to cause an increase in the bubble population due to a decrease in ΔT_{sat} in Eq. (4), and hence in the heat transfer coefficient. This is against the trend of data.

Preusser [13] also has studied the effect of $(dp/dT)_{sat}$ on boiling inception in binary mixture systems. Defining the ideal slope of the vapor pressure curve for a binary mixture as

$$(dp/dT)_{sat_I} = \bar{x}_1(dp/dT)_{sat_1} + (1 - \bar{x}_1)(dp/dT)_{sat_2} \tag{9}$$

he then compared the ideal value to the actual value. Figure 5 illustrates the ratio of the actual to the ideal values for five mixture systems at 1.0 bar. The deviation from ideality is shown by Preusser to be insignificant other than for the acetone–water system.

B. Work of Bubble Formation

The work required for reversible isothermal–isobaric vapor bubble formation in a binary mixture system has been studied by Stephan and Korner [14, 15] and approaches the nucleation problem from a point of view different from that in Section III,A. Grigor'ev [16] earlier looked at this problem but incorrectly [14] used the integral latent heat instead of

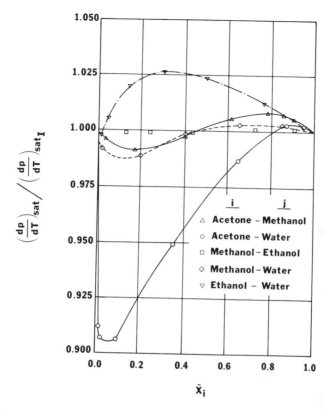

FIG. 5. Variation in the slope of the saturation curve from ideality for several mixture systems by Preusser [13].

the differential latent heat (the former quantity being the difference between the molar enthalpy of the vapor and the molar enthalpy of the liquid with the same composition as the vapor instead of the equilibrium composition, which applies to the latter). The Gibbs potential $\Delta \tilde{g}$, which is interpreted as the work of formation of a vapor nucleus in an infinite liquid phase, has been derived for single-component liquids or azeotropic mixtures to be

$$\Delta \tilde{g} = \frac{16\pi\sigma^3 \tilde{v}_g^2}{3 \, \Delta T_{\text{sat}}^2 \{\Delta \bar{h}/T_{\text{sat}}\}^2} \tag{10}$$

Extending the analysis to binary mixtures using the Gibbs–Duhem equation, Stephan and Korner arrived at

$$\Delta \tilde{g} = \frac{16\pi\sigma^3 \tilde{v}_g^2}{3 \, \Delta T_{\text{sat}}^2 \{\Delta \bar{h}/T_{\text{sat}} + [(\tilde{y} - \tilde{x})(\partial^2 \tilde{g}/\partial \tilde{x}^2)(\Delta \tilde{x}/\Delta T_{\text{sat}})]\}^2} \tag{11}$$

Later, Stephan and Preusser [17] extended Eq. (11) to the general case of multicomponent mixtures and obtained the following equation:

$$\Delta\tilde{g} = \frac{16\pi\sigma^3\tilde{v}_g^2}{3\,\Delta T_{sat}^2\left\{\dfrac{\Delta\tilde{h}}{T_{sat}} + \left(\dfrac{\Delta\tilde{x}/\Delta T_{sat}}{\tilde{y}_n - \tilde{x}_n}\right)\displaystyle\sum_{i=1}^{n-1}\sum_{j=1}^{n-1}\left(\dfrac{\partial^2\tilde{g}}{\partial\tilde{x}_i\,\partial\tilde{x}_j}\right)(\tilde{y}_i - \tilde{x}_i)(\tilde{y}_j - \tilde{x}_j)\right\}^2}$$

(12)

Then, using Eq. (7) and Konovalov's first rule [18] (which states that at constant pressure the boiling point rises with the addition of the component whose composition is lower in the vapor than in the liquid), Stephan and co-workers showed that the additional term in Eqs. (11) and (12) compared to Eq. (10) is always negative and thus the work of formation of a vapor bubble in an ideal binary mixture is greater than that for an equivalent pure fluid for which this additional term is zero. Hence their conclusion is equivalent to that of Shock in Section III,A.

C. Effect of Additional Physical Properties

The functional dependency of ΔT_{sat} and $\Delta\tilde{g}$ on composition in binary mixtures is controlled by the variation in the pertinent physical proper-

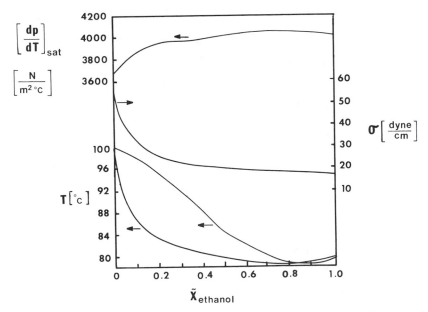

Fig. 6. Phase equilibrium diagram, surface tension, and slope of saturation curve for ethanol–water by Shock [12].

ties. Shock [12] has studied the effects of σ, $(dp/dT)_{sat}$, and contact angle β on the equilibrium superheat for ethanol–water and ethanol–benzene mixtures at 1 bar. Figure 6 shows the temperature–composition diagram and the variations in σ and $(dp/dT)_{sat}$ for the ethanol–water system. Figure 7 depicts the same for ethanol–benzene mixtures. From these figures Shock concluded that $(dp/dT)_{sat}$ will have only a small influence on ΔT_{sat} and surface tension will be important in the ethanol–water mixture because of its large nonlinear variation with composition. He then noted that the nucleation radius R in Eq. (4) is very susceptible to changes in the contact angle of the liquid–vapor interface with the solid surface. Hence

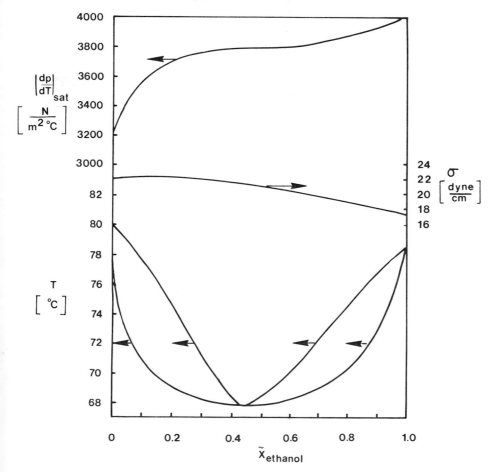

FIG. 7. Phase equilibrium diagram, surface tension, and slope of saturation curve for ethanol–benzene by Shock [12].

R may change dramatically with composition for aqueous mixtures that display a large decrease in contact angle with addition of an organic liquid to water. Figure 8 illustrates this point for two aqueous systems studied by Ponter *et al.* [19] and Fig. 9 shows later results for the ethanol–water system tested by Eddington and Kenning [20]. As the contact angle decreases, the amount of vapor trapped in cavities in the heated surface diminishes. This results in smaller nucleation radii, which require larger values of ΔT_{sat} to activate. Shock concluded that the activation of boiling in aqueous mixtures is dominated by the contact angle effect, and hence a large variation in boiling site density with composition may exist.

D. Experimental Studies on Boiling Inception

A number of researchers have measured the effect of composition on the wall superheat required for the initiation of boiling. Several of the studies are on the onset of nucleate boiling in convective boiling and the others are on the activation of the first boiling site in nucleate pool boiling.

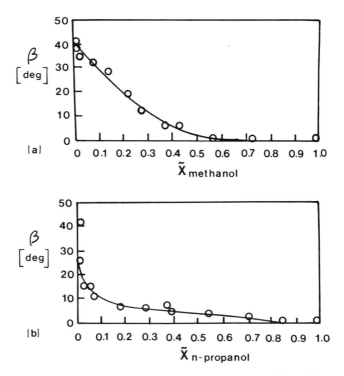

Fig. 8. Equilibrium contact angles on graphite by Ponter *et al.* [19]. (a) Methanol–water; (b) *n*-propanol–water.

TABLE I

SHOCK [21] DATA FOR THE ONSET OF NUCLEATE
BOILING IN ETHANOL–WATER MIXTURES

$\bar{x}$ (ethanol)	ΔT_{sat} (°C)	R (μm)	p (bar)
0.0	9.9	1.05	2.61
0.058	26.6	0.23	2.59
0.197	36.7	0.095	2.47

Shock [21] conducted experiments for forced convective boiling of eth-anol–water mixtures at about 2.5 bar with a 26 mm i.d. electrically heated copper tube. Table I lists his results for the onset of nucleate boiling at three compositions obtained by observing changes in the axial tempera-ture profiles. The nearly fourfold increase in ΔT_{sat} for $\bar{x} = 0.197$ compared to pure water supported his postulate that the contact angle is the most significant physical parameter governing ΔT_{sat} for aqueous mixtures since $(dp/dT)_{sat}$ has only a negligible change and σ suggests that ΔT_{sat} in the mixtures should be lower than that for pure water.

Similar experiments with a 26 mm i.d. tube were conducted by Toral

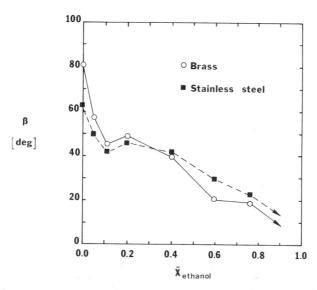

FIG. 9. Advancing contact angle for ethanol–water mixtures against nitrogen at 25°C by Eddington and Kenning [20].

TABLE II

Toral [22] Data for the Onset of Nucleate
Boiling in Ethanol–Cyclohexane Mixtures

$\bar{x}$ (ethanol)	ΔT_{sat} (°C)
0.12	23
0.54	11
0.82	15

[22] for ethanol–cyclohexane mixtures at 2.6 bar. Table II notes his values at the azeotrope ($\bar{x} = 0.54$) and at the compositions where $|\bar{y} - \bar{x}|$ has maxima ($\bar{x} = 0.12$ and 0.82). The equilibrium superheat is seen to be higher at compositions to either side of the azeotrope. It was concluded in Toral *et al.* [23] that the physical properties of the mixture and the mass transfer resistance, which is related to $|\bar{y} - \bar{x}|$, are important factors in the problem.

Thome *et al.* [24] obtained the incipient superheats for pool boiling of nitrogen–argon mixtures for the activation of apparently the same boiling site at 14 compositions in the range 0.04–1.0 mole fraction of nitrogen at 1.1 bar. Figure 10a demonstrates that no effect of composition occurred here, where care was taken not to prepressurize the system. (Prepressurization is known to cause partial or complete condensation of the trapped vapor nuclei, resulting in higher incipient superheats.) On lowering the heat flux, this site deactivated at the same wall superheat as it previously activated, as shown in Fig. 10b. The variation in the contact angle with composition for this mixture system is not known, but the values of the contact angle are very small. The solid curve in Fig. 10a represents Eq. (4) for a nucleation radius of 1 μm.

Thome and co-workers [24] have also obtained the activation and deactivation superheats over the whole composition range for ethanol–water mixtures at 1.01 bar. For a 25.4 mm diameter flat-disk heated surface facing upward without prepressurization, a maximum in the activation superheat was found at a mole fraction of 0.5 (see Fig. 11a). No explanation for the fact that there is a maximum is yet forthcoming, but the maximum does not coincide with the maximum in $|\bar{y} - \bar{x}|$. The solid curve in Fig. 11a is Eq. (4) evaluated at a constant value of R in micrometers. For instance, assuming $R = 1$ μm predicts incipient superheats of 33 and 8°C for water and ethanol, respectively, and assuming $R = 2$ μm gives incipient superheats of 16.5 and 4°C. Thus the actual activation superheat for pure ethanol is about three times that of pure water, instead of one-fourth, as predicted by Eq. (4), neglecting the effect of the wetting charac-

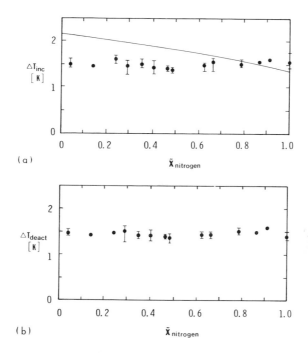

(a)

$\bar{X}_{nitrogen}$

(b)

$\bar{X}_{nitrogen}$

FIG. 10. Boiling incipience data of Thome *et al.* [24] for nitrogen–argon mixtures. (a) Activation superheats; (b) deactivation superheats.

teristics on R. The trend in the results at low ethanol compositions is similar to those of Shock in Table I for convective boiling.

The deactivation superheat of the last active site when the heat flux was lowered was also obtained [24]. These results are plotted in Fig. 11b. There is a significant variation with composition, the mixtures demonstrating higher values than the single components and the azeotrope. Two maxima are evident, neither matching the composition of the maximum in the activation superheat. However, the minimum at $\bar{x} = 0.5$ in the deactivation superheat matches the maximum in the activation superheat.

A very large hysteresis exists in Fig. 11 between the superheats at which the first bubble activates and the last bubble deactivates, except for pure water, which hardly wets. (It is unlikely that the first boiling site to activate is also the last to deactivate.) For pure water individual sites were noted to activate as the heat flux was increased. Deactivation occurred at essentially the same superheat. However, as the mole fraction of ethanol increased, activation of the first bubble almost instantaneously activated the whole boiling surface, with a high boiling site density resulting and a sharp drop occurring in the wall temperature. Thus a vapor seeding phe-

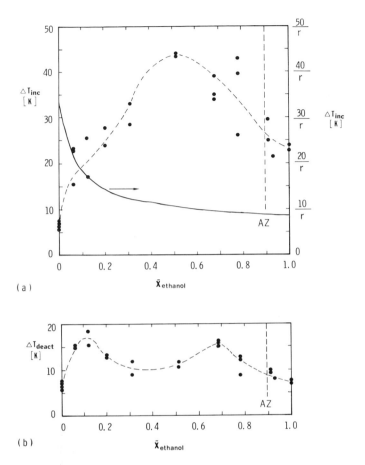

Fig. 11. Boiling incipience data of Thome *et al.* [24] for ethanol–water mixtures. (a) Activation superheats; (b) deactivation superheats.

nomenon must activate neighboring sites when a high activation super-heat is required to activate the first boiling site, and this results in a very high boiling site density.

In further experiments using prepressurization, Mercier [25] again mea-sured the activation superheats for nitrogen–argon mixtures. His new results are shown in Fig. 12. A maximum in the activation superheat is evident with prepressurization. This may be due to mass transfer effects in the mixtures, since the variation in the incipient superheat with compo-sition is similar to that of $|\bar{y} - \bar{x}|$.

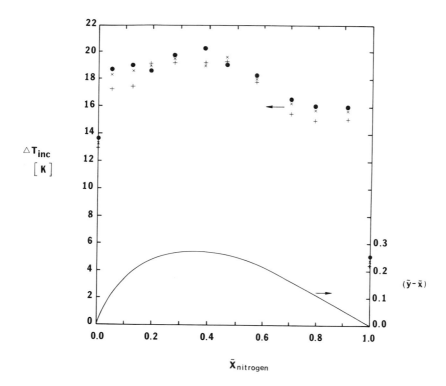

Fig. 12. The effect of prepressurization on boiling activation superheats by Mercier [25] for nitrogen–argon mixtures at 1.55 bar (legend: +, 0.2 bar; ×, 0.4 bar; ●, 0.7 bar of prepressurization).

E. Experimental Studies on Boiling Site Densities

Boiling site density is one of the most difficult boiling parameters to measure in the laboratory. Consequently, few studies have attempted to determine the variation in boiling site density with composition.

Van Stralen and Cole [26] were the first to report boiling site densities for binary mixtures. By using a horizontal thin wire of 0.2 mm diameter, the problem of one bubble obscuring the view of another bubble was essentially eliminated, but at the cost of utilizing a surface that is very uncharacteristic of a real boiling surface (e.g., a tube). Their tests covered a number of aqueous systems. Figure 13 depicts the results for water, methyl ethyl ketone (MEK), and one mixture (4.1 wt. % methyl ethyl ketone). At a heat flux of 0.3 MW/m², for instance, the number of boiling sites per square centimeter in pure water is 30 and in MEK over 200, but

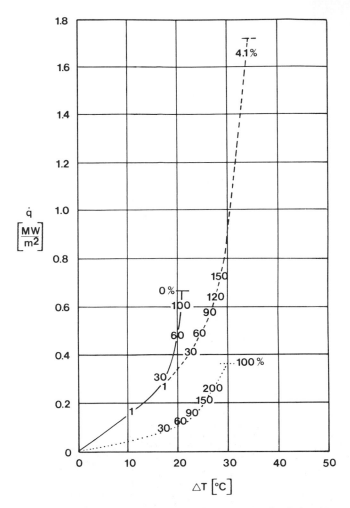

Fig. 13. Pool boiling curves with boiling site densities (sites/cm²) for the water–MEK system at 1.0 bar by Van Stralen and Cole [26].

for the 4.1% mixture only one site per unit area is active. Thus the variation in the boiling site density at constant heat flux shows a marked minimum. Looking at the data from the perspective of superheat, which is indicative of Eq. (4), at a constant wall superheat of 20 K there are about 60, 50, and 10 sites/cm² active for water, MEK, and 4.1% MEK, respectively. Again a minimum in the mixtures is evident. This is interesting because the 4.1% MEK mixture has physical properties similar to pure

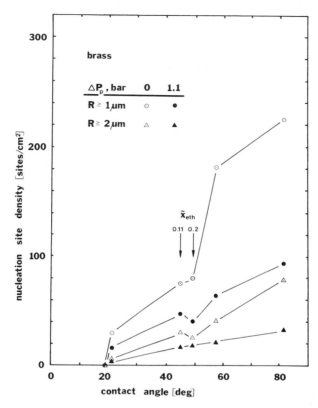

FIG. 14. Nucleation site densities for ethanol–water mixtures, with and without pre-pressurization, by Eddington and Kenning [20] for radii in two size ranges.

MEK but behaves very differently. This may be caused by the mass diffusion effect noted earlier by Toral *et al.* [23].

Eddington and Kenning [20] have studied the effect of contact angle on the bubble nucleation density in ethanol–water mixtures using a gas diffusion method. The bubble nucleation densities on two metallic surfaces (copper and brass) were obtained by the nucleation of gas bubbles from solutions supersaturated with nitrogen. The variation in nucleation density with advancing contact angle (see Fig. 9) is shown in Fig. 14. The results showed that the nucleation density increased as the contact angle increased and, hence, decreased with increasing ethanol composition.

Hui [27] has recently completed an experimental study on the effects of composition, heat flux, and subcooling on boiling site densities in ethanol–water and ethanol–benzene mixtures at 1.01 bar. The heated test surface was a 25.4 mm diameter disk mounted in a vertical orientation. The number of active boiling sites in an inscribed circle of 19 mm diameter was determined photographically. Figures 15 and 16 depict Hui's results for the variation in the boiling site density N/A with composition at a heat flux of 75 kW/m^2 with 15°C of subcooling.

Figure 15 shows that the boiling site density increased two orders of magnitude from pure water to the azeotrope. (The bubble departure diameters were also noted to be much smaller at the azeotrope composition than at that for pure water.) A large negative deviation in the boiling site density from a linear mixing law interpolation between the pure water and the azeotrope values was found. The actual minimum occurred at a com-

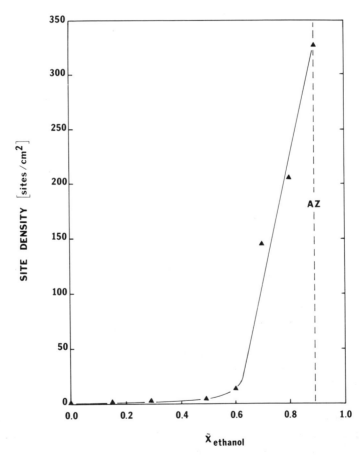

Fig. 15. Boiling site densities for ethanol–water mixtures for 15°C subcooling at $\dot{q}$ = 75 kW/m^2 by Hui [27].

position of about 30% ethanol. A sharp rise in the boiling site density above 60% ethanol was demonstrated at all subcoolings where site densities were measurable. (Low subcoolings and high heat fluxes tended to cause excessive evaporation such that individual boiling sites could no longer be identified.) These boiling results are in direct opposition to the gas diffusion results of Eddington and Kenning [20] mentioned previously. The sharp rise in the boiling site density with increasing ethanol

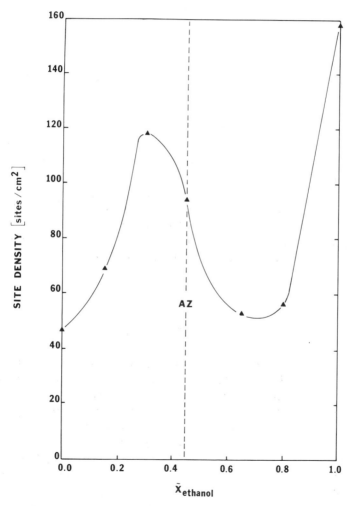

FIG. 16. Boiling site densities for ethanol–benzene mixtures for 15°C subcooling at $\dot{q}$ = 75 kW/m² by Hui [27].

composition was noted to be linked to the vapor spreading phenomenon observed by Thome *et al.* [24].

In Fig. 16 the ethanol–benzene mixture results demonstrate a maximum in the boiling site density to the left of the azeotrope but a minimum to the right. It was noted that the vapor spreading phenomenon occurred over the entire composition range from pure benzene to pure ethanol while activating the boiling surface. Thus a nonaqueous mixture system is shown also to have a very nonlinear variation in boiling site density with composition. Since the boiling site density is widely thought to play a major role in governing the heat transfer rate in boiling, these results may shed some new light on mixture boiling heat transfer coefficients.

F. Summary

In summary, the variation of the equilibrium superheat with composition that is required to maintain a vapor nucleus in a binary liquid mixture was seen to be primarily dependent on the wetting characteristics of the fluid, its surface tension, and to a lesser extent, the slope of its vapor pressure curve. The work of formation of a vapor bubble in an ideal binary or multicomponent mixture system was demonstrated to be greater than that for an equivalent pure fluid. The boiling site density in the mixtures was determined to vary drastically with composition in aqueous and nonaqueous mixtures. Further study is suggested to investigate the mechanism causing the activation of the whole boiling surface from the inception of boiling at one site, and a comprehensive explanation should be sought for the functional dependence of boiling site density on composition.

IV. Bubble Growth

The understanding of the mechanics of bubble growth has long been thought to be one of the keys to the eventual understanding of the boiling process. Consequently, much effort has been expended on the study of vapor bubble growth rates in binary liquid mixtures to further the fundamental understanding of multicomponent boiling. Early bubble growth models considered the growth of vapor bubbles remote from a wall because of the relative simplicity of the boundary conditions. Later models add the effects superimposed on the problem when bubbles grow at a heated wall.

The growth of a vapor bubble at a heated wall begins when the equilibrium superheat given by Eq. (4) is exceeded. The early stage of its growth is hydrodynamically controlled by the liquid inertia, namely, the liquid put into motion by the growth of the bubble. The inertia force decreases because of a slowing of the bubble growth rate as the bubble grows and then the growth of the bubble becomes limited by the rate at which heat can diffuse from the superheated liquid surrounding the bubble to the bubble interface to provide the latent heat of vaporization. During the bubble growth process the temperature at the interface drops to the saturation value of the bulk liquid as the pressure differential across the interface decreases with increasing radius.

The growth of a vapor bubble in a binary mixture is further complicated compared to pure fluids by the difference in the compositions of the vapor and liquid phases. For a mixture not at an azeotrope, the composition of the more volatile component in the vapor bubble $\tilde{y}$ is greater than that in the liquid phase $\tilde{x}$ at the bubble interface, assuming that phase equilibrium exists as shown in Fig. 17. Thus as the evaporation process proceeds and the bubble grows, the more volatile component in the liquid layer adjacent to the bubble diffuses to the interface to provide the additional more volatile component in the vapor and simultaneously produces a local composition gradient around the bubble. As the local value of $\tilde{x}$ decreases, the bubble point at the interface rises, as shown in Fig. 17. Eventually, the bubble point at the interface reaches its maximum value

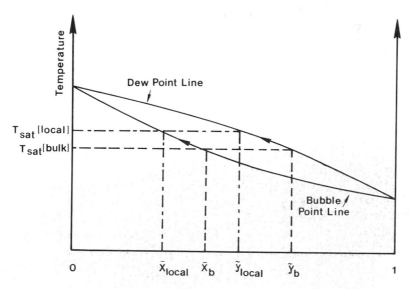

Fig. 17. Phase equilibrium diagram showing the decrease in the local mole fractions of the more volatile component.

when the rate of diffusion of the more volatile component to the interface matches the rate of its excess evaporation at the interface, which is proportional to $|\bar{y} - \bar{x}|$. This was first explained by Van Wijk *et al.* [28].

A. THEORETICAL BUBBLE GROWTH MODELS

The models for the growth of vapor bubbles in binary mixtures can be separated into two categories, those for bubbles growing homogeneously in a superheated liquid and those growing heterogeneously at a heated wall. A review of these theories will begin with the homogeneous models.

1. *The Scriven Model*

Scriven [29] developed the first analytical model for bubble growth in a binary liquid mixture. His model comprises a spherical bubble growing homogeneously in an initially uniformly superheated liquid. Looking at the asymptotic stage of bubble growth, he considered heat and mass transfer by one-dimensional radial conduction and convection to the bubble interface and perfect mixing in the vapor phase. Starting from an energy balance and a mass balance on the two components, Scriven arrived at a set of equations governing bubble growth that required numerical solution. To make the analysis of more practical importance, he then determined two simplified expressions governing bubble growth, one for small superheats and one for large superheats. For nucleate pool boiling superheats of practical interest, the large superheat expression is more valuable and is given as

$$R = \left(\frac{12}{\pi}\right)^{1/2} \left\{ \frac{\Delta T (\kappa_1 t)^{1/2}}{(\rho_g/\rho_1)(\Delta h_v/c_{pl})[1 - (y - x)(\kappa_1/\delta)^{1/2}(c_{pl}/\Delta h_v)(dT/dx)]} \right\} \quad (13a)$$

or

$$R = R_{pure}[1 - (y - x)(\kappa_1/\delta)^{1/2}(c_{pl}/\Delta h_v)(dT/dx)]^{-1} \quad (13b)$$

where dT/dx is the slope of the bubble point line and the bracketed term is a simplification of his original equation [30]. This equation reverts to the Plesset and Zwick [31] solution, that is, R_{pure}, for pure fluids when $(y - x)$ equals zero.

Since Konovalov's first rule [see Section III,B] requires that $(y - x)$ and (dT/dx) always be of opposite signs, Eq. (13a) shows that the rate of bubble growth dR/dt in a binary mixture is always slower than that of an equivalent pure fluid with the same physical properties as the mixture but with $y = x$. Rewriting Eq. (13a) in the form

$$R = 2\beta\sqrt{\kappa_1 t} \quad (14)$$

Scriven evaluated β (the bubble growth coefficient) for the binary system water–ethylene glycol (water being the more volatile component). A minimum in β for this system is shown in Fig. 18.

Bruijn [32] also studied the growth of a spherical vapor bubble in a binary liquid mixture but for the special case where the densities of the two phases are identical. Physically, this means that there is no radial motion in the liquid caused by the growth of the bubble. This is in fact a special case of Scriven's theory. Bruijn also suggested that the bubble growth rate can be slowed additionally by competitive consumption of the more volatile component by neighboring bubbles.

2. *The Van Stralen Model*

Van Stralen [33] extended the bubble growth model of Plesset and Zwick [31] for a spherical bubble growing remote from a wall in an initially

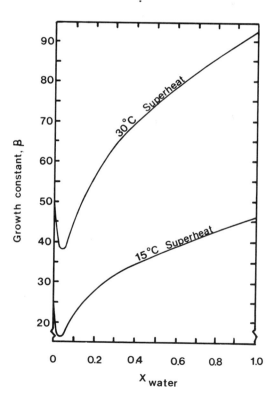

FIG. 18. Variation in bubble growth coefficient for water–ethylene glycol at 1.0 bar from Scriven [29].

uniformly superheated single-component liquid to binary liquid mixtures. His derivation is presented below.

Considering only the asymptotic stage of bubble growth illustrated in Fig. 19, Van Stralen begins with a component mass balance on the more volatile component. As shown in the diagram, the mass fraction of the more volatile component y in the vapor is greater than that of the liquid x and causes a composition gradient in the liquid surrounding the bubble to

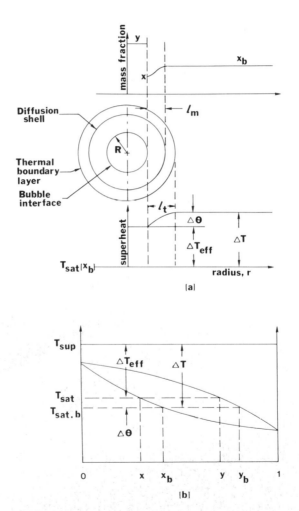

FIG. 19. Bubble growth model of Van Stralen [33] for a spherical vapor bubble growing in a superheated binary mixture. (a) Temperature and composition profiles; (b) process illustrated on phase diagram.

form. Equating the rate of accumulation of the more volatile component in the bubble to its rate of loss by the liquid yields

$$\rho_g y (dR/dt) = \rho_g (dR/dt) + \rho_l \, \delta(\partial x / \partial r)_{r=R} \tag{15a}$$

or

$$\rho_g (y - x)(dR/dt) = \rho_l \, \delta(\partial x / \partial r)_{r=R} \tag{15b}$$

Note that the initial state of growth, which is inertia controlled, is ignored because it is assumed to be relatively short. The asymptotic growth state (with the composition gradient already established before any evaporation takes place) is assumed to begin at $t = 0$. If the liquid mass fraction drops from its bulk value x_b to x at the interface across a diffusion shell of thickness l_m, then the derivative $(\partial x / \partial r)_{r=R}$ can be approximated as

$$(\partial x / \partial r)_{r=R} = (x_b - x)/l_m \tag{16}$$

where the thickness of the one-dimensional radial transient mass diffusion shell is

$$l_m = [(\pi/3) \, \delta t]^{1/2} \tag{17}$$

This is consistent with the one-dimensional radial transient thermal boundary layer thickness used by Plesset and Zwick as

$$l_t = [(\pi/3)\kappa_l t]^{1/2} \tag{18}$$

Since κ_l is an order of magnitude greater than δ for most liquid mixtures, the thermal boundary layer ($l_t \sim 100 \ \mu m$) is much thicker than the diffusion shell ($l_m \sim 10 \ \mu m$). Substituting Eqs. (16) and (17) into Eq. (15b) gives the bubble growth rate based on the mass balance as

$$\frac{dR}{dt} = \left(\frac{\rho_l}{\rho_g}\right)\left(\frac{x_b - x}{y - x}\right) \frac{\delta}{[(\pi/3)\delta t]^{1/2}} \tag{19}$$

Van Stralen defines the ratio of the composition drop in the liquid to the vapor–liquid composition difference as the "vaporized mass diffusion fraction," G_d. Thus Eq. (19) becomes

$$\frac{dR}{dt} = \left(\frac{\rho_l}{\rho_g}\right) G_d \frac{\delta}{[(\pi/3)\delta t]^{1/2}} \tag{20}$$

An energy balance on the growth of a vapor bubble in a single-component liquid led Plesset and Zwick [31] to the following expression for bubble growth, which arises from Eq. (18):

$$R = (12/\pi)^{1/2} Ja(\kappa_l t)^{1/2} \tag{21}$$

where the Jakob number, a dimensionless superheat, is defined as

$$\text{Ja} \equiv (\rho_l c_{Pl}/\rho_g \, \Delta h_v) \, \Delta T \tag{22}$$

and

$$\Delta T = T_{sup} - T_{sat.b} \tag{23}$$

As Van Wijk et al. [28] earlier noted, the actual superheat for a bubble growing in a binary mixture is less than that given by Eq. (23) because of the rise in the saturation temperature at the bubble interface compared to the original bulk saturation temperature (see Fig. 19). Thus the effective superheat ΔT_{eff} is reduced by the increase in the local boiling point $\Delta \theta$ at the bubble interface such that

$$\Delta T_{eff} = \Delta T - \Delta \theta \tag{24}$$

The modified Jakob number becomes

$$\text{Ja}_m \equiv (\rho_l c_{Pl}/\rho_g \, \Delta h_v)(\Delta T - \Delta \theta) \tag{25}$$

Consequently, the bubble growth rate equation using this corrected temperature difference as the driving force for heat conduction is given as

$$\frac{dR}{dt} = \frac{\rho_l c_{Pl}}{\rho_g \, \Delta h_v} (\Delta T - \Delta \theta) \frac{\kappa_l}{[(\pi/3)\kappa_l t]^{1/2}} \tag{26}$$

Then, equating the expression obtained from the mass balance, Eq. (20), to that for the energy balance, Eq. (26), since they both describe the growth of the same bubble, yields

$$\frac{\Delta \theta}{G_d} = \left[\left(\frac{\Delta h_v}{c_{pl}} \right) \left(\frac{\delta}{\kappa_l} \right)^{1/2} \left(\frac{\Delta T}{\Delta \theta} - 1 \right) \right]^{-1} \tag{27}$$

Substituting from Eq. (26) using Eq. (27) results in

$$\frac{dR}{dt} = \frac{\Delta T}{\rho_g - \rho_l[\Delta h_v/c_{pl} + (\delta/\kappa_l)^{1/2}\Delta \theta/G_d]} \left(\frac{3\kappa_l}{\pi t} \right)^{1/2} \tag{28}$$

Van Stralen [34] obtained an approximate graphic solution to the ratio of $\Delta \theta / G_d$ using a temperature–composition diagram such that

$$\frac{\Delta \theta}{G_d} = -x_b(K_b - 1)(dT/dx)_{x=x_b} \tag{29}$$

where $\Delta \theta / G_d$ is evaluated at the bulk composition x_b rather than at the unknown interfacial composition. For an ideal binary mixture system $(dT/dx)_{x=x_b}$ can be evaluated using Eq. (8), making the appropriate changes from mole fraction to mass fraction. Also, the right-hand side of Eq. (29) is seen to be always positive; thus the bubble growth rate in a binary mixture is seen to be less than that of an equivalent pure fluid. Substituting Eq. (29) into Eq. (28) and integrating gives the bubble radius as a function of time as

$$ R = \left(\frac{12\kappa_l}{\pi}\right)^{1/2} \frac{\rho_l c_{pl} \Delta T}{\rho_g \Delta h_v} \left[\frac{1}{1 - (y - x)(\kappa_l/\delta)^{1/2}(c_{pl}/\Delta h_v)(dT/dx)}\right] t^{1/2} \quad (30) $$

This expression is identical to Scriven's equation for large superheats, Eq. (13a). Equation (3) differs from the Plesset and Zwick solution, Eq. (21), by the bracketed term.

3. Other Models

Skinner and Bankoff have extended their theory for bubble growth in single-component liquids [35] to binary mixtures [36]. They solved for the asymptotic spherical bubble growth rate with initial arbitrary spherically symmetric temperature and composition profiles. The Scriven solution is a special case of their derivation when uniform initial conditions exist.

Van Ouwerkerk [37] presented a theoretical model for the growth of a bubble in a binary mixture at a solid surface. He considered a hemispherical bubble growing in a liquid with initially uniform temperature and concentration. He allowed for evaporation over the hemispherical surface of the bubble and from a microlayer underneath the bubble. The heat and mass diffusion in the microlayer were assumed to be one-dimensional. Also, the thicknesses of the thermal and mass diffusion layers were assumed to be less than the thickness of the microlayer. This latter assumption is physically justifiable, since by the time the thermal and mass diffusion wave fronts travel to the heated wall and are reflected back to the bubble interface, the bubble has become much bigger. Since most of the microlayer contribution to bubble growth is at its outer perimeter, the interaction of the diffusion processes with the wall at this smaller radial position is no longer important. Not surprisingly, Van Ouwerkerk arrived at a solution similar to Scriven's, differing by the growth constant β containing $(1 + \sqrt{3})$ instead of $\sqrt{3}$.

Van Stralen [38] extended his theory for a bubble growing in an initially uniformly superheated liquid remote from a wall, Eq. (30), to a bubble growing at a heated wall with an exponential radial temperature profile in the liquid. Evaporation takes place at the segment of the bubble surface covered by the superheated thermal boundary layer (i.e., Van Stralen's relaxation microlayer). A parameter b is introduced for the ratio of the height of the bubble covered by the superheated liquid layer δ_t to the height of the bubble (see Fig. 20). This leads to

$$R = \left(\frac{12\kappa_1}{\pi}\right)^{1/2} b \frac{\rho_1 c_{pl}}{\rho_g \, \Delta h_v} \left[\frac{\Delta T \exp(t/t_g)^{1/2}}{1 - (y - x)(\kappa_1/\delta)^{1/2}(c_{pl}/\Delta h_v)(dT/dx)}\right] t^{1/2} \quad (31)$$

Since both l_t and R grow with respect to $\sqrt{t}$, Van Stralen assumes b to be independent of time. However, the zero time for the growth of the thermal layer is at the instant of departure of the previous bubble and, after a waiting time of t_w, the bubble begins to grow. Since the initial times are separated by t_w, b cannot be independent of time. This is also borne out from physical intuition; namely, for nucleation to occur, the vapor nucleus must be covered by the thermal boundary layer and then in Van Stralen's model it grows to a diameter greater than the thickness of the thermal boundary layer.

Van Stralen *et al.* [39] later derived a comprehensive bubble growth model to describe the entire growth stage of hydrodynamic and asymptotic growth for the combined effect of the relaxation and evaporation microlayers. The evaporation microlayer is the thin liquid layer trapped between the bubble and the solid heating surface during rapid bubble growth. The bubble growth equation is given as

$$R(t) = \frac{R_1(t)R_2(t)}{R_1(t) + R_2(t)} \quad (32)$$

where the hydrodynamically controlled growth solution (i.e., the modified

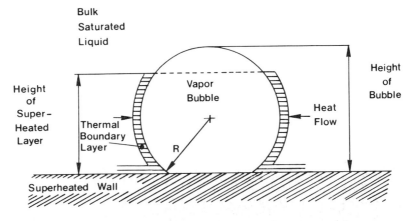

FIG. 20. Relaxation microlayer model for bubble growth at a heated wall by Van Stralen [38].

Rayleigh solution) is

$$R_1(t) = \left[\frac{2\rho_g\,\Delta h_v\,\Delta T\,\exp-(t/t_g)^{1/2}}{3\rho_l T_{sat}} \right]^{1/2} t \qquad (33)$$

and the asymptotic combined relaxation and evaporation microlayer diffusion solution is

$$R_2(t) = \left(\frac{12}{\pi}\right)^{1/2} \frac{C_{1,m}}{C_{1,p}} \left[|b^* \exp-(t/t_g)^{1/2}| + \frac{\Delta T_{sat.b}}{\Delta T} \right] Ja(\kappa_1 t)^{1/2}$$

$$+\ 0.373 \left(\frac{C_{1,m}}{C_{1,p}}\right)^2 Pr_l^{-1/6} |exp-(t/t_g)^{1/2}| Ja(\kappa_1 t)^{1/2} \qquad (34)$$

The ratio of the bubble growth constants (as in all the Van Stralen models) is

$$\frac{C_{1,m}}{C_{1,p}} = \left[1 + \frac{c_{Pl}}{\Delta h_v}\left(\frac{\kappa_1}{\delta}\right)^{1/2}\frac{\Delta\theta}{G_d} \right]^{-1} \le 1 \qquad (35)$$

Thus the binary mixture bubble growth rate in the asymptotic stage is slowed down by both the relaxation and evaporation microlayer contributions in this model.

Thome [40] has suggested that the effective superheat for the growth of sequential and neighboring bubbles can be modeled as

$$\Delta T_{eff} = \Delta T - \Delta\theta_1 - \Delta\theta_2 \qquad (36)$$

where $\Delta\theta_1$ is the steady-state local rise in the boiling point of the liquid adjacent to the heated wall due to the competitive consumption of the more volatile component by neighboring bubbles and $\Delta\theta_2$ is the rise in the local boiling point due to the growth of an individual bubble in this local liquid layer [i.e., $\Delta\theta$ in Eq. (24)]. Assuming that $\Delta\theta_1$ is equal to the $\Delta\theta$ for the growth of a single bubble, he showed that the effective wall superheat is

$$\Delta T_{eff} = Sn^2\,\Delta T \qquad (37)$$

where Sn is the dimensionless mixture bubble growth parameter written in terms of mole fractions and, from Eq. (13b), is defined as

$$Sn \equiv \left[1 - (\bar{y} - \bar{x})\left(\frac{\kappa_1}{\delta}\right)^{1/2}\left(\frac{c_{Pl}}{\Delta h_v}\right)^{1/2}\left(\frac{dT}{d\bar{x}}\right) \right]^{-1} \le 1 \qquad (38)$$

Thus the bubble growth rate in a binary mixture at a heated wall is predicted to be further reduced in comparison to the other bubble growth rate models.

B. EXPERIMENTAL STUDIES

Using high-speed cinematography, the growth of bubbles in binary liquids has been studied experimentally by a number of researchers. Both freely rising bubbles in superheated bulk liquid and bubbles growing at heated walls have been examined.

Experimental bubble growth data were obtained by Florschuetz and Khan [41] for ethanol–water and isopropanol–water mixtures. Freely rising vapor bubbles in a superheated bulk liquid were studied. Uniform superheating of the test liquid was produced by a sudden reduction in the system pressure. Good agreement with the Scriven theory was found, except at large growth times, as shown in Fig. 21.

Bubble growth rates in mixtures of water and ethylene glycol at 1.01 bar were measured by Benjamin and Westwater [42]. Bubbles grew from a 102 μm diameter artificial reentrant cavity in a vertically oriented heated surface and at some compositions were so large that they were bigger than the motion picture film's frame. Experimental values of β from Eq. (14) were obtained over the whole composition range and compared to Scriven's theory, Eq. (13). Figure 22 shows that there is qualitative agreement at a superheat of 8°C. The reduction in the bubble growth is seen to be more severe than that predicted by Scriven's theory. This may be due

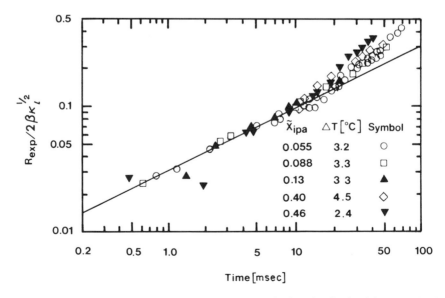

FIG. 21. Dimensionless bubble radius versus growth time for freely rising bubbles in superheated isopropanol–water mixtures by Florschuetz and Khan [41].

to the fact that the actual bubbles are growing at a heated wall rather than remote from a wall, as modeled by Scriven. Thus the experimental conditions may be closer to the model of Van Stralen, which includes microlayer evaporation, Eq. (34). Also, as noted by Thome [43], some bubbles in their study were measured up to departure from the surface while others had only the first one-third of their growth period measured. Since early growth is inertia controlled ($n = 1$), this may have distorted their results.

Yatabe and Westwater [44] later tested ethanol–water and ethanol–isopropanol mixtures using the same experimental facility as Benjamin and Westwater [42]. Whereas the previous study found a pronounced

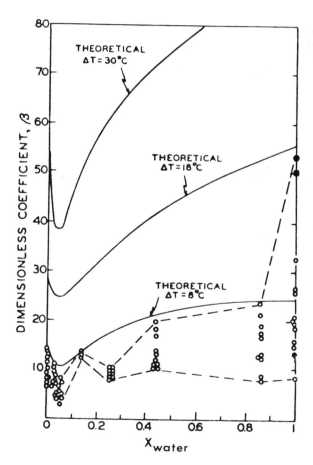

FIG. 22. Comparison of Eq. (13) to experimentally obtained bubble growth coefficients for water–ethylene glycol mixtures by Benjamin and Westwater [42].

minimum in β, the expected minima were not observed experimentally in the later tests. There was no explanation for this unexpected result.

Van Stralen and co-workers [34, 45–47] have studied the growth of vapor bubbles from electrically heated wires of 0.2 mm diameter for several aqueous binary mixtures. The very thin wires were apparently used to simplify photography of the vapor bubbles. They found that a small addition of solvent to water greatly reduced the bubble growth rate, as expected from theory; the maximum in $\Delta T/G_d$ just happens to occur at low values of $\bar{x}$ for the mixtures they tested. The maximum in $\Delta T/G_d$ was seen to coincide with the minimum in the experimental bubble growth rate.

Thome and Davey [48] measured growth rates for bubbles growing from cylindrically shaped artificial cavities in a horizontal heated surface in nitrogen–argon mixtures at 1.3 bar. They found quite a wide variation in the growth rates of sequential bubbles from a boiling site and hence determined the statistically mean bubble growth rate for the sequence measured. Figure 23 shows the bubble growth rate in terms of the bubble diameter plotted versus the vapor–liquid molar composition difference. A reduction similar to that found in the previous studies was obtained as well as qualitative agreement with the Scriven and Van Stralen homogeneous bubble growth models.

For a bubble growing in a binary liquid, the local rise in the saturation temperature due to preferential evaporation of the more volatile compo-

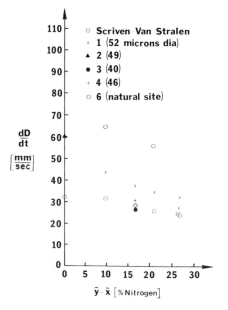

Fig. 23. Comparison of Eq. (30) to measured bubble growth rates in nitrogen–argon mixtures by Thome and Davey [48].

nent (in this case, nitrogen) must be zero at time zero and then increase until it reaches its asymptotic value assumed in the preceding theories. Figure 24 illustrates this effect on the bubble growth rate for this particular case. At small times the growth rates in the mixtures are slightly higher than those for the pure components because the tests were done at constant heat flux, and hence ΔT was slightly higher in the mixtures than for the single components. Clearly, the effect of mass diffusion becomes dominant as growth proceeds. Thus it is only in the asymptotic stage ($t \doteq$ 0.005 sec) that agreement with theory can be expected.

Cooper and Stone [49] have recently reported bubble growth data for n-hexane and n-octane mixtures. Individual bubbles were grown at a wall in an initially stagnant and uniformly superheated liquid by passage of a small current through a resistance heater on the wall. They found that the bubble growth behavior in the mixtures was the same as that for the pure components if the local interfacial boiling point was used for evaluating the superheating rather than the bulk liquid saturation temperature. This is in agreement with Van Ouwerkerk's theory for bubbles growing at a heated wall.

Cooper and Stone [49] also measured the vapor temperature inside a bubble as it grew using a microthermocouple probe. Figure 25 shows a typical result where the temperature fell close to the calculated initial interfacial temperature and then rose as the bubble continued to grow. The rise in the vapor temperature was consistent with the rise in the local

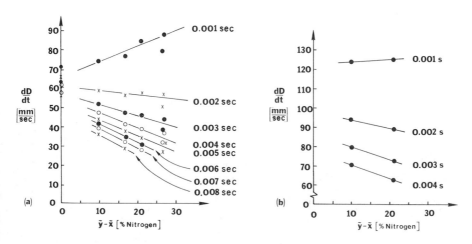

Fig. 24. Measured variation in the bubble growth rate with vapor–liquid composition difference and growth time for nitrogen–argon mixtures by Thome and Davey [48]. (a) Site 1; (b) site 6.

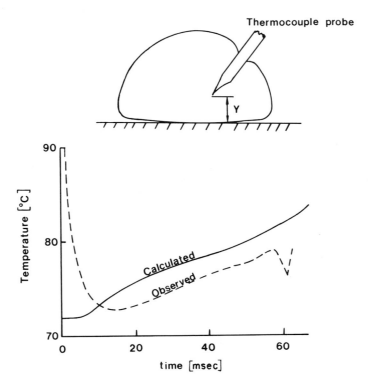

FIG. 25. Small thermocouple probe at height Y above heated surface and comparison of measured and predicted temperature–time history during growth of a bubble in a superheated mixture by Cooper and Stone [49].

boiling point due to depletion of the more volatile component and lack of mixing in the vapor; this runs counter to the assumption made by Scriven and others in evolving their bubble growth models (see Section IV,A).

C. SUMMARY

Bubble growth rates in binary liquids have been analytically predicted for freeing rising bubbles and bubbles attached to a heated surface. Theory predicts a slowing down in the bubble growth rates in mixtures compared to an equivalent pure liquid resulting from the rise in the local saturation temperature caused by preferential evaporation of the more volatile component. Experimental studies have qualitatively confirmed these theories.

V. Bubble Departure

Bubble departure diameter and frequency are two parameters used in the development of many of the single-component nucleate pool boiling correlations available in the literature (see, e.g., Rohsenow [50], Mikic and Rohsenow [51], and Stephen and Abdelsalam [52]). This is because the departure diameter and frequency of vapor bubbles from a heated surface are intrinsic to all heat transport mechanisms used to explain the large increase in the heat transfer coefficient with the transition from single-phase natural convection to nucleate pool boiling. Hence the effect of composition on these two parameters is important to the overall understanding of multicomponent mixture boiling.

A. BUBBLE DEPARTURE DIAMETER

Van Wijk *et al.* [28] in their experimental work on bubble growth rates in binary liquid mixtures noted visually that the departure diameters of the bubbles from their thin heated wire decreased drastically with the addition of a small percentage of organic liquid to water. They attributed the smaller bubble departure diameters in the mixtures to their slower growth rates. At about the same time, Tolubinskiy and Ostrovskiy [53] completed a comprehensive study on bubble departure diameter over the whole composition range for ethanol–water and methanol–water mixtures. They later also studied ethanol–butanol and ethanol–benzene [54] and water–glycerine mixtures [55]. Their data for ethanol–water and ethanol–butanol are shown in Fig. 26. A minimum in the departure diameter was observed for each mixture system at the maximum in $|y - x|$ except for water–glycerine (see Fig. 27). The conflicting result for water–glycerine is probably best explained by the large decrease in the surface tension over the 200°C rise in saturation temperature from pure water to pure glycerine.

Isshiki and Nikai [56] have also measured bubble departure diameters for the ethanol–water system, which are also shown in Fig. 26. Their study found a minimum in the bubble departure diameter at $x_{eth} = 0.02$ and a maximum at $x_{eth} = 0.5$, strikingly different results from those of Tolubinskiy and Ostrovskiy.

Thome [43] obtained bubble departure diameters for nitrogen–argon mixtures at 1.3 bar for bubbles growing from artificial and natural sites on a heated surface. An average bubble departure diameter D_d at a boiling site was obtained by measuring a sequence of bubble departures. Figure 28 shows the results for a constant heat flux of 2.1 kW/m². The minimum in

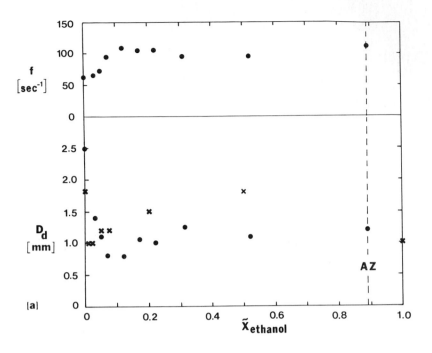

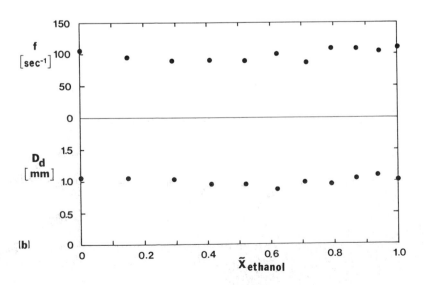

FIG. 26. Bubble departure diameter and frequency in binary mixtures (●, Tolubinskiy and Ostrovskiy [53, 54]; ×, Isshiki and Nikai [56]). (a) Ethanol–water; (b) ethanol–butanol.

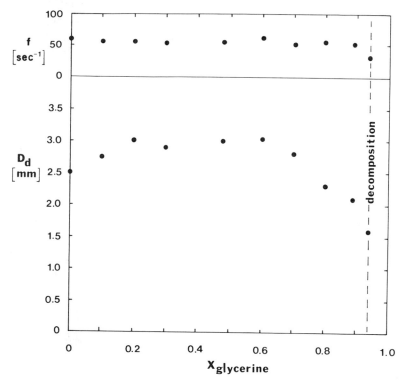

FIG. 27. Bubble departure diameter and frequency in water–glycerine mixtures at 1.0 bar by Tolubinskiy *et al.* [55].

D_d is very close to the composition at which $|\bar{y} - \bar{x}|$ is a maximum, that is, 35% $\bar{x}_{N_2}$.

Thome and Davey [57] sought a physical explanation for the minimum in the departure diameter in the mixtures. A dynamic force balance for bubble departure in a single-component liquid developed by Keshock and Seigel [58] was selected as the best model available. It balances the buoyancy force against a combination of the surface tension, inertia, drag, and excess pressure forces. (Note that some researchers believe that the inertia force assists bubble departure rather than resists it.) The slower bubble growth rates in the mixtures were seen to cause a reduction in their liquid inertia and drag forces. There is thus less resistance to bubble departure and smaller diameters result. The inertia force can be shown to be equal to

$$F_i = \frac{33\pi\rho_l}{8g_c}\left[a^4\,\frac{n(4n-1)}{3}\,t^{4n-2}\right] \tag{39}$$

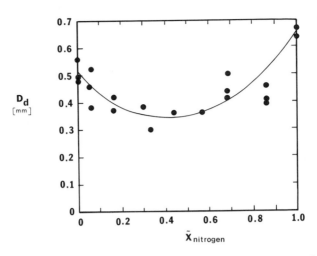

FIG. 28. Average bubble departure diameter at various boiling sites for nitrogen–argon mixtures at $\dot{q} = 2.1$ kW/m^2 by Thome [43].

where a and n are from the bubble growth expression $R = at^n$. Figure 29 does indeed show that the bubble departure diameter in the mixtures decreases as the inertia force decreases. (Experimental values from Thome and Davey [43, 48] were used for a and n.) This is in agreement with the single-component bubble departure correlations of Staniszewski [59] and Cole and Shuman [60] of the form

$$D_d \sim \left[\frac{\sigma}{g_n(\rho_l - \rho_g)} \right]^{1/2} \left[1 + c_1 \left(\frac{dD}{dt} \right)^{c_2} \right] \tag{40}$$

which implies that the inertia force does indeed resist bubble departure.

Thome [40] later reworked the Keshock and Siegel [58] bubble departure equation to include the effect of the binary mixture bubble growth rate. He excluded the excess pressure term because it is determined using the bubble contact radius in the Laplace equation, Eq. (2), rather than the bubble radius itself. Thus the Keshock and Siegel model implies that a pressure gradient exists in the vapor bubble between its base, which has a small radius, and the "spherical" segment of the bubble, which has a relatively large radius. It seems unlikely that such a pressure gradient could exist in the vapor phase.

The resulting ratio of the actual bubble departure diameter D_d to the ideal, linear mixing law bubble departure diameter D_{dl} is shown to be

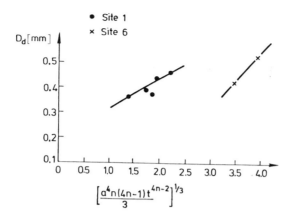

FIG. 29. Variation in bubble departure diameter with bubble inertia force term in nitrogen–argon mixtures by Thome and Davey [57] for two boiling sites.

$$\frac{D_d}{D_{d_I}} = \left\{ \frac{[(11\pi\rho_l/24g_c)(Sn^2 c_a \, \Delta T)^4 + (45/4g_c)\eta_l(Sn^2 c_a \, \Delta T)^2 + \pi D_b \sigma \sin \beta]}{[(11\pi\rho_l/24g_c)(c_{a_I} \, \Delta T_I)^4 + (45/4g_c)\eta_{l_I}(c_{a_I} \, \Delta T_I)^2 + \pi D_{b_I} \sigma_I \sin \beta_I]} \right\}^{1/3}$$

(41)

where Sn is defined by Eq. (38), Eq. (37) is used for the effective wall superheat, and ΔT_I is the ideal linear mixing law wall superheat originally defined by Stephan and Korner [14] as

$$\Delta T_I = \tilde{x}_1 \, \Delta T_1 + \tilde{x}_2 \, \Delta T_2 \qquad (42)$$

For inertia-controlled growth Thome [40] obtained the ratio of the superheats to be

$$\frac{\Delta T}{\Delta T_I} = Sn^{-7/5} \qquad (43)$$

and therefore Eq. (41) reduces to

$$\frac{D_d}{D_{d_I}} = Sn^{4/5} \qquad (44)$$

Since $Sn < 1.0$ for mixtures other than the azeotrope, Eq. (44) predicts a minimum in the departure diameter at the minimum in Sn, which is near the composition of the maximum in $|\tilde{y} - \tilde{x}|$. Thus Eq. (44) agrees qualitatively with the trends shown in Figs. 26 and 28.

For surface tension controlled departure, which is characteristic of low wall superheats, Eq. (41) reduces to

$$\frac{D_d}{D_{d_I}} = \left\{ \frac{D_b \sigma \sin \beta}{D_{b_I} \sigma_I \sin \beta_I} \right\}^{1/3} \tag{45}$$

Van Stralen et al. [39] have derived an expression for the radius of the dry area underneath a growing bubble on a heated wall for both single-component liquids and binary mixtures. The expression for single-component liquids is

$$R_{d,p}^*(t) \doteq 0.156 \left(\frac{\lambda \, \Delta T}{\eta_l \, \Delta \bar{h}_v} \right)^{1/2} \mathrm{Pr}_l^{1/6} \, \mathrm{Ja}(\kappa_l t)^{1/2} \tag{46}$$

and the equivalent equation for binary mixtures, including the effect of mass diffusion, is

$$R_{d,m}^*(t) \doteq 0.156 \left(\frac{C_{1,m}}{C_{1,p}} \right)^{7/2} \left(\frac{\lambda \, \Delta T}{\eta_l \, \Delta \bar{h}_v} \right)^{1/2} \mathrm{Pr}_l^{1/6} \, \mathrm{Ja}(\kappa_l t)^{1/2} \tag{47}$$

Therefore at departure the ratio of D_b/D_{b_I} is obtained using Eqs. (29), (35), and (38) as

$$\frac{D_b}{D_{b_I}} = \mathrm{Sn}^{7/2} \left(\frac{\Delta T}{\Delta T_I} \right) \left(\frac{t_g}{t_{gI}} \right) \tag{48}$$

neglecting the nonlinearity in the physical properties other than that for Sn. Thome [40] has shown that the ratio of the actual wall superheat to the ideal superheat is given as

$$\frac{\Delta T}{\Delta T_I} = \frac{1}{\mathrm{Sn}} \left(\frac{D_{d_I}}{D_d} \right)^{1/2} \tag{49}$$

when the contribution of natural convection to the overall boiling heat flux is small and the cyclic thermal boundary layer stripping mechanism is dominant. It can then be shown that for surface tension controlled departure that

$$\frac{D_d}{D_{d_I}} = \mathrm{Sn} \left\{ \frac{\sigma \sin \beta}{\sigma_I \sin \beta_I} \right\}^2 \tag{50}$$

Comparing Eq. (50) to Eq. (44) for $\sigma \doteq \sigma_I$ and $\beta \doteq \beta_I$, the reduction in the bubble departure diameter ratio is very similar whether inertia or surface tension controls departure. For a system such as ethanol–water, the surface tension–contact angle term in Eq. (50) is quite significant, as illustrated in Figs. 6 and 9 for σ and β, respectively. Figure 30 demonstrates that the Tolubinskiy and Ostrovskiy experimental data for ethanol–water (Fig. 26) are bounded by the inertia and the surface tension controlled departure expressions, Eqs. (44) and (50), respectively. However, this is not true for all the data of Isshiki and Nikai (Fig. 26).

In summary, the smaller bubble departure diameter in a mixture may be due not only to its slower growth rate, which reduces the bubble inertia force, but also to the retardation of its microlayer evaporation rate, which is caused by preferential evaporation of the more volatile component, as noted by Van Ouwerkerk [37], Zeugin *et al.* [61], and Van Stralen *et al.* [39]. In addition, the negative deviations from linearity in σ and β can also partially explain the smaller departure diameters in aqueous mixtures.

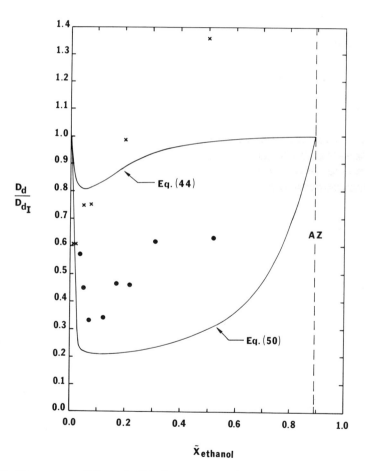

FIG. 30. Comparison of Thome bubble departure equations to data of (●) Tolubinskiy and Ostrovskiy [53, 54] and (×) Isshiki and Nikai [56] for ethanol–water system.

B. BUBBLE DEPARTURE FREQUENCY

The frequency of bubble departure f is defined as

$$f \equiv 1/(t_g + t_w) \tag{51}$$

where t_g is the time period when the bubble grows from its nucleation size to its departure diameter and t_w is the waiting time during which the vapor nucleus left behind waits to be activated again. The effect of composition on the bubble growth rate and the waiting time determines the overall variation in the bubble departure frequency.

Figure 26 shows the experimentally measured departure frequencies for ethanol–water and ethanol–butanol mixtures studied by Tolubinskiy and Ostrovskiy [53]. The ethanol–water results have an intermediate maximum at 30% ethanol composition. The ethanol–butanol data exhibit relatively no change. Figure 31 depicts the departure frequency data of Thome and Davey [43, 57] for liquid nitrogen–argon mixtures. There is a large degree of scatter in the data but general trend is a maximum in the departure frequency.

The length of the bubble growth time t_g in the mixtures is affected by two factors. First, the bubbles do not have to grow as large as in the single components in order to depart. Second, their rate of growth is slower, as was shown earlier. The variation in the ratio of the growth times for inertia-controlled growth can be shown to be

$$t_g/t_{g_I} = Sn^{2/5} \tag{52}$$

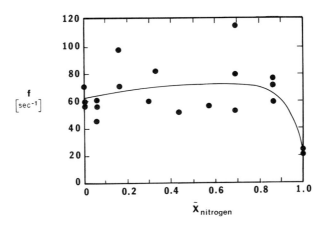

FIG. 31. Average bubble departure frequency at various boiling sites for nitrogen–argon mixtures by Thome [43] and Thome and Davey [57].

Thus the growth time is predicted to decrease at the more volatile compositions. The five mean experimental bubble growth times from a sequence of bubble departures available in Thome [43] for boiling site 1 at 2.1 kW/m² heat flux show that a pronounced minimum does indeed occur in t_g, which qualitatively agrees with Eq. (52), as shown in Fig. 32.

For surface tension controlled growth the ratio of (t_g/t_{gI}) is obtained as

$$t_g/t_{gI} = Sn[\sigma \sin \beta/\sigma_I \sin \beta_I]^6 \qquad (53)$$

Here there is more effect by Sn than in Eq. (52) and the potential of large effects by the surface tension and contact angle. The exponent of 6 seems suspiciously high and probably is the result of the simplifying assumptions used in developing the microlayer evaporation model that yielded Eq. (47).

The bubble waiting time is determined by the rate at which the thermal boundary layer reforms to sufficient thickness and superheat for the vapor nucleus left behind in the cavity to be activated. The waiting time is therefore affected by the equilibrium superheat needed to activate the

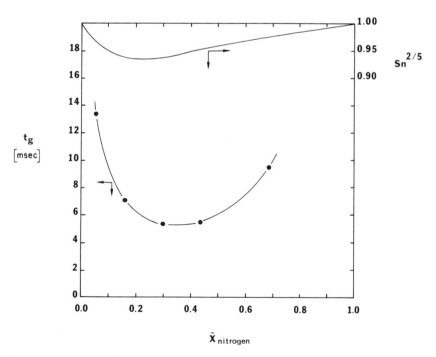

FIG. 32. Average bubble growth time for boiling site 1 [43] and variation in $Sn^{2/5}$ for nitrogen–argon mixtures.

boiling site, Eq. (4). Consequently, the surface tension's variation with composition in aqueous mixtures could significantly alter t_w. Also, the formation of the new thermal boundary layer is controlled by the liquid thermal diffusivity and the wall superheat. The waiting time could be shorter in the mixtures since the wall superheat in the mixtures is higher than for the ideal linear mixing law superheat. In addition, Toral et al. [23] suggest that a complex heat and mass transfer process may be important since the vapor left behind by the departing bubble may not correspond to the equilibrium composition of the liquid that rushes in and contacts the vapor nucleus.

Van Stralen and Cole [62] have considered the effect of surface tension on bubble waiting time through the equilibrium superheat equation. Assuming the bubble waiting time to be a constant multiple of the bubble growth time, they derived the ratio of the bubble departure frequency in a mixture to that of an equivalent pure fluid as

$$f/f_1(t_{g_1}/t_g)(\sigma_1/\sigma)^2 \tag{54}$$

They thus concluded that the bubble departure frequency will be increased in aqueous mixtures since $(\sigma_1/\sigma) > 1$. In addition, Eqs. (52) and (53) give $(t_{g_1}/t_g) > 1$ such that the frequency will also tend to increase because of the shorter bubble growth time in the mixtures.

The assumption that the bubble departure time t_d is a constant multiple of the bubble growth time t_g (equivalent to Van Stralen and Cole's assumption) with variation in composition is not borne out experimentally, although it is convenient for analytical considerations. Figure 33 shows a plot of k_2, which is defined as

$$k_2 \equiv t_d/t_g \tag{55}$$

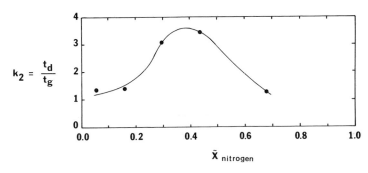

Fig. 33. Variation in ratio of bubble departure time to bubble growth time for nitrogen–argon mixtures for site 1.

for the experimental results of Thome [43] for boiling site 1. The increase in k_2 by a factor of 2 at the more volatile compositions offsets the 50% reduction in t_g seen in Fig. 32. Thus the bubble waiting time in this instance is found to be relatively independent of composition.

In summary, the bubble departure frequency in a binary mixture is predicted to be higher than that for its equivalent pure fluid. Experimental results on aqueous, organic, and cryogenic mixtures tend to support this fact. However, further theoretical and experimental work in this area is recommended.

VI. Nucleate Pool Boiling Heat Transfer

The pool boiling curve is a curve showing the heat flux $\dot{q}$ as a function of the temperature difference ΔT between the heating surface and the bulk liquid. Figure 34 shows a schematic representation of the pool boiling curve when ΔT is the independent variable. Four distinguishable heat

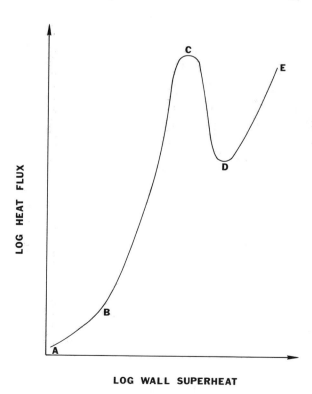

FIG. 34. Pool boiling curve.

transfer regions are noted. Curve *AB* represents single-phase natural convection where point *B* represents the ΔT at which boiling sites activate on
the heated surface. Curve *BC* is called the nucleate pool boiling curve
where boiling sites nucleate to form vapor bubbles on the heated surface.
Point *C* is the peak nucleate heat flux. Region *CD* is a transition region in
which the heated surface is gradually covered by a coherent layer of
vapor. The insulating effect of the vapor causes a drop in the heat flux.
Curve *DE* represents the film boiling region where the heated surface is
completely blanketed by a layer of vapor. In this section, the nucleate
pool boiling region will be discussed. Sections VIII and IX will review the
peak nucleate heat flux and film boiling, respectively.

Figure 35a depicts typical nucleate pool boiling curves for two single-
component liquids and some of their mixtures obtained by Clements and
Colver [63] for the propane–*n*-butane mixture system. The first observation is that the pool boiling curves of the mixtures are not necessarily
bounded by the single-component curves as they were in Fig. 13 for
water–MEK mixtures. Figure 35b shows the variation in the wall super-
heat, ΔT, defined as $T_w - T_{sat.b.}$, at four heat fluxes. The problem, then, is

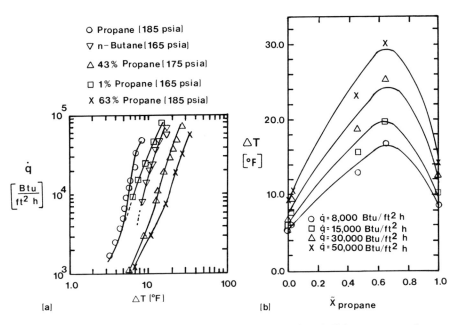

FIG. 35. Nucleate pool boiling data of Clements and Colver [63] for propane–*n*-butane
mixtures. (a) Pool boiling curves; (b) variation in wall superheat with composition at fixed
heat flux.

to explain the increase in ΔT in the mixtures above the value predicted by a simple linear interpolation between their single-component values as a function of composition.

A. PHYSICAL EXPLANATIONS FOR THE DECREASE IN THE HEAT TRANSFER COEFFICIENT

Van Wijk *et al.* [28] presented the first physical explanation for the reduction in the mixture boiling heat transfer coefficient. They suggested that preferential evaporation of the more volatile component takes place at the heated surface such that the local boiling point rises. This process is shown schematically in Fig. 17.

Sternling and Tichacek [64] cited three possible explanations: (1) the change in the mixture physical properties, notably viscosity, with composition; (2) the change in the rate of bubble growth caused by the resistance to the mass transfer of the more volatile component in diffusing into the growing bubble; and (3) changes in the rate of nucleation of new boiling sites on the surface. They suggested that the rate of nucleation will decrease because of the clogging of the boiling cavities by accumulation of the less volatile liquid caused by preferential evaporation of the more volatile component. This last physical mechanism was suggested to be dominant for explaining decreases in the boiling heat transfer coefficient of more than twofold. It is not clear as to how clogging of the cavities occurs, since the mass diffusion boundary layer is very thin (on the order of 0.01 mm) and is probably removed by the departing bubble.

Stephan and Korner [14] (see Section III) showed that the work of formation of bubbles in a mixture is greater than that for an equivalent pure fluid and concluded that the mixture heat transfer coefficient for a given heat flux is lower than that for the equivalent pure fluid because of the projected decrease in the bubble population. Stephan and Preusser [17] later extended this conclusion to multicomponent liquids and also demonstrated [65] conclusively that some of the negative deviation of the heat transfer coefficient calculated from the linear mixing law prediction [Eq. (42)] is due to the nonlinear variation in the physical properties of the mixture with composition. For example, Fig. 36a shows their comparison of the heat transfer coefficients predicted by four different single-component correlations with their measured values for the acetone–water system. Figure 36b depicts a similar, although less significant, deviation between a single-component correlation and a linear mixing law for a nonaqueous mixture system ethanol–benzene, as calculated by Shock [66].

Thome [67] sought an explanation of the lower mixture heat transfer

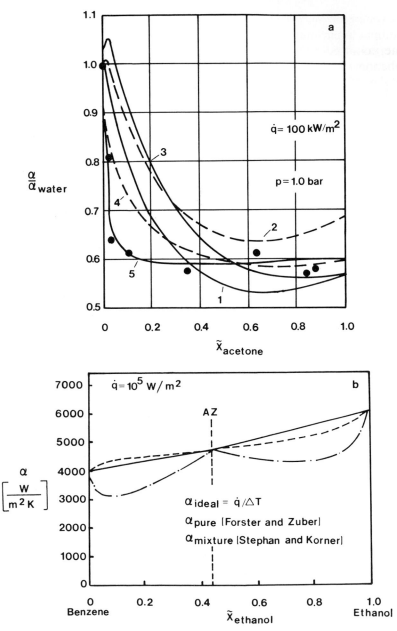

FIG. 36. Mixture heat transfer coefficients predicted by single-component correlations. (a) Stephan and Preusser [65] for acetone–water at 1.0 bar. 1, Calculated according to Labuntsov; 2, according to Stephan; 3, according to Kutateladze; 4, according to Stephan and Preusser; 5, experimental results. (b) Shock [66] for ethanol–benzene at 1.0 bar.

coefficients by investigating the heat transfer mechanisms responsible for the majority of the heat transfer removal from the heated surface in nucleate pool boiling because the previous studies on the individual boiling phenomenon only indirectly infer the lower heat transfer rates. He looked at the effect of composition on the bubble evaporation mechanism and the thermal boundary layer stripping mechanism at an isolated boiling site. The bubble evaporation mechanism refers to the latent heat transported from the heated surface by departing bubbles, estimated as

$$Q_{\Delta h_v} = \rho_g \, \Delta h_v \, V_d f \tag{56}$$

The thermal boundary layer stripping mechanism refers to the cyclic removal of the thermal boundary layer from the heated surface by departing vapor bubbles. Assuming the area of influence of a departing bubble to be twice its departure diameter as shown in Fig. 37, the heat transport rate due to this mechanism is estimated as

$$Q_{c_p} = \tfrac{1}{2}\pi^{3/2}(\rho_l c_{pl} \kappa_l f)^{1/2} D_d^2 \, \Delta T \tag{57}$$

The experimental bubble departure diameters and frequencies available in Thome [43] were used to evaluate Eqs. (56) and (57). Figure 38 shows that a minimum in these two heat transfer mechanisms at an individual boiling site coincides with the minimum in the heat transfer coefficient for the nitrogen–argon system.

It is seen that the decrease in the boiling heat transfer coefficient is a combined result of mixture effects on bubble growth rate and departure, bubble nucleation and boiling site density, and nonlinear variation of the pertinent physical properties.

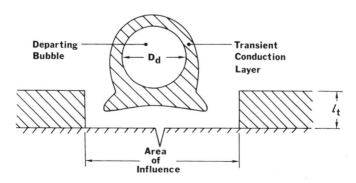

FIG. 37. Thermal boundary layer stripping model.

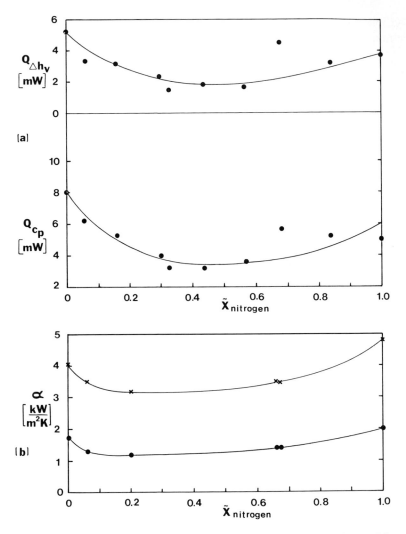

Fig. 38. Nitrogen–argon mixture boiling. (a) Heat transfer rates at 1.3 bar and $\dot{q}$ = 2.1 kW/m² [67]; (b) heat transfer coefficient at 1.0 bar [43] at two heat fluxes.

B. Subcooled Boiling of Mixtures

Sterman *et al.* [68] presented the first experimental results for the effect of subcooling on the nucleate pool boiling of binary mixtures. They investigated the system benzene–diphenyl at 3.5 and 8.0 bars with subcoolings of 0, 10, 30, and 80°C for boiling from a thin-walled stainless steel tube of 5 mm diameter. Nitrogen gas was used to obtain the subcooling. As in

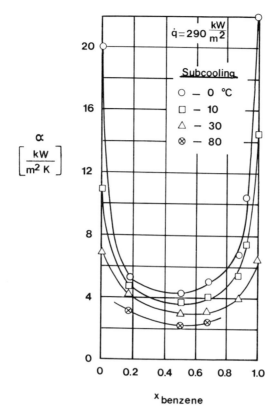

Fig. 39. Effect of subcooling on nucleate pool boiling of benzene–diphenyl mixtures by Sterman *et al.* [68] at 8.0 bar.

pure component boiling, they found that the heat transfer coefficient based on $(T_w - T_b)$ decreased as subcooling increased (see Fig. 39). The heat transfer coefficients in the mixtures are seen to drop less with subcooling than those for the two single components. No explanation for this behavior was presented.

Recently, Hui [27] studied the effect of subcooling on the heat transfer coefficient and the boiling site density in ethanol–water and ethanol–benzene. Figures 40a and b demonstrate the effect of subcooling on the heat transfer coefficient at 1.0 bar. Hui obtained a so far inexplicable maximum in α for ethanol–water at $x_{eth} = 0.7$. The relative decreases in the heat transfer coefficient with subcooling were similar in magnitude here for both mixtures and single components. However, more work is required for modeling and correlating the effect of subcooling on the heat transfer coefficient.

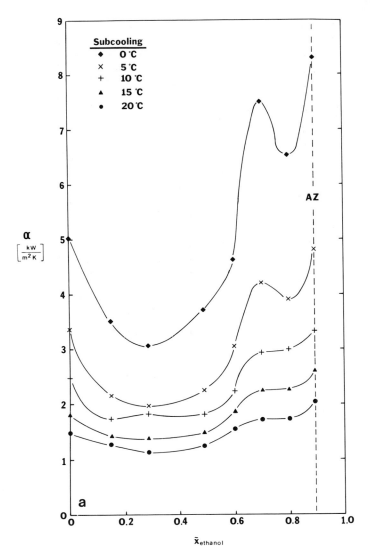

FIG. 40. Effect of subcooling on boiling heat transfer coefficient by Hui [27] at a heat flux of 75 kW/m². (a) Ethanol–water; (b) ethanol–benzene.

C. Effect of Pressure

As in single-component boiling, the heat transfer coefficients for mixtures tend to increase as the pressure increases. That is, the nucleate pool boiling curves move to the left with increasing pressure. The increase in α

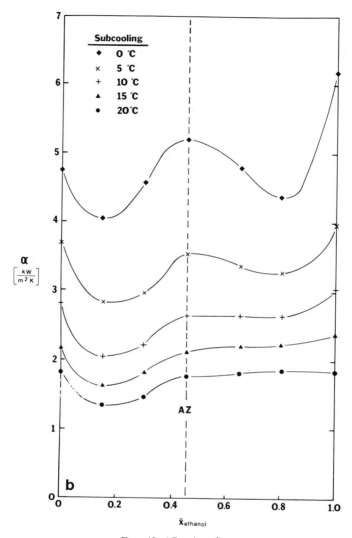

Fig. 40. (*Continued*)

is caused by the increase in the boiling site density, which is due to a decrease in the equilibrium superheat required for boiling site activation with increasing pressure. Furthermore, the shape of the temperature–composition phase equilibrium diagram can change significantly with pressure.

The effect of pressure on mixture boiling has been investigated experimentally for aqueous [69–71], organic [68, 69, 72–76], refrigerant [77], and cryogenic [78, 79] systems. Figure 41 shows the variations in the heat transfer coefficient and $\tilde{y} - \tilde{x}$ with composition at several pressures for R-23–R-13 mixtures obtained by Jungnickel *et al.* [77]. The minima in α are seen to be more pronounced with increasing pressure, even though the maxima in $|\tilde{y} - \tilde{x}|$ decrease.

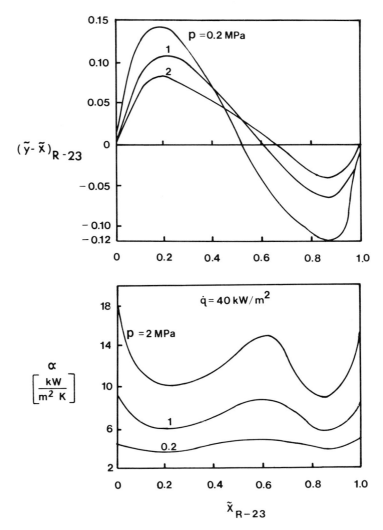

FIG. 41. Variation in vapor–liquid composition difference and boiling heat transfer coefficient at three pressures in R-23–R-13 mixtures by Jungnickel *et al.* [77].

Bier *et al.* [80] have also recently noted the same behavior for SF_6–CF_3Br mixtures near the critical point. They suggest that the rise in the local saturation temperature in the boundary layer at the heated wall resulting from preferential evaporation of the more volatile component continues to be an important factor, since the overall wall superheat decreases rapidly with increasing pressure. This idea is demonstrated here in Fig. 42, assuming that the local rise in the boiling point due to all the various effects, $\Delta\theta$, remains constant while the linear mixing law wall superheat from Eq. (42), ΔT_I, decreases linearly with pressure. The overall wall superheat ΔT hence decreases with increasing pressure but the relative heat transfer performance, given by α/α_I or $\Delta T_I/\Delta T$, still decreases. Therefore the minimum in α/α_I becomes more profound with

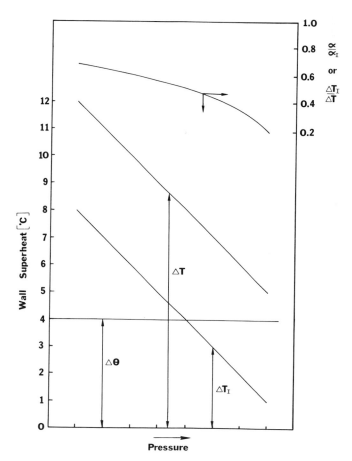

FIG. 42. Decrease in relative heat transfer coefficient with increasing pressure.

increasing pressure in this example, with particulars quite similar to the ethanol–water system tested by Tolubinskiy *et al.* [71], whose results also demonstrated decreasing relative heat transfer performance with increasing pressure.

For a particular mixture system, the variation in the heat transfer performance will depend on the changes in $\Delta\theta$ and ΔT_1 with pressure, and a decrease in performance will not necessarily always occur.

D. Effect of a Third Component

Nearly all the studies of mixture boiling have dealt with binary mixtures. The study of multicomponent mixture boiling is much more tedious because of the large number of experiments required to cover the composition range of all the components; yet multicomponents are encountered much more frequently than binary mixtures in industrial practice. There is no apparent reason for the characteristics of binary mixture boiling not being similar to those of multicomponent boiling. To test this hypothesis, two studies covering a wide range of composition have been completed for several ternary mixture systems.

Grigor'ev *et al.* [81] have tested the ternary mixture system of acetone–methanol–water and acetone–ethanol–water at 1.0 bar for boiling on a 7.7 mm diameter horizontal stainless steel tube. They plotted their experimental data as isoheat transfer coefficients at various heat flux levels on triangular composition diagrams. Figure 43 shows that the minimum in the heat transfer coefficient in the ternary mixtures was lower than the minima for the three binary mixture systems. It should be noted, however, that the additional decrease in α in the ternary mixtures is not much more than the decrease in α from single-component boiling to binary mixture boiling.

Stephan and Preusser [65] have studied the boiling of acetone–methanol–water mixtures and methanol–ethanol–water mixtures. They reported data contradictory to those of Grigor'ev in that the reduction in the heat transfer coefficients for ternary mixtures was less than that for the binary mixtures. They suggest that the local rise in the saturation temperature is smaller in the ternary mixtures because of a flattening of the bubble point curve on addition of a third component. Hence the ternary mixture perhaps experiences less degradation in the effective wall superheat than the corresponding binary mixture systems.

More experimental work is required on ternary mixtures, especially nonaqueous systems, to resolve the conflicting conclusions of these two studies.

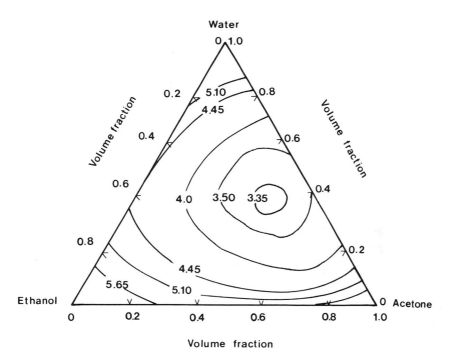

FIG. 43. Iso-heat transfer coefficients [kW/(m² °C)] for the boiling of acetone–ethanol–water mixtures at $\dot{q}$ = 116 kW/m² by Grigor'ev *et al.* [81].

VII. Prediction of Nucleate Boiling Heat Transfer Coefficients

Numerous methods for predicting the nucleate pool boiling curve for binary and multicomponent mixtures have been proposed [14, 17, 30, 40, 49, 54, 65, 73, 75, 80, 82–84]. They attempt to predict the variation in the heat transfer coefficient or wall superheat with composition at a constant heat flux. Figures 35b, 38b, 40, and 41 are typical of the variation of the heat transfer coefficient or wall superheat with composition. Obviously, a linear mixing law heat transfer coefficient can grossly overestimate the actual value. Figure 36 shows that even evaluating a single-component nucleate pool boiling correlation allowing for the nonlinear variation in physical properties is not sufficiently accurate for predicting mixture boiling heat transfer coefficients. A selection of the predictive methods is presented below to demonstrate the spectrum of approaches used.

A. Predictive Methods

Palen and Small [84] published perhaps the earliest correlation in this field for predicting the boiling heat transfer coefficients for multicomponent mixtures with reboiler applications in mind. They give the following relation for mixtures with a wide boiling range

$$\alpha/\alpha_I = \exp[-0.027(T_{b0} - T_{bi})] \tag{58}$$

The ideal heat transfer coefficient α_I is calculated from the McNelly [85] correlation using average physical properties. The term $(T_{b0} - T_{bi})$ is the temperature difference between the vapor leaving a kettle reboiler and the liquid feed to the reboiler, and this is equal to the temperature difference between the dew point and the bubble point at the liquid feed composition if all the liquid feed is evaporated.

Stephan and Korner [14] later developed a simple method for correlating binary mixture boiling heat transfer coefficients using an excess function approach common to the prediction of mixture physical properties. They noted that the maximum in the wall superheat in the mixtures occurred at about the same composition as the maximum in $|\tilde{y} - \tilde{x}|$. Hence they correlated the wall superheat at a prescribed heat flux as

$$T_w - T_{sat.b} = \Delta T = \Delta T_I + \Delta T^E \tag{59}$$

where the ideal superheat ΔT_I is defined from a linear molar mixing law at constant heat flux as

$$\Delta T_I = \tilde{x}_1 \Delta T_1 + \tilde{x}_2 \Delta T_2 \tag{60}$$

and the excess superheat ΔT^E is calculated from

$$\Delta T^E = A|\tilde{y} - \tilde{x}| \Delta T_I \tag{61}$$

The empirical quantity A still contains the influence of the specific characteristics of the particular mixture system. A value of A is determined at an intermediate heat flux level for each mixture system by fitting Eqs. (59)–(61) to experimental data. The values of ΔT_1 and ΔT_2 for the two single-component liquids comprising the mixture are obtained from suitable experimental data or correlations. To account for the pressure dependence, Stephan and Korner presented a further empirical equation valid over the range of 1–10 bar as

$$A = A_0(0.88 + 0.12p) \tag{62}$$

with p in bars and A_0 being the value of A at 1.0 bar. They list recommended values of A_0 for 17 binary mixture systems. See Table III for a list of values. (They suggest a value of 1.53 for A_0 if it is unknown.)

TABLE III

NUMERICAL VALUES OF A_0

Mixture	Numerical values, A_0
Acetone–ethanol	0.75
Acetone–butanol	1.18
Acetone–water	1.40
Ethanol–benzene	0.42
Ethanol–cyclohexane	1.31
Ethanol–water	1.21
Benzene–toluene	1.44
Heptane–methylcyclohexane	1.95
Isopropanol–water	2.04
Methanol–benzene	1.08
Methanol–amyl alcohol	0.80
Methylethyl ketone–toluene	1.32
Methylethyl ketone–water	1.21
Propanol–water	3.29
Water–glycol	1.47
Water–glycerol	1.50
Water–pyridine	3.56

Stephan and Preusser [17] have extended this technique to multicomponents (n) by changing Eq. (60) to

$$\Delta T_{\mathrm{I}} = \sum_{i=1}^{n} \tilde{x}_i \, \Delta T_i \tag{63}$$

and Eq. (61) to

$$\Delta T^{\mathrm{E}} = \left| \sum_{i=1}^{n-1} K_{\mathrm{in}}(\tilde{y}_i - \tilde{x}_i) \right| \tag{64}$$

where K_{in} are the values of A for the constituent binary mixtures. (See the appendix of the paper by Stephan [86] for further discussion of K_{in}.)

A semiempirical equation based on the assumption that the reduction in the bubble growth rate is directly proportional to the reduction in the heat transfer coefficient in binary mixtures was developed by Calus and Rice [30]. Using the Scriven and Van Stralen bubble growth equation, Eq. (30), for spherical bubble growth remote from a wall in binary mixtures, they obtained, from data for boiling of aqueous systems on wires and tubes,

$$\alpha/\alpha_{\mathrm{I}} = [1 + |y - x|(\kappa_{\mathrm{I}}/\delta)^{1/2}]^{-0.7} \tag{65}$$

where the terms (dT/dx) and $(c_{\mathrm{pl}}/\Delta h_{\mathrm{v}})$ have been dropped with no particular justification other than that the resulting expression correlated their data. The ideal heat transfer coefficient α_I is evaluated using the Borishanskii–Minchenko (cited in Kutateladze [87]) correlation for single-component liquids. The exponent of -0.7 was found empirically.

Calus and Leonidopoulos [82] derived the first completely analytical expression for predicting the variation in the wall superheat with composition. Integrating Eq. (26) and equating it to its equivalent expression, Eq. (30), they solved for the value of $\Delta\theta$. Apparently assuming this to be equal to the rise in the local saturation temperature due to the multibubble evaporation process at a heated wall, they obtained the following expression for the wall superheat:

$$\Delta T = (x_1 \, \Delta T_1 + x_2 \, \Delta T_2) \left[1 - (y - x) \left(\frac{\kappa_l}{\delta} \right)^{1/2} \left(\frac{c_{pl}}{\Delta h_v} \right) \left(\frac{dT}{dx} \right) \right] \tag{66}$$

This equation incorporates a linear mixing law based on mass fractions rather than mole fractions and is also equivalent to

$$\frac{\Delta T}{\Delta T_I} = Sn^{-1} \tag{67}$$

with the substitution of Eq. (38) in mass fractions.

Stephan and Preusser [65] give a new correlation for multicomponent boiling that was formulated specifically to allow for nonlinear variations in the physical properties with composition. Their correlation is

$$\frac{\alpha D_d}{\lambda} = 0.100 \left(\frac{\dot{q} D_d}{\lambda T_{sat}} \right)^{0.674} \left(\frac{\rho_g}{\rho_l} \right)^{0.156} \left(\frac{\Delta h_v \, D_d^2}{2\kappa_l} \right)^{0.371} \left(\frac{\kappa_l^2}{\sigma D_d} \right)^{0.350}$$

$$\times \left(\frac{\eta_l c_{Pl}}{\lambda} \right)^{-0.162} \left[1 + \left| \sum_{i=1}^{n-1} (\tilde{y}_i - \tilde{x}_i) \left(\frac{\partial \tilde{y}_i}{\partial \tilde{x}_i} \right)_{\tilde{x}_{i,p}} \right| \right]^{-0.0733} \tag{68}$$

where D_d is calculated from

$$D_d = 0.0146\beta\{2\sigma/[g_n(\rho_l - \rho_g)]\}^{1/2} \tag{69}$$

The contact angle β is taken as $45°$ for water and $35°$ for organics and their aqueous mixtures. However, using values of β plotted in Figs. 8 and 9 shows that Eq. (69) predicts organic liquids to have departure diameters approaching zero.

Thome [40] has developed a nucleate pool boiling model for binary mixtures based on the cyclic thermal boundary layer stripping mechanism as depicted schematically in Fig. 44. The enthalpic heat transfer areas are those affected by the cyclic thermal boundary layer stripping mechanism. The unaffected area is under the influence of single-phase natural convec-

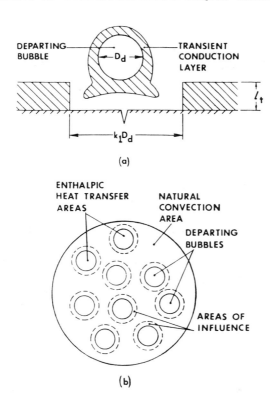

FIG. 44. Nucleate pool boiling model of Thome [40]. (a) Thermal boundary layer removal by departing bubble; (b) heat transfer regions.

tion. He assumed that the enthalpic heat transport rate is given by the expression

$$Q_{c_p} = \frac{\pi}{2} \rho_l c_{Pl} l_t k_1^2 D_d^2 fNA \, \Delta T \tag{70}$$

using one-dimensional transient heat conduction from the heated wall into the liquid. He then derived an expression for the energy balance at the heated wall. For the region of the nucleate pool boiling curve where natural convection is insignificant and assuming that the area of influence, the ratio of the bubble departure time to the bubble growth time, and the boiling site density all vary linearly with composition, Thome arrived at the following expression for predicting the variation in the boiling heat transfer coefficient for inertia-controlled bubble departure:

$$\alpha/\alpha_1 = \Delta T_1/\Delta T = Sn^{7/5} \tag{71}$$

This expression is similar to Eq. (67) of Calus and Leonidopoulos, but the model is more general, in that it incorporates the features of bubble departure (besides just the growth rate) into an energy balance. ΔT_I is defined by Eq. (60) and Sn by Eq. (38).

A corresponding equation for surface-tension-controlled departure is obtained by substituting Eq. (50) into Eq. (49), giving

$$\frac{\alpha}{\alpha_I} = \frac{\Delta T_I}{\Delta T} = \mathrm{Sn}^{3/2} \left\{ \frac{\sigma \sin \beta}{\sigma_I \sin \beta_I} \right\} \tag{72}$$

Equation (72) is essentially equivalent to Eq. (71) for a mixture following a linear mixing law in σ and β. The effects of σ and β in aqueous solutions are seen to be very important.

Using a new approach, Thome [83] postulated that the rise in the local saturation temperature for a mixture boiling on a smooth tube or plate is controlled by the total rate of evaporation at the heated surface. Thus the rise in the local saturation temperature $\Delta\theta$ is zero for single-phase natural convection at low heat fluxes and increases to its highest value at the peak heat flux. All the liquid arriving at the heated surface is assumed to be evaporated at the peak heat flux. Therefore for steady-state conditions at the peak heat flux the composition of the liquid arriving at the heated wall, $\tilde{x}_{\mathrm{bulk}}$, must equal its composition in the vapor, $\tilde{y}_{\mathrm{local}}$, leaving the wall in order to satisfy a conservation of molecular species. Figure 45 shows that the rise in the local bubble point temperature ΔT_{bp} is given as

$$\Delta T_{\mathrm{bp}} = T_{\mathrm{local}} - T_{\mathrm{bulk}} \tag{73}$$

where both T_{local} and T_{bulk} are evaluated at $\tilde{x}_{\mathrm{bulk}}$. Consequently, this is the maximum rise in the local bubble point temperature for nucleate pool

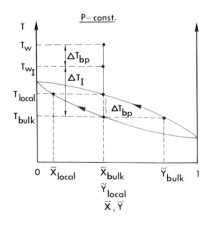

FIG. 45. Maximum rise in local boiling point temperature by Thome [83].

boiling. ΔT_{bp} is obtainable from phase equilibrium data as the temperature difference between the dew point and the bubble point at the bulk liquid mole fraction.

To calculate the heat transfer coefficient as a function of composition, ΔT_{bp} is substituted for ΔT^E in Eq. (59) such that

$$\frac{\alpha}{\alpha_I} = \frac{\Delta T_I}{\Delta T} = \frac{\Delta T_I}{\Delta T_I + \Delta T_{bp}} \tag{74}$$

where α_I is defined as

$$\alpha_I = 1/[(\bar{x}_1/\alpha_1) + (\bar{x}_2/\alpha_2)] \tag{75}$$

which is an expression equivalent to Eq. (60). This ΔT_{bp} method is also applicable to multicomponent boiling if Eq. (63) is used instead of Eq. (60).

B. COMPARISONS AND COMMENTS

Before comparing the relative accuracy of the various predictive methods, a number of points require further discussion.

First of all, the ultimate method for predicting mixture boiling heat transfer coefficients is one that (1) incorporates all the pertinent physical mechanisms into the model, (2) can be applied universally to all binary mixture systems over a wide pressure range, (3) accounts for the effect of heat flux on the level of degradation in the heat transfer coefficient [e.g., see Fig. 35(b)], (4) is extendable to multicomponent boiling, and (5) is easily implemented. Applying these criteria, we see that all the models reviewed in some manner incorporate a facet of the heat and mass diffusion process. Only the Stephan and Korner and Thome ΔT_{bp} equations have been tested for numerous mixture systems over a range of pressures. All the methods that are not already generalized to handle multicomponents could be, using Eq. (63) with appropriate modifications. The methods that do not require a specific empirical constant for each binary system are most easily used since the number of empirical constants required for all the binary pairs in a multicomponent mixture becomes prohibitive. Yet the calculation of Sn is also difficult because it involves the mass diffusivity as a function of composition and temperature. The Palen and Small correlation, the second Stephan and Preusser correlation, Eq. (68), and the Thome ΔT_{bp} equation are therefore attractive because they only require phase equilibrium data to evaluate the mixture term.

Another point concerning the various predictive methods is the definition of the ideal wall superheat or ideal heat transfer coefficient. Some methods use a particular single-component correlation; others use a linear mixing law. The problem associated with utilizing a single-component correlation is that it may work well for one component (e.g., water) but poorly for the second (e.g., an organic fluid). This is alleviated by use of a linear molar mixing law, obtaining ΔT_1 and ΔT_2, utilizing perhaps two different single-component correlations but at the expense of losing the effect of nonlinear variations in the physical properties with composition. Also, there is not uniformity in the application of the linear mixing law to mixtures forming an azeotrope. As illustrated in Fig. 46, Stephan and co-workers do not necessarily match the wall superheat at the azeotrope to ΔT_1, while Thome does. For instance, to the left of the azeotrope the ideal wall superheat is prorated by Thome as

$$\Delta T_1 = \left(\frac{\tilde{x}_1}{\tilde{x}_{az}}\right)\Delta T_{az} + \left(\frac{\tilde{x}_{az} - \tilde{x}_1}{\tilde{x}_{az}}\right)\Delta T_{\tilde{x}_1 = 0} \qquad (76)$$

and to the right a similar expression is used.

The system of ethanol–water mixtures at 1.0 bar has been studied experimentally by many different researchers and is an obvious choice to use to test the accuracies of the various methods presented, although it is not a typical industrial mixture. Also, aqueous mixtures tend to be the most difficult ones to predict.

Figure 47 demonstrates a comparison of the normalized experimental heat transfer coefficients α/α_1 to four of the predictive methods. The Stephan and Preusser correlation is seen to overpredict α/α_1 dramatically, while the surface-tension-controlled departure equation of Thome drastically underpredicts the experimental values. The inertia-controlled departure equation of Thome is an improvement, but the correlation of Palen

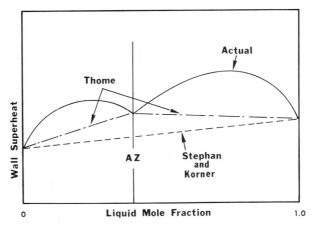

FIG. 46. Differing definitions of a linear mixing law for an azeotropic mixture system.

and Small handles the intermediate compositions better with an accuracy of about $\pm 30\%$. The Calus and Rice and the Calus and Leonidopoulos methods are not shown, but neither has the accuracy of the Palen and Small correlation.

Figure 48a shows a comparison of the Thome ΔT_{bp} equation to the data in Fig. 47. (The data of Stephan and Preusser are discarded because they deviate substantially from the other data sets.) Almost all the remaining data are predicted to within $\pm 15\%$. In Figure 48b the Stephan and Korner correlation is tested using their published value of A equal to 1.21. The scatter is broader than $\pm 15\%$, actually deviating up to about 30%.

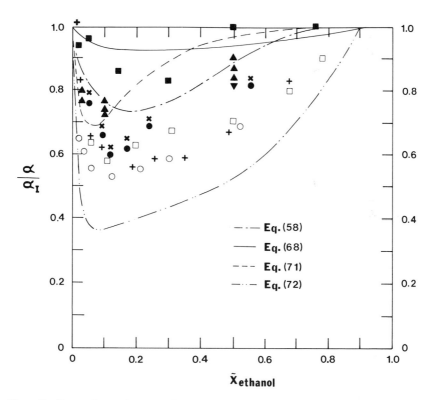

FIG. 47. Comparison of ethanol–water data at 1.0 bar with four predictive methods. Legend: (—) Stephan and Preusser [65], Eq. (68); (—··—) Palen and Small [84], Eq. (58); (– – –) Thome [40], Eq. (71); (—·—) Thome, Eq. (72); (●) Valent and Afgan [87a] at 300 kW/m²; (×) Valent and Afgan [87a] at 190 kW/m²; (○) Tolubinskiy and Ostrovskiy [53] at 116 kW/m²; (+) Grigor'ev *et al.* [81] at 232 kW/m²; (▲) Bonilla and Perry [87b] at 95 kW/m²; (▼) Cichelli and Bonilla [69] at 221 kW/m²; (□) Shakir [87c] at 200 kW/m²; (■) Preusser [13] at 200 kW/m².

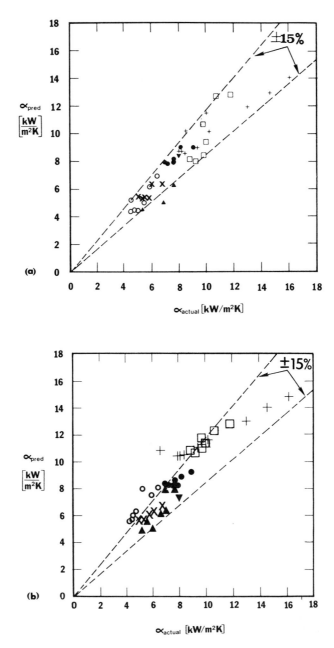

FIG. 48. Comparison of ethanol–water data at 1.0 bar with two predictive methods. (a) Thome [83], Eq. (74); (b) Stephan and Korner [14], Eq. (59). (Legend same as for Fig. 47.)

In summary, the Thome ΔT_{bp} method appears to be one of the easiest methods to apply (only phase equilibrium data are required) and also the most accurate. It was shown in Thome [83] that this method predicted boiling heat transfer coefficients quite well at medium and high heat fluxes for the following additional systems: ethanol–benzene, acetone–water, nitrogen–argon, nitrogen–oxygen, and nitrogen–methane. Even so, the method needs to include the effect of heat flux in order to perform adequately at low heat fluxes where reboilers tend to be designed to reduce operating costs.

VIII. Peak Nucleate Heat Flux

The peak nucleate heat flux or critical heat flux $\dot{q}_{cr}$ (point C, Fig. 34) denotes the highest heat flux that can be obtained in the nucleate pool boiling regime. The prediction of $\dot{q}_{cr}$ is therefore very important in the design of heat exchange equipment operating in the nucleate pool boiling regime, since it limits the upper range of operation. For mixtures there is no consistent method for predicting $\dot{q}_{cr}$ because of the many conflicting experimental results. Some of these experimental results are presented here to illustrate the dilemma. Then a number of physical phenomena related to the peak heat flux will be reviewed.

A. Experimental Investigations of $\dot{q}_{cr}$

Experimental studies on the peak nucleate heat flux have covered both a wide range of mixture systems and a wide variety of heater geometries. Mixture systems studied include aqueous mixtures, organic–organic mixtures, refrigerants, and cryogens. Geometries tested range from very small diameter wires up to flat plates and tubes of sizes typical to engineering practice. The ambiguity in the experimental results arises in that some experimentalists have obtained a pronounced maximum in $\dot{q}_{cr}$ while others have observed a minimum for the same mixture system.

Figure 49 from Preusser [13] shows a comparison of the results of three independent studies on the peak nucleate heat flux for the ethanol–water system. The data of Van Wijk et al. [28] for boiling from a 0.2 mm diameter wire form a maximum at all four pressures tested. The experiments of Isshiki and Nikai [56] for a 0.3 mm diameter wire gave similar results. Bobrovich [88], instead, found a minimum in $\dot{q}_{cr}$ for boiling off of horizontal and vertical plates while obtaining a maximum in $\dot{q}_{cr}$ for a wire. Figure 50 depicts the results of three additional studies on the ethanol–water system by Matorin [89], Tobilevich et al. [90], and Preusser [13] for heat-

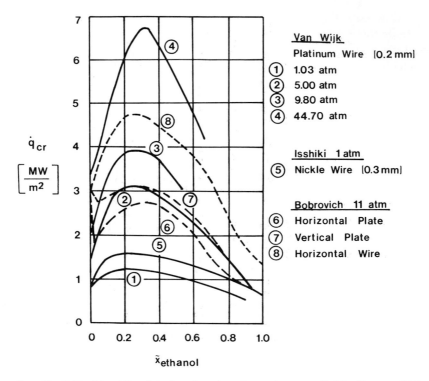

FIG. 49. Critical heat flux data for ethanol–water system compiled by Preusser [13].

ing surfaces of industrial dimensions. These show minima in qualitative agreement to the Bobrovich data. Thus both Bobrovich and Preusser concluded that the pronounced maximum in $\dot{q}_{cr}$ only occurs with heating surfaces of small dimensions. In addition, Bobrovich [88] also observed this trend with butyl alcohol–water mixtures, as shown in Fig. 51.

Pitts and Leppert [91], however, obtained results for methyl ethyl ketone–water mixtures that conflict with the preceding conclusion. Figure 52 shows that they obtained minima for wires of 0.0087 and 0.13 cm diameter but a maximum for a wire of 0.0254 cm diameter.

Afgan [92] obtained a maximum to each side of the azeotrope composition for ethanol–benzene mixtures boiling from a 5.1 mm diameter tube as shown in Fig. 53. Sterman *et al.* [68] also found an increase in $\dot{q}_{cr}$ in the mixtures of diphenyl–benzene compared to their single-component values for boiling off of a vertical plate (see Fig. 54). Thus large heating surfaces have also been found to result in a maximum in $\dot{q}_{cr}$ rather than in a minimum.

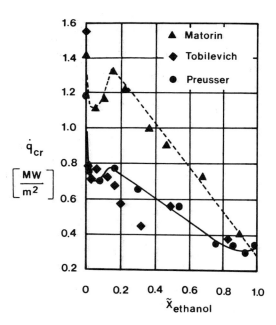

FIG. 50. Additional critical heat flux data for ethanol–water compiled by Preusser [13].

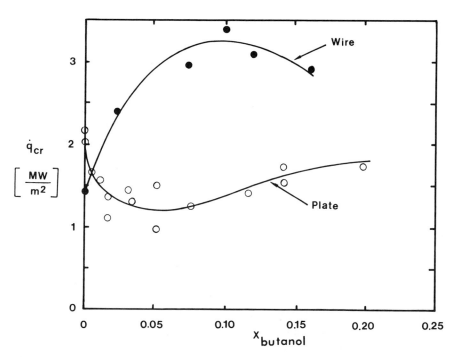

FIG. 51. Dependence of $\dot{q}_{cr}$ on surface geometry in butyl alcohol–water mixtures at 9.9 bar (wire) and 11.1 bar (plate) from Bobrovich et al. [88].

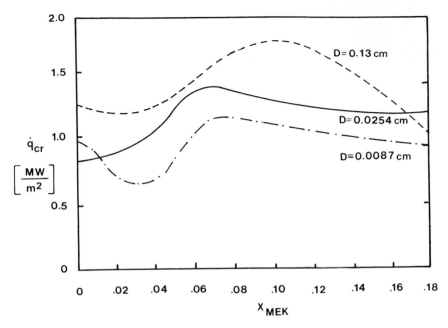

FIG. 52. Dependence of $\dot{q}_{cr}$ on wire diameter for MEK–water mixtures by Pitts and Leppert [91].

In conclusion, it is clear that additional experimental work is required to determine the cause of these seemingly contradictory results.

B. PHYSICAL PHENOMENA AND MECHANISMS

The Zuber [93] hydrodynamic instability model works quite well for predicting the peak nucleate heat flux for boiling many different single-component liquids. Basically, the model assumes that vapor jets emanating from the heated surface will become unstable at some point and trigger a Helmholtz instability. This causes a coherent vapor film to cover the heated surface. Assuming that the addition of a second component does not affect the physical phenomena, Zuber's analytical solution for $\dot{q}_{cr}$, given for a large flat heater as

$$\dot{q}_{cr} = 0.149\rho_g^{1/2}\,\Delta h_v[g_n(\rho_l - \rho_g)\sigma]^{1/4} \tag{77}$$

with further development by Lienhard and Dhir [94], in assuming that the unstable Helmholz wavelength was equal to the one-dimensional Taylor wavelength, can be used for predicting the values for mixtures. As Fig. 50, Preusser [13] shows that the nonlinear variation in the physical properties results in Eq. (77) predicting a minimum in $\dot{q}_{cr}$ with respect to composition. This, then, is in qualitative agreement with some of the

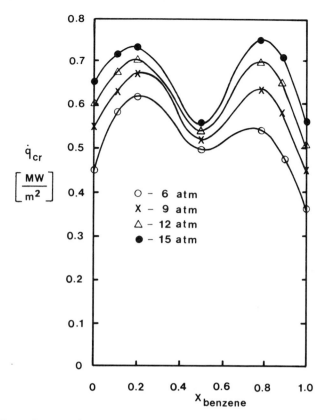

FIG. 53. Dependence of $\dot{q}_{cr}$ on pressure for ethanol–benzene mixtures by Afgan [92].

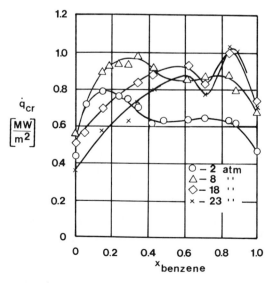

FIG. 54. Dependence of $\dot{q}_{cr}$ on pressure for diphenyl–benzene mixtures by Sterman *et al.* [68].

experimental studies. However, it is not at all clear what the effect of mass diffusion will be on the formation of vapor jets and their stability.

Van Stralen [33] has approached the study of the boiling crisis from a perspective different from Zuber's. He considers the boiling crisis to be a surface-related phenomenon caused by bubble packing. From his experiments on pool boiling and bubble growth from thin wires he observed that the maximum in the peak nucleate heat flux occurred at the same composition as the minimum bubble growth rate. Since the bubble departure diameter decreases to a minimum in the same neighborhood, Van Stralen surmised that the increase he observed in $\dot{q}_{cr}$ was due to the higher bubble-packing density possible in the mixtures. The peak heat flux was thought to occur when the microlayers of neighboring bubbles touched. The increase in the peak heat flux is then explained by the summation of the contributions of direct vaporization at the heated surface and the convective heat transfer caused by the bubble motion. Van Stralen showed that while the evaporation heat flux decreased with the addition of a solvent to water, the convective heat flux increased by a greater margin. Thus $\dot{q}_{cr}$ experienced an increase in the mixture in comparison to pure water.

Hovestreijdt [95] also based his physical interpretation of the transition of stable nucleate boiling to unstable film boiling by the coalescence of neighboring bubbles on the heated surface. However, he explained the maximum in $\dot{q}_{cr}$ as being due to two phenomena: preferential evaporation of the more volatile component, as previously noted by Van Stralen, and the Marangoni effect. The flow of liquid as a result of a surface tension gradient along a vapor–liquid interface is known as the Marangoni effect. For a positive system (a system whose more volatile component has the lower surface tension) the preferential evaporation of the more volatile component causes the surface tension to be greater near the heated wall than in the bulk. Thus liquid is drawn in between neighboring bubbles, and this tends to inhibit their coalescence. Consequently, the density of boiling sites can be higher before transition to film boiling occurs. Presumably, this results in a higher peak nucleate heat flux, although no thermal model is presented by Hovestreijdt to explain how this occurs.

An aspect of the influence of surface tension on bubble coalescence at the peak heat flux neglected by Hovenstreijdt is the effect of the temperature gradient in the liquid near the heated wall. Since the surface tension for a good many fluids decreases with increasing temperature on the order of 0.1 dyne/(cm °C), see the table compiled by Jasper [96], the surface tension of the liquid near the wall tends to be lower than that in the bulk liquid. Thus the temperature gradient works to oppose liquid circulation caused by the preferential evaporation of the more volatile component for a positive system.

C. Summary

No general model or predictive method can currently explain the seemingly conflicting experimental results. Further experimental and analytical work is suggested. A study on the effects of mass diffusion and surface tension gradients on the formation of vapor jets, which are assumed to exist in the Zuber hydrodynamic instability model, may provide some fruitful insight into the problem.

IX. Film Boiling

Film boiling occurs when heat is being transferred to a two-phase mixture and the surface is in contact only with vapor. Heat passes through the vapor to the vapor–liquid interface and is at least partly used there to generate further vapor. For most pool boiling situations the interface is intensely wavy and streams of bubbles are produced at the nodes of the waves; in other cases (e.g., vertical plates or horizontal plates with parallel flow of fluid), the interface is assumed to be planar. Several studies have been concerned with the effect on film boiling of the addition of a second component to a pure fluid coolant (though very few discuss mixtures of three or more components), and some interesting and important conclusions have emerged. The most important of these are first, that, in contrast to nucleate boiling, the second component increases the heat transfer coefficient in the established film boiling region and, second, that it increases the heat flux at the minimum in the boiling curve (Liedenfrost point). It is, however, found both from theory and practice that for a given mixture relative volatility, $\bar{\alpha}_{AB}$, the effects are smaller in film boiling than in nucleate boiling.

In this section we examine first film boiling on flat plates or tubular heaters of large diameters, then film boiling on thin wires, and finally heat transfer at the Liedenfrost point.

A. Film Boiling on Flat Plates and Tubes

Perhaps the clearest illustration of the mixture effect is given in the work of Kautsky and Westwater [97], illustrated in Fig. 55. They boiled CCl_4–R-113 mixtures on a horizontal flat plate immersed in a pool of saturated liquid $[T_b = T_{sat}(x_b)]$. The figure shows that at several fixed superheats $(T_w - T_{sat})$, the heat flux was higher for mixtures than for the pure fluids; large increases, out of proportion to any changes in transport properties, can be seen when small quantities of CCl_4 are added to R-113 or vice versa. Similar behavior is also found, for example, for acetone–cyclohexanol [98] and for small additions of high molar mass polymers to

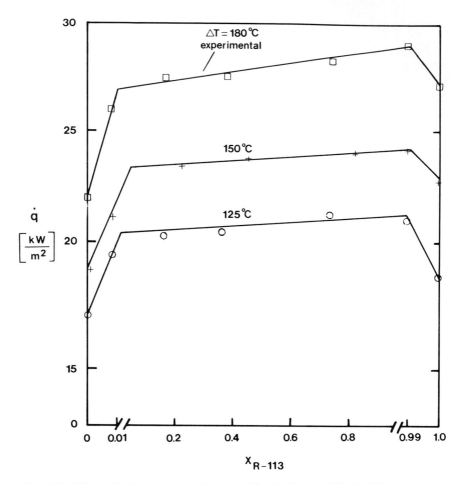

Fig. 55. Effect of mixture composition on film boiling of CCl₄–R-113 mixtures by Kautsky and Westwater [97].

cyclohexane [99]. Other workers have reported that there is no special mixture effect. For example, Wright *et al.* [76] experimented with ethane–ethylene mixtures on a horizontal tube and found a smooth transition in the heat transfer coefficient in going from one pure fluid to the other, although they found a marked deterioration in the nucleate boiling heat transfer coefficient. Brown and Colver [100] boiled LNG on a horizontal tube and found its film boiling, but not its nucleate boiling, behavior identical to that for pure methane.

The explanation for these effects, or lack of them, is as follows. For film boiling, as for nucleate boiling, the fact that the vapor is generated from

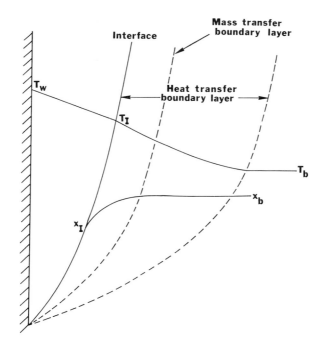

FIG. 56. Temperature and concentration profiles for film boiling of mixtures from Yue and Weber [98].

the liquid at the interface and that one component is more volatile than the other means that the interface will be preferentially stripped of the light component. Since the interface will thus be at $T_{sat}(x_I)$ it will be hotter than the bulk and hence heat will be conducted from the interface into the liquid. The temperature and concentration profiles now resemble those shown in Fig. 56, which is taken from the work of Yue and Weber [98] (see following discussion). Thus the conventional wisdom argues that there is an extra path for heat removal and $\dot{q}$ is increased for a given $T_w - T_{sat}(x_b)$. We note that the heat transfer resistance in the liquid is in series with that in the vapor and ask the question "For a given $T_w - T_{sat}(x_b)$, an increase in T_I will mean a decrease in $T_w - T_I$ (i.e., in the temperature drop across the vapor film). As the extra heat also has to pass through the vapor film, how can it do so at a reduced ΔT across the vapor?" The answer may be that if some heat arriving at the interface is being conducted into the liquid, less heat is being used to generate vapor; the vapor resistance is actually reduced because the film is thinner than it would be for a pure fluid. Hence we can manage the extra heat transfer at reduced $T_w - T_I$. These points are illustrated in Fig. 57.

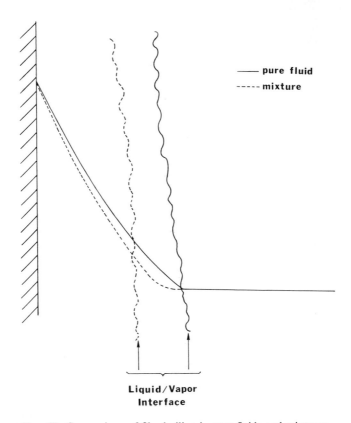

pure fluid
---- mixture

Liquid/Vapor
Interface

Fig. 57. Comparison of film boiling in pure fluids and mixtures.

The fact that a temperature gradient exists from the interface into the bulk liquid means that there are many similarities between this situation and subcooled film boiling in pure fluids (where the heat flux is known to increase with bulk subcooling). The latter situation has been modeled by Nishikawa and Ito [101], and their model was extended to saturated film boiling of mixtures by Yue and Weber [98], by Marschall and Moresco [102] for vertical plates and horizontal cylinders, and, with some differences, by Prakash and Seetharamu [103] for horizontal plates with parallel flow. Figure 56 shows the configuration used by Yue and Weber. Both phases have velocity, temperature, and concentration gradients; there is no slip in temperature or velocity at the interface (in contrast to the work of Sparrow and Cess [104]). Momentum, heat, and mass balances and rate equations are written for each phase; equilibrium at the interface is assumed. (We would question the suggestion by Yue and Weber [98] that the vapor phase composition at the interface is necessarily the composi-

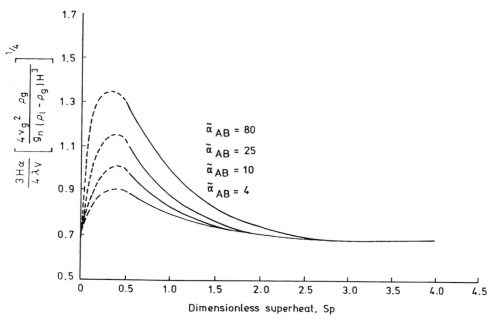

FIG. 58. Effect of relative volatility on film boiling of mixtures on vertical flat plates by Yue and Weber [98].

tion weighted average of the mass fluxes through the interface; Colburn and Drew [105] analyzed the similar problem for condensation in a binary mixture and showed that the interface conditions and condensation rate are interrelated.) The system of equations was solved with the aid of similarity transforms identical to those used by Nishikawa and Ito.

The main illustration of the mixture effect is shown in Fig. 58. The vertical axis shows a dimensionless heat transfer coefficient for film boiling on a vertical cylinder [98]; a further simple similarity transform gives the coefficient for a horizontal cylinder. On the horizontal axis is plotted dimensionless wall superheat, Sp $[= c_{pg}(T_w - T_{sat}(x_b)]/Pr_g \Delta h_v)$. We can see for a given wall superheat that the coefficient increases with volatility (i.e., with the tendency to strip the light component from the interface). We can now suggest that Wright et al. [76] found no effect for ethane–ethylene mixtures because $\bar{\alpha}_{AB}$, the relative volatility, is not more than 2, which (see Fig. 58) does not give significant conduction into the liquid. This case is an excellent illustration of the relative weakness of mixture effects in film boiling compared to nucleate boiling.

The amelioration is also found to increase with liquid Schmidt number; low Sc produces high diffusion rates in the liquid, rapid replenishment of the light component at the interface, and reduction of the driving force for heat conduction into the liquid.

Figure 58 shows that for a given $\bar{\alpha}_{AB}$, its influence on the heat transfer coefficient diminishes with superheat and is not felt at Sp > 3. This is because as the superheat increases, the stripping effect is at its maximum when there is none of the light component at the interface. The conduction into the liquid consequently reaches a maximum and, with further increases in wall superheat, is a decreasing proportion of the total. We note that this logic would suggest that there might be a maximum in the heat transfer coefficient at some point in Fig. 58 as the situation built up to this maximum in conduction; this is illustrated by the dashed line in the figure. No such turning point is seen in Yue and Weber's theoretical results, but they do not show Sp < 0.5. It is perhaps worth noting that Sp for the results in Fig. 55 is 0.33 for $\Delta T = 125$ K and 0.48 for $\Delta T = 180$ K.

The mathematical model discussed above was tested against data by Yue and Weber [98] for a horizontal carbon cylinder 8 mm in diameter. Photographs of the boiling showed that the vapor bubbles were released at the top of the tube, but everywhere else the vapor film was fairly smooth and conditions were close to those assumed in the model. The results confirmed the predictions that at low $\bar{\alpha}_{AB}$ (e.g., n-hexane–toluene and ethanol–benzene), the fluid behaves like a pure fluid and the appropriate prediction method [106] is adequate, whereas at high $\bar{\alpha}_{AB}$ (e.g., acetone–cyclohexanol), the additional heat transfer due to the mixture effects must be taken into account.

The work carried out so far in this area is restricted to the case where a flat plate is sufficiently short or a tube is of sufficiently small diameter to ensure that the vapor flow is always laminar. The case where turbulence is significant has been studied by Hsu and Westwater [107] for pure fluids but has not yet been extended to mixtures.

B. Film Boiling on Thin Wires

On thin wires the bubbles that are emitted in regular streams dominate the heat transfer. The spacing of the streams is determined by the wavelengths of the disturbances in an otherwise smooth film, and the analysis of these waves is central to the calculation of heat transfer in this situation.

Van Stralen et al. [108] studied film boiling of mixtures of water and methyl ethyl ketone (MEK) on a platinum wire of diameter 0.2 mm. The composition $x_{MEK} = 0.041$ was chosen as that which gives the minimum bubble growth rate in nucleate boiling; the corresponding $\bar{\alpha}_{AB}$ is 35. Tests over a large range of superheats again showed an increase in the heat transfer coefficient—up to 80% for $T_w - T_{sat}(x_b) = 1100°C$.

The situation was modeled by treating the "bubbles" growing at the nodes of the waves in a similar way to those extensively treated by Van Stralen and co-workers for nucleate boiling (see Section IV). For pure water the measured growth rates suggest an effective superheat of 6°C and this increased to 11°C for the mixtures at the highest superheat; the increase was interpreted as a complete exhaustion of the more volatile component at the liquid–vapor interface. Application of the standard bubble growth model by Van Stralen et al. suggests that for water 95% of the heat input is used to generate vapor directly at the interface; for the mixtures this value drops to 53% (in line with the conclusions for nucleate boiling). The rest of the heat is held to pass into the liquid, presumably by conduction, and subsequently to the rising vapor bubbles, which grew further. The model also predicts a reduction in the bubble growth rate for mixtures, and this was again confirmed by photographic evidence. It is interesting to note that for the value of Sp under these conditions of 1.45 and for $\bar{\alpha}_{AB}$ equal to 35, Figure 58 suggests that for flat plates there indeed would be a beneficial effect of the second component on the heat transfer.

C. FILM BOILING AT THE LEIDENFROST POINT

So far we have been concerned with established film boiling; here we discuss the location of the minimum point in the Nukiyama boiling curve, also known as the Leidenfrost point. Figure 59 shows data obtained by Yue and Weber for several mixtures on a carbon tube, o.d. 8 mm. Note the large increase for small additions of cyclohexanol to acetone and the lack of any marked effect for CCl_4–R-113. In examinations of film boiling of pure fluids, the minimum is always found by determining the wavelengths of the Taylor instabilities [109–112]. The minimum heat flux is then found from

$$\dot{q}_{min} = \left(\begin{array}{c} \text{energy transport} \\ \text{per bubble} \end{array} \right) \cdot \left(\begin{array}{c} \text{bubbles per unit} \\ \text{heater area in one} \\ \text{oscillation} \end{array} \right)$$
$$\cdot \left(\begin{array}{c} \text{minimum number of} \\ \text{oscillations per} \\ \text{unit time} \end{array} \right) \quad (78)$$

A complete model for mixtures (or subcooled fluids) would involve rectifying this relationship to allow for the simultaneous conduction into the liquid. For wires Van Stralen et al. [108] have shown from photographs that the second term is unaffected by the addition of the second compo-

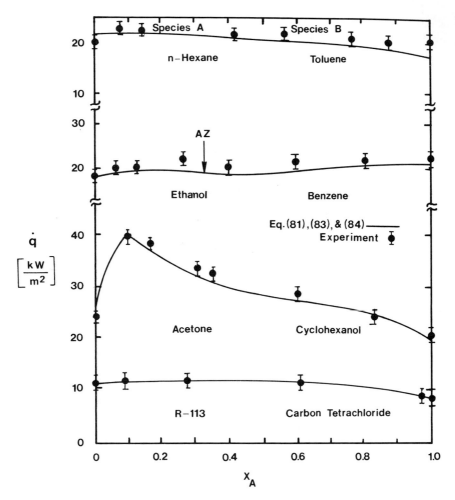

Fig. 59. Minimum film boiling heat flux of mixtures by Yue and Weber [113].

nent, whereas the rate of energy transport from a given "node" (the product of the first and third terms) is reduced. For flat plates Yue and Weber [113] found that the bubble spacing is given by the most dangerous wavelength and that the frequency of oscillation is the same as for pure fluids. A comprehensive model would also have to consider Marangoni forces that act along an interface when a gradient of surface tension is set up by one of concentration. (We can assume that the wave structure may cause such gradients.)

A form of modification of the preceding relationship was carried out by

Yue and Weber [113] and applied to the results in Fig. 59. They took the equation for $\dot{q}_{min}$

$$\dot{q}_{min} = 0.15\rho_g\,\Delta h_v\left[\frac{\sigma g_n(\rho_l - \rho_g)}{(\rho_l + \rho_g)^2}\right]^{1/4} \tag{79}$$

which arises [109] from (1) Eq. (78) and (2) modeling of the Taylor instability for plane horizontal surfaces, and they combined this equation with Bromley's [106] relation for the film boiling heat transfer coefficient on a cylinder

$$\alpha = 0.62\left[\frac{\lambda_g^2\rho_g(\rho_l - \rho_g)\,\Delta h_v}{\eta_g D\,\Delta T}\right]^{1/4} \tag{80}$$

After combining these two equations and rearranging they obtained

$$\Delta T_{min} = 0.334\left[\frac{D}{\lambda_g}\right]^{1/3}\left[\frac{\rho_g\,\Delta h_v}{\lambda_g}\right]\left[\frac{g_n(\rho_l - \rho_g)}{(\rho_l + \rho_g)}\right]^{2/3}$$

$$\left[\frac{\sigma}{g_n(\rho_l - \rho_g)}\right]^{1/2}\left[\frac{\eta_g}{g_n(\rho_l - \rho_g)}\right]^{1/3} \tag{81}$$

They found that this equation was satisfactory for pure fluids and mixtures of low volatility ($\bar{\alpha}_{AB} < 4$) on a carbon tube (o.d. 8 mm) if the constant 0.334 were replaced by 0.42, but it underpredicted ΔT_{min} for mixtures of higher volatility. In order to predict these results they proposed that the preceding equation should apply to the temperature difference across the film (which determines the onset of critical instability in the wave pattern). Hence they wrote

$$\Delta T_{min} = T_w - T_{sat}(x_b) \tag{82}$$

$$= (T_w - T_1) + [(T_1 - T_{sat}(x_b)] \tag{83}$$

The first term in Eq. (83) is given by Eq. (81) and the second, following a suggestion of Kutateladze *et al.* [114] for the prediction of the *maximum* heat flux, is correlated with $(y^* - x_b)/(y^* - x_b)_{max}$, where $(y^* - x_b)_{max}$ is the maximum value of this difference for all possible concentrations of the binary pair. An empirical correlation obtained from the four mixtures tested gave

$$T_1 - T_{sat}(x_b) = \frac{y^* - x_b}{(y^* - x_b)_{max}}[T^* - T_{sat}(x_b)] \tag{84}$$

where the temperature difference on the right-hand side is the boiling range of the mixture of composition x_b. Figure 59 shows that good agreement is attained with this relationship, but, as always, the real test is with some mixture that was not included in the data base. Yue and Weber comment that they have searched for, but not found, a theoretical reason for this simple relationship; this correlation has the advantage that in use it requires only the vapor–liquid equilibrium data of the mixture.

X. Convective Boiling

The feedstocks to many boilers in the chemical and petrochemical industries are complex mixtures of many components and have large boiling ranges. Thus the problem of predicting heat transfer coefficients in evaporation of mixtures flowing through heat exchangers is a very real and important one for designers of chemical plant and related equipment. In spite of the importance of the problem, much less work has, however, been carried out on convective boiling than on pool boiling of mixtures; Butterworth [115] has included this problem along with the major unresolved problems in heat exchanger design. This section summarizes the work that has been carried out on convective boiling of mixtures. The general topic of convective boiling of pure fluids, used as a basis for much of the mixture work, is reviewed elsewhere [116, 117].

In the boiling of a flowing mixture there are several general features to consider in addition to those discussed in earlier sections, and these are illustrated in Fig. 60, which shows the flow patterns and temperature distributions from the single-phase liquid to beyond the dryout point. Near the entry to a boiler tube with little vapor present (low quality) the heat transfer is generally assumed to be dominated by nucleate boiling, whereas further downstream, at higher qualities, two-phase convection becomes increasingly more important as nucleate boiling is suppressed. The onset of nucleate boiling in mixtures is discussed in Section III. In the nucleation region we may reasonably expect effects similar to those discussed in Sections III–VI, and this is indeed found (see following discussion); in the two-phase convection region, usually annular flow (a thin liquid film being dragged up the walls of the tube by a droplet-laden gas core), we may also expect concentration gradients to be set up as the system adjusts toward the axially changing equilibrium concentrations. However, a further point to note is that not only are the concentrations changing but so is the equilibrium temperature as the lower boiling components are progressively boiled off; this effect can often overcome, as shown in Fig. 60, the tendency of the saturation temperature to fall with the pressure. The increase in mean fluid temperature means that some of the heat input is required for sensible heating; the variation of the mean flash temperature with heat load must be calculated carefully for boiler applications.

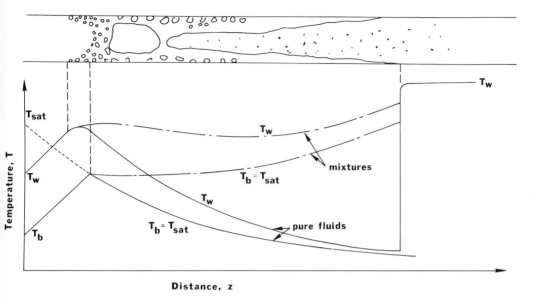

FIG. 60. Temperature profiles and boiling regimes for convective evaporation.

The few predictive methods that have been suggested for mixtures are usually based on modifications of pure fluid methods. There are many such methods. We discuss briefly here that of Chen [118] because (1) it is one of the most commonly used for pure fluids and (2) it has been suggested as a basis for mixture methods by application of suitable corrections. It is an "additive" method, summing nucleate and convective contributions, and is expressed in basic form as

$$\alpha = \alpha_l F + \alpha_p S \tag{85}$$

where α_l is the heat transfer coefficient that would be found for the liquid phase flowing alone, F is a multiplier (empirically related to hydrodynamic parameters) that accounts for the apparent increase in velocity due to the presence of the vapor, α_p is the pool boiling coefficient at the local wall superheat, and S is a "suppression factor" that accounts for the fact that α_p is found from pool boiling correlations that overpredict the nucleate boiling coefficient [117].

A correction to this method for mixtures was proposed by Bennett and Chen [119], who derived the following equation for comparing with data on ethylene glycol–water—a rather special case where only the water is volatile:

$$\alpha = \alpha_l F'C + \alpha_p SC' \tag{86}$$

where F' is a modified version of F accounting for Prandtl number effects (having no relationship to mixture effects) and C and C' are correction factors.

The factor C allows for mass transfer effects (in the liquid phase only) on the convective heat transfer coefficient and is given by

$$C = 1 - \frac{1 - y}{\rho_l \, \Delta h_v \beta (T_w - T_{sat})} \frac{\partial T_{sat}}{\partial x} \tag{87}$$

In deriving this equation it is assumed that the mass transfer affects the driving force for heat transfer but not the heat transfer coefficient itself—an assumption also implicitly made in all the pool boiling work discussed in this review. It is also assumed that all the vapor that is generated is in equilibrium with the liquid and that all the heat input results in evaporation at the interface. (As noted at the beginning of this section, this is not so because some of the heat is required as sensible heating following the increase in equilibrium temperature.) Bennett and Chen evaluate a correction to the apparent latent heat to allow for this effect.

The factor C' allows for mass transfer limiting effects on the nucleate boiling heat transfer and is modeled in a similar way to the methods shown in Section IV,B giving

$$C' = \frac{1}{1 - \dfrac{c_{pl}}{\Delta h_v} (y - x) \dfrac{dT}{dx} \left(\dfrac{\kappa_l}{\delta} \right)^{1/2}} \tag{88}$$

Tests were carried out on Eq. (86) using a data base dominated by convective heat transfer and reasonable agreement was found. It was concluded that application of the correction factors produced a significant improvement in accuracy.

The conclusion that nucleate boiling in flow situations is reduced in a similar way to that found in pool boiling is supported by data of Toral et al. [23] (see Fig. 61) for in-tube boiling of ethanol–cyclohexane mixtures. They plotted the nucleate boiling portion of the total heat flux against the wall superheat. None of the current established correlations are particularly successful, especially since they all implicitly assume that the boiling curves for all the mixtures and pure components are straight parallel lines. This is clearly not the case here, and indeed is not generally so (see Fig. 35). Several of these correlations were indeed formulated from data at constant heat flux. Similar reduction in nucleate boiling heat transfer coefficients was also found by Kadi [120] for ethylene glycol–water mixtures, by Fink et al. [121] for R-11–R-113 mixtures, and by Rose et al. [122] for aqueous binary mixtures of ethanol, methanol, and n-butanol, all for boiling on a horizontal tube with a flow stream normal to it.

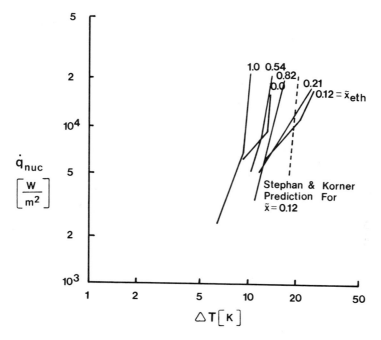

FIG. 61. Boiling curve for ethanol–cyclohexane mixtures derived from flow boiling data from Toral *et al.* [23].

The convective heat transfer region was investigated, theoretically, in detail by Shock [123]. He analyzed heat and mass transfer in the liquid and vapor in the annular region in the absence of nucleate boiling and assuming that all the liquid flows in the film. Liquid phase mass transfer coefficients were found from a mass transfer analogy to the heat transfer study carried out by Hewitt [124]; the diffusional mass transfer within the liquid film was found to affect the heat transfer coefficient (although usually only to a small extent). Vapor phase mass transfer is analyzed with the Chilton–Colburn analogy ignoring the presence of the liquid film.

Mass balances give the change in component flows over an element as the sum of contributions from diffusive and net flow (normal to the liquid–vapor interface):

$$-d(W_1 x_b) = 2\pi r_I j_{l,I} \, dz - \frac{dW_l}{dz} x_I \, dz \tag{89}$$

$$d(W_g y_b) = 2\pi r_I j_{g,I} \, dz + \frac{dW_g}{dz} y_I \, dz \tag{90}$$

These were combined with heat balances (which effectively neglect the heat absorbed by the liquid film) and the resultant equations were integrated to give the axial variation of the liquid and vapor concentrations and temperatures (both in the "bulk" flow and at the interface). It was assumed that the vapor phase temperature was always the dew point corresponding to the mean concentration. This is a constraint on the calculation, an effect that requires closer future examination.

The model was applied to ethanol–water, which has a lower boiling range (~15°C maximum) than many mixtures of interest in the chemical industry. A typical axial profile is shown in Fig. 62. We can see the effect of allowing the removal of either or both of the resistances to mass transfer. This result, typical of those found, shows that the effect of the mass transfer resistances on the interface (and hence wall) temperature is small and that any change that is produced is mainly due to the vapor phase resistance. This work is a useful starting point as shown but requires repetition for mixtures of wider boiling range and removal of some of the assumptions in the method.

The material discussed so far is an almost complete summary of the

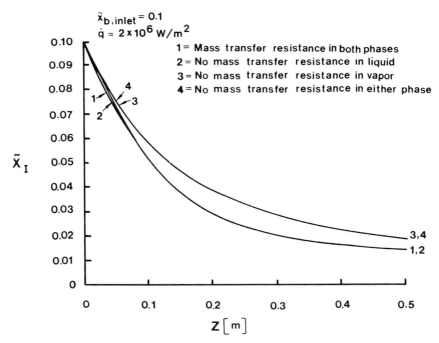

FIG. 62. Axial variation of interface concentration for annular flow heat transfer to ethanol–water from Shock [123].

published work in the prediction of heat transfer coefficients for convective boiling of mixtures. Perhaps this paucity of progress goes hand in hand with the real problem, which is the lack of data. The production of such data is a first priority for research workers.

All the theoretical studies carried out so far have tended to be specific to particular cases or to be too complex for design use. Since studies of condensation of mixtures have produced both reliable data and well-verified generalized calculation methods, we may be able to learn from this field, which does not, of course, suffer from the added complexity of nucleation effects.

The case of condensation of binary mixtures has been studied by Colburn and Hougen [125] for one condensable component and one noncondensable and by Colburn and Drew [105] for both condensable components. Both studies were primarily concerned with the resistance to heat and mass transfer in the vapor phase.

Temperature and concentration gradients build up in both phases because of the resistance to heat and mass transfer. For the case of two condensables the total heat flux at the liquid–vapor interface is given by

$$\dot{q} = \dot{q}_c + \dot{q}_{lat} \tag{91}$$

$$= \alpha_{\dot{g}}(T_g - T_l) + \Delta h_v \dot{n}_T \tag{92}$$

where

$$\dot{n}_T = \beta^{\cdot} \frac{\tilde{y}_b - \tilde{y}_l}{z - \tilde{y}_b} \tag{93}$$

and $\alpha_{\dot{g}}$ and $\beta^{\cdot}$ are, respectively, the heat and mass transfer coefficients corrected for the effect of mass transfer [126]. These equations can be solved stepwise along a channel (usually making the assumption that the bulk vapor temperature follows the dew point corresponding to the bulk concentration). The solution techniques for this and other problems in the condensation of mixtures are described by Webb and McNaught [127]. In principle, the same methods can be applied to evaporation (with the addition of a nucleate boiling contribution). This has not been done to our knowledge.

Modern sophisticated methods of analysis of multicomponent heat and mass transfer rates can be applied to model the condensation of mixture systems of any number of components [128], producing results similar in form to Eqs. (91)–(93). The full analysis of multicomponent mass transfer requires considerable matrix manipulation during solution and is too great a step to consider *yet* for evaporation; a method that is far simpler and in many cases not much less accurate is that originally devised by Silver

[129] and later, independently, by Bell and Ghaly [130]. The derivation of an equation for evaporation making the same assumptions as the model for condensation but allowing for the simultaneous occurrence of nucleate boiling has been given by Sardesai *et al.* [128] and is as follows.

The total heat flux is given by

$$\dot{q} = d\dot{Q}/dA \tag{94}$$

and the sensible heat flux in the vapor phase by

$$\dot{q}_g = d\dot{Q}_g/dA \tag{95}$$

Combining these equations we get

$$\dot{q}_g/\dot{q} = d\dot{Q}_g/d\dot{Q} \tag{96}$$

If we make the assumption, usually minor, that all the sensible heat for the liquid-phase heating crosses all the liquid film, we get

$$\dot{q} = \alpha_c(T_w - T_I) + \alpha_n(T_w - T_{sat}) \tag{97}$$

The sensible heat flux in the vapor can also be described by

$$\dot{q}_g = \alpha_g(T_I - T_g) \tag{98}$$

The value of T_{sat} needs to be carefully defined. The most "correct" value would be T_{sat} at x_b (α_n will naturally require correction for mixture effects as discussed in Section VIII). One of the assumptions in the original Silver–Bell–Ghaly method is that the concentrations are uniform within the two phases and are equivalent to flash vaporization values at the local pressure and enthalpy. It is also assumed that the vapor is at its saturation temperature, hence T_{sat} can be identified with T_g. Alternatively, a rather more conservative assumption would be to assume that $T_{sat} = T_I$. Making the first assumption and combining Eqs. (97) and (98) we get

$$\dot{q} = \frac{1 + \alpha_n/\alpha_c}{1/\alpha_c + (\dot{q}/\dot{q}_g)/\alpha_g} (T_w - T_g) \tag{99}$$

Alternatively, the second assumption gives

$$\dot{q} = \frac{1}{\dfrac{1}{\alpha_c + \alpha_n} + (\dot{q}/\dot{q}_g)/\alpha_g} (T_w - T_g) \tag{100}$$

In each case $\dot{q}/\dot{q}_g$, which is given by Eq. (96), can be shown to be equal to $x_g c_{pg}(dT_{sat}/dh)$ (usually known as Z) which can be evaluated before carrying out heat transfer calculations for the particular mixture since it is a function of physical properties and local average equilibrium quality. In the basic Silver–Bell–Ghaly method two further assumptions which are made and have not yet been noted are:

(1) α_g is calculated as a value corresponding to the vapor flowing alone and is not corrected for any effects due to the presence of the liquid or the waves on the film.

(2) α_g is not corrected for mass transfer effects (i.e., we use α_g not $\alpha_{\dot{g}}$).

Using either Eq. (99) or Eq. (100), it is possible to evaluate $\dot{q}$ for a given T_w (or vice versa) where the variables on the right-hand side are given as follows:

$$\alpha_n = \alpha_p SC'$$

$$\alpha_c = \alpha_l F'C$$

$$\dot{q}/\dot{q}_g = Z = xc_{pg}(dT_{sat}/dh)$$

and T_g = mean temperature assuming equilibrium flash vaporization. In the absence of nucleate boiling both Eqs. (99) and (100) reduce to

$$\dot{q} = \frac{1}{1/\alpha_c + Z/\alpha_g}(T_w - T_g) \tag{101}$$

which is the simple evaporation analog of the original derivation for condensation.

A suggestion that the method be used in this form was made by Palen *et al.* [131], who derived an equation similar to Eq. (100). They compared this with data in air–water systems. Figure 63 shows a comparison with the data; curve 1 shows the prediction of Eq. (85) (i.e., in the absence of any mixture correction), and curve 2 shows the prediction of Eq. (100) (i.e., including the effect of the correction). It can be seen that without the correction the data are always overpredicted, whereas with it a conservative prediction is achieved; perhaps a closer match of prediction to data would be achieved if either assumption (1) or (2) above were relaxed; in each case this would cause the value of α_g in Eqs. (99)–(101) to increase. This conclusion is also reached for air–benzene and air–toluene mixtures (with a nucleate boiling contribution included). The method has not yet been tested on a mixture where all components are present in each phase.

In conclusion, it can be seen that some ideas have been suggested for convective evaporation of mixtures. Several have combined methods

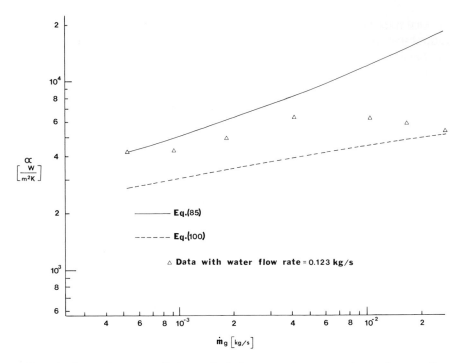

FIG. 63. Comparison of prediction methods for convective evaporation in air–water flow from Palen *et al.* [131].

originally evaluated for condensation of mixtures with those for evaporation of pure fluids. There is a great need for data with which to check these ideas and so produce methods that can be used by designers.

XI. Recommendations

Additional work is suggested to focus primarily on convective boiling where few experimental studies have been completed and heat transfer predictive methods require refinement. For nucleate pool boiling, future studies should focus on the effect of composition on boiling incipience and site densities and the effect of heat flux on the deterioration in the heat transfer coefficient. Experimental work on the boiling of mixtures from "enhanced" boiling surfaces (not dealt with here because of insufficient published work on the subject) is recommended, since it has recently been suggested that these surfaces will be ineffective for mixtures with wide boiling ranges [132, 133]. Furthermore, the measurement of boiling heat transfer coefficients for nonaqueous ternary mixtures and several multicomponent mixtures (with many components) should be obtained to test the various models developed for such cases.

NOMENCLATURE

A	empirical coefficient	$j_{l,I}$	liquid mass flux at interface [kg/(m² sec)]
A	area [m²]		
A_0	empirical coefficient at 1.0 bar	K_b	equilibrium constant
a	bubble growth constant [m/sec$^{1/2}$]	K_{in}	empirical coefficient
		k_1	multiple of bubble departure diameter
b	bubble growth parameter		
$b*$	microlayer parameter	k_2	multiple of bubble growth time
C	mass transfer correction factor for convection	l_m	mass diffusion shell thickness [m]
C'	mass transfer correction factor for nucleate boiling	l_t	thermal boundary layer thickness [m]
$C_{1,m}$	mixture bubble growth constant [m/(K sec$^{1/2}$)]	N	boiling site density [m^{-2}]
		n	exponent
$C_{1,p}$	pure component bubble growth constant [m/(K sec$^{1/2}$)]	n	number of components
		$\dot{n}_T$	evaporating mass flux corrected for mass transfer effects [kg/(m² sec)]
c_a	bubble growth coefficient [m/(K sec$^{1/2}$)]		
c_p	specific heat [J/(kg K)]	p	pressure [N/m²]
c_1	empirical coefficient [sec/m]	p_g	vapor pressure [N/m²]
c_2	empirical exponent	$p_{g\infty}$	vapor pressure for planar interface [N/m²]
D	bubble diameter [m]		
D	tube diameter [m]	Pr	Prandtl number
D_b	bubble base diameter	$(dP/dT)_{sat}$	slope of saturation curve [N/(m² K)]
D_d	bubble departure diameter [m]		
F	empirical multiplier	W	heat flow rate [W]
F'	modified empirical multiplier	Q_{c_p}	sensible heat transport rate [W]
F_i	inertia force [N]		
f	bubble departure frequency [sec^{-1}]	$Q_{\Delta h_v}$	latent heat transport rate [W]
		$\dot{q}$	heat flux [W/m²]
G_d	vaporized mass fraction	$\dot{q}_c$	convective heat flux [W/m²]
$\bar{g}$	free enthalpy, Gibbs thermodynamic potential [J/kmol]	$\dot{q}_{cr}$	critical heat flux [W/m²]
		$\dot{q}_g$	sensible heat flux in vapor phase [W/m²]
g_c	gravitational conversion factor [kg m/N sec²]		
		$\dot{q}_{lat}$	latent heat flux [W/m²]
g_n	gravitational acceleration [m/sec²]	$\dot{q}_{min}$	minimum heat flux [W/m²]
		R	bubble radius [m]
h	enthalpy [J/kg]	R	bubble nucleus radius [m]
$\Delta \bar{h}$	molar transfer enthalpy [J/kmol]	$\bar{R}$	universal gas constant [Nm/kmol]
Δh_v	latent heat of vaporization [J/kg]	$R_{d,m}^*$	radius of dry spot for mixture [m]
Ja	Jakob number	$R_{d,p}^*$	radius of dry spot for pure fluid [m]
Ja$_m$	modified Jakob number		
$j_{g,I}$	vapor mass flux at interface [kg/(m² sec)]	R_1	hydrodynamically controlled bubble radius [m]

R_2	asymptotic bubble radius [m]	ΔT^{E}	maximum rise in local bubble point temperature [K]
r	radial dimension [m]		
r_1	interfacial radius [m]	ΔT_{eff}	effective superheat [K]
S	entropy [J/k]	$\Delta T_{\mathrm{local}}$	local superheat [K]
Sc	Schmidt number	ΔT_{min}	minimum superheat [K]
Sn	dimensionless mixture bubble growth parameter	ΔT_{sat}	equilibrium superheat [K]
		V_{d}	bubble departure volume [m³]
Sp	dimensionless superheat	$\bar{v}$	specific volume [m³/kmol]
$\bar{s}$	molar entropy [J/(K kmol)]	W_{g}	mass flow rate of vapor [kg/sec]
T	temperature [K]		
T^*	dew point temperature [K]	W_1	mass flow rate of liquid [kg/sec]
T_{bi}	bubble point temperature of inlet mixture [K]		
		x	vapor quality
T_{bo}	bubble point temperature of mixture at outlet [K]	x	mass fraction in liquid
		$\tilde{x}$	mole fraction in liquid
T_{bulk}	bulk temperature [K]	x_1	mass fraction at interface
T_{g}	vapor temperature [K]	$\dfrac{d\tilde{x}}{dT}$	slope of bubble line [1/K]
T_1	interfacial temperature [K]		
T_{sat}	saturation temperature [K]	y	mass fraction in vapor
T_{sup}	superheated liquid temperature [K]	y^*	equilibrium mass fraction in vapor
T_{w}	wall temperature [K]	$\tilde{y}$	mole fraction in vapor
t	time [sec]	y_{b}	bulk mass fraction in vapor
t_{g}	bubble growth period [sec]	y_1	mass fraction in vapor at interface
t_{w}	bubble waiting period [sec]		
ΔT	$= T_{\mathrm{sup}} - T_{\mathrm{sat}}$, superheat [K]	Z	vapor heat storage term
		z	z-axis coordinate [m]

Greek Symbols

α	heat transfer coefficient [W/(m² K)]	$\beta^{\cdot}$	mass transfer coefficient [m/sec]
$\bar{\alpha}_{\mathrm{AB}}$	relative volatility	δ	mass diffusion coefficient [m²/sec]
α_{c}	convective heat transfer coefficient [W/(m² K)]		
		η	dynamic viscosity [Nsec/m²]
α_{g}	heat transfer coefficient in vapor phase [W/(m² K)]	$\Delta\theta$	rise in local boiling point temperature [K]
$\alpha_{\mathrm{g}}^{\cdot}$	α_{g} corrected for effect of mass transfer [W/(m² K)]	$\Delta\theta_1$	steady-state rise in local boiling point [K]
α_1	heat transfer coefficient for liquid phase flowing alone [W/(m² K)]	$\Delta\theta_2$	rise in local boiling point due to growth of a bubble [K]
		κ_1	liquid thermal diffusivity [m²/sec]
α_{n}	nucleate boiling heat transfer coefficient [W/(m² K)]	λ	liquid thermal conductivity [W/(m K)]
α_{p}	pool boiling heat transfer coefficient [W/(m² K)]	ρ	density [kg/m³]
β	contact angle [rad]	σ	surface tension [N/m]
β	bubble growth coefficient		
β	mass transfer coefficient [m/sec]		

Superscripts

~   molar quantity

Subscripts

az	azeotrope	l	liquid
b	bulk	sat	saturation
g	vapor	sat.b	saturated bulk
I	ideal	1	component number one
I	interface	2	component number two
i, j, n	components		

References

1. W. M. Rohsenow, *in* "Handbook of Heat Transfer" (W. M. Rohsenow and J. P. Hartnett, eds.), Chap. 13. McGraw-Hill, New York, 1973.
2. E. Kirschbaum, "Distillation and Rectification." Chem. Publ. Co., New York, 1948.
3. J. C. Chu, "Distillation Equilibrium Data." Reinhold, New York, 1950.
4. E. Hala *et al.*, "Vapor–Liquid Equilibrium Data at Normal Pressures." Pergamon, Oxford, 1968.
5. J. M. Prausnitz, "Molecular Thermodynamics of Fluid-Phase Equilibria." Prentice-Hall, Englewood Cliffs, New Jersey, 1969.
6. C. Corty and A. S. Foust, *Chem. Eng. Prog., Symp. Ser.* **17**, 51 (1955).
7. S. G. Bankoff, *Proc. Heat Transfer Fluid Mech. Inst.* (1956).
8. J. W. Westwater and P. H. Strenge, *Chem. Eng. Prog., Symp. Ser.* **29**, 95 (1959).
9. R. Haase, "Mixed-Phase Thermodynamics." Springer-Verlag, Berlin and New York, 1956.
10. H. N. Stein, *in* "Boiling Phenomena" (S. J. D. Van Stralen and R. Cole, eds.), Chap. 17. Hemisphere, Washington, D.C., 1978.
11. W. Malesinsky, "Azeotrophy and Other Theoretical Problems of Vapor–Liquid Equilibrium." Wiley (Interscience), New York, 1965.
12. R. A. W. Shock, *Int. J. Heat Mass Transfer* **20**, 701 (1977).
13. P. Preusser, Ph.D. Thesis. Ruhr-Univ., Bochum, Fed. Rep. Ger., 1978.
14. K. Stephan and M. Korner, *Chem.-Ing.-Tech.* **41**, 409 (1969).
15. M. Korner, Ph.D. Thesis. Aachen Univ., Aachen, Fed. Rep. Ger., 1967.
16. L. N. Grigor'ev, *Teplo Massoperenos* **2**, 120 (1962).
17. K. Stephan and P. Preusser, *Heat Transfer, Int. Conf., 6th, Toronto* Pap. PB-13 (1978).
18. I. P. Basarow, "Thermodynamik," p. 284. Berlin, 1964.
19. A. B. Ponter, E. A. Davies, W. Beaton, and T. K. Ross, *Int. J. Heat Mass Transfer* **10**, 733 (1967).
20. R. I. Eddington and D. B. R. Kenning, *Int. J. Heat Mass Transfer* **22**, 1231 (1979).
21. R. A. W. Shock, *U.K., At. Energy Authority Rep.* **AERE-R7593** (1973).
22. H. Toral, Ph.D. Thesis. Oxford Univ., 1979.
23. H. Toral, D. B. R. Kenning, and R. A. W. Shock, *Proc. Int. Heat Transfer Conf., 7th, Munich* Pap. FB-14 (1982).
24. J. R. Thome, S. Shakir, and C. Mercier, *Proc. Int. Heat Transfer Conf., 7th, Munich* Pap. PB-14 (1982).

25. C. Mercier, Personal communication. Oxford Univ., 1983.
26. S. J. D. Van Stralen and R. Cole, "Boiling Phenomena," p. 38. Hemisphere, Washington, D.C., 1978.
27. T.-O. Hui, M. S. Thesis, Michigan State Univ., East Lansing, 1983.
28. W. R. Van Wijk, A. S. Vos, and S. J. D. Van Stralen, *Chem. Eng. Sci.* **5,** 65 (1956).
29. L. E. Scriven, *Chem. Eng. Sci.* **10,** 1 (1959).
30. W. F. Calus and P. Rice, *Chem. Eng. Sci.* **27,** 1687 (1972).
31. M. S. Plesset and S. A. Zwick, *J. Appl. Phys.* **25,** 493 (1954).
32. P. J. Bruijn, *Physica (Utrecht)* **26,** 326 (1960).
33. S. J. D. Van Stralen, *Br. Chem. Eng.* **4,** 8 (1959); **4,** 78 (1959); **7,** 80 (1962).
34. S. J. D. Van Stralen, *Int. J. Heat Mass Transfer* **9,** 995 (1966); **9,** 1021 (1966); **10,** 1469 (1967); **10,** 1485 (1967).
35. L. A. Skinner and S. G. Bankoff, *Phys. Fluids* **7,** 1 (1964).
36. L. A. Skinner and S. G. Bankoff, *Phys. Fluids* **7,** 643 (1964).
37. H. J. Van Ouwerkerk, *Chem. Eng. Sci.* **27,** 1957 (1972).
38. S. J. D. Van Stralen, *Proc. Int. Heat Transfer Conf., 4th, Paris* Pap. B7.6 (1970).
39. S. J. D. Van Stralen, M. S. Sohal, R. Cole, and W. M. Sluyter, *Int. J. Heat Mass Transfer* **18,** 453 (1975).
40. J. R. Thome, *AIChE Symp. Ser. No.* 208, 238 (1981).
41. L. W. Florscheutz and A. R. Khan, *Proc. Int. Heat Transfer Conf., 4th, Paris* Pap. B7.3 (1970).
42. J. E. Benjamin and J. W. Westwater, *Int. Dev. Heat Transfer, Proc. Heat Transfer Conf., Boulder, Colo., 1961* p. 212 (1963).
43. J. R. Thome, D.Phil. Thesis, Oxford University, 1978.
44. J. M. Yatabe and J. W. Westwater, *Chem. Eng. Prog., Symp. Ser.* **62,** 17 (1966).
45. W. R. Van Wijk and S. J. D. Van Stralen, *Physica (Utrecht)* **28,** 150 (1962); *Chem. Eng. Tech.* **37,** 509 (1965).
46. S. J. D. Van Stralen, *Physica (Utrecht)* **29,** 602 (1963).
47. S. J. D. Van Stralen, "High Speed Motion Picture." Agric. Univ., Wageningen, Netherlands, 1960.
48. J. R. Thome and G. Davey, *Int. J. Heat Mass Transfer* **24,** 89 (1981).
49. M. G. Cooper and C. R. Stone, *Int. J. Heat Mass Transfer* **24,** 1937 (1981).
50. W. M. Rohsenow, *Trans. ASME* **74,** 969 (1952).
51. B. B. Mikic and W. M. Rohsenow, *J. Heat Transfer* **91,** 245 (1969).
52. K. Stephan and M. A. Abdelsalam, *Int. J. Heat Mass Transfer* **23,** 73 (1980).
53. V. J. Tolubinskiy and J. N. Ostrovskiy, *Int. J. Heat Mass Transfer* **9,** 1463 (1966).
54. V. J. Tolubinskiy and J. N. Ostrovskiy, *Heat Transfer—Sov. Res.* **1**(6), 6 (1969).
55. V. J. Tolubinskiy, J. N. Ostrovskiy, and A. A. Kriveshko, *Heat Transfer—Sov. Res.* **2**(1), 22 (1970).
56. N. Isshiki and I. Nikai, *Heat Transfer—Jpn. Res.* **1**(4), 56 (1973).
57. J. R. Thome and G. Davey, *Proc. Int. Cryog. Eng. Conf.* **8,** 243 (1980).
58. E. G. Keshock and R. Siegel, *NASA Tech. Note* **NASA TN D-2299** (1964).
59. B. E. Staniszewski, Rep. No. 16. Div. Sponsored Res., MIT, Cambridge, Mass., 1959.
60. R. Cole and H. L. Shuman, *Chem. Eng. Prog., Symp. Ser.* **62** 6 (1966).
61. L. Zeugin, J. Donovan, and R. B. Mesler, *Chem. Eng. Sci.* **30,** 679 (1975).
62. S. J. D. Van Stralen and R. Cole, "Boiling Phenomena," p. 314. Hemisphere, Washington, D.C., 1978.
63. L. D. Clements and C. P. Colver, *Proc. Heat Transfer Fluid Mech. Inst.* p. 417 (1972).
64. C. V. Sternling and L. J. Tichacek, *Chem. Eng. Sci.* **16,** 297 (1961).
65. K. Stephan and P. Preusser, *Ger. Chem. Eng. (Engl. Transl.)* **2,** 161 (1979).

66. R. A. W. Shock, Personal communication, 1982.
67. J. R. Thome, *J. Heat Transfer* **104**, 474 (1982).
68. L. S. Sterman, J. V. Vilemas, and A. I. Abramov, *Proc. Int. Heat Transfer Conf., 3rd, Chicago* p. 258 (1966).
69. M. T. Cichelli and C. F. Bonilla, *Trans. Am. Inst. Chem. Eng.* **41**, 755 (1946); **42**, 411 (1946).
70. C. F. Bonilla and A. A. Eisenberg, *Ind. Eng. Chem.* **40**, 1113 (1948).
71. V. I. Tolubinskiy, A. A. Krivesko, J. N. Ostrovskiy, and V. Y. Pisarev, *Heat Transfer—Sov. Res.* **5**(3), 66 (1973).
72. D. A. Huber and J. C. Hoehne, *J. Heat Transfer* **85**, 215 (1963).
73. O. Happel and K. Stephan, *Heat Transfer, Proc. Int. Heat Transfer Conf., 5th, Tokyo* Pap. B7.8 (1974).
74. V. J. Tolubinskiy, J. N. Ostrovskiy, V. Y. Pisarev, A. A. Kriveshko, and D. M. Konstanchuk, *Heat Transfer—Sov. Res.* **7**(1), 118 (1975).
75. N. H. Afgan, *Proc. Int. Heat Transfer Conf., 3rd, Chicago* **3**, 175 (1966).
76. R. D. Wright, L. D. Clements, and C. P. Colver, *AIChE J.* **17**, 626 (1971).
77. H. Jungnickel, P. Wassilew, and W. E. Kraus, *Int. J. Refrig.* **3**, 129 (1980).
78. H. Ackermann, L. Bewilogua, R. Knoner, B. Kretzchmar, I. P. Usyugin, and H. Vinzelberg, *Cryogenics* **15**, 657 (1975).
79. P. G. Kosky and D. N. Lyon, *AIChE J.* **14**, 383 (1968).
80. K. Bier, J. Schmadl, and D. Gorenflo, *Proc. Int. Heat Transfer Conf., 7th, Munich* Pap. PB6 (1982).
81. L. N. Grigor'ev, L. A. Sakisyan, and A. G. Usmanov, *Int. Chem. Eng.* **8**, 76 (1968).
82. W. F. Calus and D. J. Leonidopoulos, *Int. J. Heat Mass Transfer* **17**, 249 (1974).
83. J. R. Thome, *Int. J. Heat Mass Transfer* **26**, 965 (1982).
84. J. W. Palen and W. Small, *Hydrocarbon Process.* **43**(11), 199 (1964).
85. M. J. McNelly, *J. Imp. Coll. Chem. Eng. Soc.* **7**, 19 (1953).
86. K. Stephan, *Proc. Int. Heat Transfer Conf., 7th, Munich* Pap. RK14 (1982).
87. S. S. Kutateladze, "Fundamentals of Heat Transfer" (Engl. transl.), p. 362. Academic Press, New York, 1963.
87a. V. Valent and N. H. Afgan, *Waerme- Stoffuebertrag.* **6**, 235 (1973).
87b. C. F. Bonilla and C. W. Perry, *Trans. Am. Inst. Chem. Eng.* **37**, 685 (1941).
87c. S. Shakir, Ph.D. Thesis, Michigan State Univ., East Lansing, 1984.
88. G. I. Bobrovich *et al., PMTF, Zh. Prikl. Mekh. Tekh. Fiz.* No. 4, 108 (1962).
89. A. S. Matorin, *Heat Transfer—Sov. Res.* **5**(1), 85 (1973).
90. N. Yu. Tobilevich, I. I. Sagan, and N. A. Pryadko, *Inzh.-Fiz. Zh.* **16**(4), 610 (1969).
91. C. C. Pitts and G. Leppert, *Int. J. Heat Mass Transfer* **9**, 365 (1966).
92. N. H. Afgan, *Heat Mass Transfer Boundary Layers, Proc. Int. Summer Sch., Hercegnovi, Yugosl.* (1968).
93. N. Zuber, AEC Rep. No. AECU-4439. Phys. Math., Oak Ridge, Tenn., 1959.
94. J. H. Lienhard and V. K. Dhir. *NASA [Contract. Rep.] CR* **NASA-CR-2270** (1973).
95. J. Hovestreijdt, *Chem. Eng. Sci.* **18**, 631 (1963).
96. J. J. Jasper, *J. Phys. Chem. Ref. Data* **1**, 841 (1972).
97. D. E. Kautsky and J. W. Westwater, *Int. J. Heat Mass Transfer* **10**, 253 (1967).
98. P. L. Yue and M. E. Weber, *Int. J. Heat Mass Transfer* **16**, 1877 (1973).
99. H. J. Gannett and M. C. Williams, *Int. J. Heat Mass Transfer* **14**, 1001 (1971).
100. L. E. Brown and C. P. Colver, *Adv. Cryog. Eng.* **13**, 647 (1968).
101. K. Nishikawa and T. Ito, *Int. J. Heat Mass Transfer* **9**, 103 (1966).
102. E. Marschall and L. L. Moresco, *Int. J. Heat Mass Transfer* **20**, 1013 (1977).

103. M. V. L. Prakash and K. N. Seetharamu, *Int. Semin. Momentum Heat Mass Transfer Two-Phase Energy Chem. Syst., Dubrovnik, 1978.*
104. E. M. Sparrow and R. D. Cess, *J. Heat Transfer* **84,** 149 (1962).
105. A. P. Colburn and T. B. Drew, *Trans. Am. Inst. Chem. Eng.* **39,** 197 (1937).
106. L. A. Bromley, *Chem. Eng. Prog.* **46,** 221 (1950).
107. Y. Y. Hsu and J. W. Westwater, *Chem. Eng. Prog., Symp. Ser.* **56,** 15 (1960).
108. S. J. D. Van Stralen, C. J. J. Joosen, and W. M. Sluyter, *Int. J. Heat Mass Transfer* **15,** 2427 (1972).
109. P. J. Berenson, *J. Heat Transfer* **83,** 351 (1961).
110. J. H. Lienhard and P. T. Y. Wong, *J. Heat Transfer* **86,** 220 (1964).
111. N. Zuber, *Trans. ASME* **80,** 711 (1958).
112. N. Zuber and M. Tribus, Rep. No. 58-5. Dep. Eng., Univ. California, Los Angeles, 1958.
113. P. L. Yue and M. E. Weber, *Trans. Inst. Chem. Eng.* **52,** 217 (1974).
114. S. S. Kutateladze, G. I. Bobrovich, I. I. Gogonin, and N. N. Manontova, *Proc. Int. Heat Transfer Conf., 3rd, Chicago* **3,** 149 (1966).
115. D. Butterworth, *Inst. Chem. Eng. Symp. Ser.* No. 60, 232 (1980).
116. R. A. W. Shock, *in* "Two Phase Flow and Heat Transfer" (D. Butterworth and G. F. Hewitt, eds.), 2nd Ed., Ch. 2. Oxford Univ. Press, London and New York, 1979.
117. D. Butterworth and R. A. W. Shock, *Proc. Int. Heat Transfer Conf., 7th, Munich* Pap. RK15 (1982).
118. J. C. Chen, *In. Eng. Chem. Process Des. Dev.* **5,** 332 (1966).
119. D. L. Bennett and J. C. Chen, *AIChE J.* **26,** 454 (1980).
120. F. J. Kadi, Ph.D. Thesis, Lehigh Univ., Bethlehem, Pennsylvania, 1976.
121. J. Fink, E. S. Gaddis, and A. Vogelpohl, *Proc. Int. Heat Transfer Conf., 7th, Munich* Pap. FB5 (1982).
122. W. J. Rose, H. L. Gilles, and U. W. Uhl, *Chem. Eng. Prog., Symp. Ser.* **59,** 62 (1963).
123. R. A. W. Shock, *Int. J. Multiphase Flow* **2,** 411 (1976).
124. G. F. Hewitt, *U.K., At. Energy Authority Rep.* **AERE R3185.**
125. A. P. Colburn and O. A. Hougen, *Ind. Eng. Chem.* **26,** 1178 (1934).
126. R. B. Bird, W. E. Stewart, and E. N. Lightfoot, "Transport Phenomena." Wiley, New York, 1960.
127. D. R. Webb and J. M. McNaught, *in* "Developments in Heat Exchanger Technology" (D. Chisholm, ed.). Appl. Sci., London, 1981.
128. R. G. Sardesai, R. A. W. Shock, and D. Butterworth, *Heat Transfer Eng.* **5,** 104 (1982).
129. L. Silver, *Trans. Inst. Chem. Eng.* **25,** 30 (1947).
130. K. J. Bell and M. A. Ghaly, *AIChE Symp. Ser.* **69,** 72 (1972).
131. J. W. Palen, C. C. Yang, and J. Taborek, *AIChE Symp. Ser.* **76,** 282 (1980).
132. J. Arshad and J. R. Thome, *Proc. ASME–JSME Therm. Eng. Jt. Conf., Honolulu* **1,** 191 (1983).
133. J. Taborek, *in* "Two-Phase Flow and Heat Transfer" (S. Kakac and M. Ishii, eds.), Vol. 2, p. 815. Nijhoff, The Hague, 1983.

Heat Flow Rates in Saturated Nucleate Pool Boiling—A Wide-Ranging Examination Using Reduced Properties

M. G. COOPER

Department of Engineering Science, Oxford University, England

I. Introduction

Boiling is of widespread industrial importance, but there is no agreement on the problem of relating heat flow to driving temperature difference. That is partly because the related experiments are difficult, and the results show wide scatter, which in turn may be a reflection of insufficient understanding of crucial physical phenomena. Those underlying physical phenomena are many and complex, so there is no shortage of theories. Many theories are backed by or based on experiments aimed at one particular underlying phenomenon, but they are often limited to a single fluid and/or single operating pressure. In fitting the experimental results to the chosen theory, several disposable constants are used, but the scatter in experimental data has so far prevented a clear decision as to which theory is best.

The method used here is the reverse. The experimental data are examined with no particular theory in mind, to search for significant effects of physical parameters. That has been done before, but the present method gains power, scope, and speed by avoiding redundancies among the fluid properties, which otherwise lead to intolerable and unhelpful complexity of analysis and to a confusing multiplicity of correlations that look different but are often numerically rather similar.

We start by examining existing correlations, finding a reason for the fact that they are often numerically similar, despite using very different properties. That leads to the improved method of analysis.

A. EXISTING CORRELATIONS

Many correlations of heat flow rates in saturated nucleate pool boiling were originally derived in the form

$$\text{Nu} = c \times \text{Re}^{c'} \times \text{Pr}^{c''} \times \ldots \times \ldots \tag{1}$$

The dimensionless groups Nu and Re (and others) were adapted or devised by the originating author to reflect his view of the important phenomena underlying boiling; they may involve lengths like cavity radius or

$\sqrt{[\sigma/(\rho_f - \rho_g)g]}$. The disposable constants $c, c', c'', \ldots$ were obtained by comparison with some experimental data. The result is often algebraically identical to the form

$$h = (q/A)^m \times \Pi \, (\text{property})_i^{\alpha i} \times (\text{a constant}) \qquad (2)$$

though in some cases the right-hand side is also multiplied by positive or negative powers of local gravity and/or a length from the apparatus (e.g., heater diameter or roughness). The properties used in Eq. (2) are a selection of properties of saturated liquid and vapor and interface and some difference properties, usually p, T_{sat}, ρ_f, ρ_g, $(\rho_f - \rho_g)$, h_{fg}, σ, k_f, μ_f, c_{pf}.

Experimental results are often expressed in these forms, or in the related form

$$h/(q/A)^m = F(p_r) \qquad (3)$$

in which m is generally in the range 0.6–0.8. In this form, $F(p_r)$ generally shows a steady rise with p_r, perhaps as $p_r^{0.2}$ or $p_r^{0.3}$, and a sharper rise near the thermodynamic critical point, perhaps a rise by a factor of 6 as p_r rises from 0.2 to 0.9. Various algebraic expressions have been proposed for $F(p_r)$, frequently using a normalized form with the left-hand side divided by its experimental value at p_{r0}, so $F(p_{r0}) = 1.0$ (p_{r0} being usually 0.1 or 0.3).

A recent suggestion is that the form of $F(p_r)$ is affected by roughness of the heater surface, and this will be considered.

It has apparently been generally accepted that the form of $F(p_r)$ should not vary between fluids, but this will be considered in Section VII,C,4.

1. Comparison of Correlations

Competing correlations contain such a variety of properties that it may seem very strange that experiments have not decided between them. The reason is partly that the experimental data on boiling are so scattered that they can be applied only with a very broad brush, of width $+40\%$, -30% (Section VII,D), but there are additional reasons.

As discussed in a previous paper [1] and described briefly here, most correlations predict much the same variation of heat flow as pressure is varied for any one fluid, despite containing different properties. Briefly, since there is always a disposable constant, it is difficult or impossible to distinguish correlations containing, say, $\sigma^{0.4}$ or h_{fg} since both of these vary roughly as $(1 - T_r)^{0.4}$. The variation of individual properties with p_r has long been known to be broadly similar for different fluids, because of the "law of corresponding states." It is useful to recognize that fact and quantify it in a new way [1], since that leads to correlations of much simpler form, avoiding such redundancies among properties of a given fluid, and simplifying the data processing. It also shows that some statistical analyses are incapable of finding a unique solution, as discussed in Cooper [1] and briefly here (Section III,B,1).

The present paper carries the argument a stage further, since it shows (in Section VI) why most correlations tell much the same story in respect to different fluids. Briefly, most correlations predict that pool boiling heat transfer is lower for fluids with high molecular weight, roughly as $M^{-0.5}$. That occurs because most properties of most fluids vary pretty much with M. It therefore remains difficult or impossible to distinguish whether heat flow is affected by variations in individual properties, even when results for many fluids are considered, since the boiling data, when applied with a brush of width $+40\%$, -30%, blur into a general trend with M.

It is useful to recognize that fact as well, since, by examining data in terms of M (or similar simple quantity) instead of using all properties, a broad pattern of variation can be tentatively seen, avoiding unhelpful confusion from the multiplicity of properties. Given precise data, including precise properties, we could expect to resolve that broad variation back into a variation dependent on properties, or at least see the effect of some further molecular parameters, such as the Pitzer acentric factor ω. However, it may be a long time before that is possible; meanwhile, there is much to be said for a simple correlation based on reduced properties and M or p_{cr}, T_{cr}.

Correlations of the more complex form of Eqs. (1) and (2) are far less convenient to use, but it might be supposed nevertheless that such correlations are more "respectable" and more likely to fit a range of conditions and fluids. That supposition is now seen to be very doubtful. In fact, correlations may as well be expressed in terms of reduced pressure and perhaps molecular weight; the experimental data do not justify the use of all properties.

The practical man will welcome the simplicity of the new correlation, and the researcher may find it focuses his ideas and experiments into productive channels, such as experiments designed more closely and realistically, to distinguish effects of different fluids or conditions.

2. Exceptional Fluids

The commonest fluid—water—appears at first glance to be exceptional in a way that is likely to affect correlations in the form of Eq. (1), because of variation of fluid properties. That arises because it has two properties (σ, k_f) that differ appreciably (factor 3) from the general run. But, as will be seen, the observed heat flow in boiling water is not very different from the general run; hence either those two properties must each have weak effect or their effects must cancel. Most correlations back both horses, since σ and k_f occur to powers of about -0.4, 0.4, so they predict effects that are both weak and canceling. Thus they reach the right answer. Whether they reach it by the right method is another question, which probably cannot be answered by comparing fit of the correlations to experimental data.

The liquid metals also have exceptional properties, notably k_f, higher by a factor of 1000 or so, and it is remarkable that, as reported in Cooper [1], they do not give exceptional boiling heat transfer coefficient at given p_r. However, there may be an interaction here with effect of roughness (nucleation), since the data for liquid metals are generally at low p_r, and a recent study, considered in Section VII,B,1, suggests effects of roughness are greater at low p_r.

B. DATA REDUCTION

It has always required a major effort of data reduction to compare experimental data with correlations that involve many properties, as in Eqs. (1) and (2). Comparisons between competing claims of different correlations are confused by uncertainties and indeed errors in tabulated property values, especially for uncommon fluids or old tabulations. These problems are greatly reduced by using a simpler correlation (e.g., in terms of reduced pressure), where the only tabulated property required for each fluid is critical pressure. Taking advantage of that simplicity, correlations of a more subtle type can be considered without undue labor in data reduction.

II. Reformulation of Correlations

In Cooper [1], many correlations of form (1) or (2) are evaluated for water and plotted against p_r. It is clear that most of them follow the usual trend of steady rise followed by sharper rise near the thermodynamic critical point (CP). In one way, that similarity is not surprising, because they were each devised by adjusting several disposable parameters, such as c, c', c'' in Eq. (1), to suit some experimental data. In another way, the similarity is surprising, because the correlations look very dissimilar when rewritten as Eq. (2), having widely differing selections of properties and powers α_i. On further consideration, it is seen that a rise near the CP can be obtained by having a suitable positive power of c_{pf} (which goes to infinity at the CP) or by having a suitable negative power of a difference property h_{fg} or $(\rho_f - \rho_g)$ or of σ, which all go to zero at the CP.

A. Properties in Terms of p_r, T_r, $(1 - T_r)$

It is well known that for many fluids, σ goes to zero at the CP approximately as $(1 - T_r) \times$ constant, and it is fairly well known that h_{fg} goes approximately as $(1 - T_r)^{0.4} \times$ constant. If these approximations were exact, there would be no difference between having in the correlation the term $\sigma^{0.4}$ or the term h_{fg} (apart from the value of a disposable multiplying constant), since each is proportional to $(1 - T_r)^{0.4}$. Of course, it is not possible to approximate all properties in this way; in fact, for ρ_g the rough approximation $\rho_g \approx p/RT$ comes to mind. That suggests investigating whether each of the relevant properties $(\rho_f, \rho_g, (\rho_f - \rho_g), h_{fg}, \sigma, k_f, \mu_f, c_{pf})$ can be represented as

$$p_r^a \times T_r^b \times (1 - T_r)^c \times \text{constant} \tag{4}$$

for a given fluid, with sufficient accuracy for the current purpose. As described in [1], a simple computer program was written to do that, by least squares fit among log(property), $\log(p_r)$, $\log(T_r)$, $\log(1 - T_r)$ and 1.0 (for the constant, regarded as 10^e). Investigations, described in [1], showed that the accuracy of the representation depends on the range covered. For water in the very wide range from 25°C to 365°C ($p_r = 0.00017$–0.90) the representation is within a few percent, almost invariably less than 2%. In a more normal range, 100–350°C, that discrepancy is halved. Each of these is less than the reproducibility error in experimental data for boiling heat flow rates, even for the most careful experiments, as discussed in Section V. For fluids other than water, it is difficult to obtain all the properties with sufficient accuracy over a range wide enough to provide a good test of the representation.

This recognizes and quantifies the fact that the many properties that are used in common boiling correlations do not vary independently with pressure but are numerically interdependent. There is a form of redundancy among all properties. The properties of a given fluid can therefore be replaced by a much smaller set of quantities, which are numerically independent in an appropriate sense. Clearly, $(1 - T_r)$ and T_r are not mathematically independent, nor indeed are T_r and p_r, as discussed in Section VI. However, in the present analysis, the convenient set of quantities is that which enables properties (and heat flows) to be expressed as a product of powers.

From the set of quantities p_r, T_r, $(1 - T_r)$, or alternatives discussed later, we can select a "set of tools" appropriate to the present task. The ideal set of tools is the smallest and simplest set that, by product of powers, can replace with sufficient accuracy the full set of 10 or more properties of the given fluid.

B. CORRELATIONS IN TERMS OF p_r, T_r, $(1 - T_r)$

By substituting properties expressed by Eq. (4) into Eq. (2), we see that any correlation that is expressible in the form of Eq. (2) with modest powers α_i is necessarily expressible in the form

$$h/(q/A)^m = p_r^A \times T_r^B \times (1 - T_r)^C \times C_E \qquad (5)$$

(where $A = \alpha_1 \times a_1 + \alpha_2 \times a_2 + \ldots$; similarly, B, C, E, and $C_E = 10^E$) within a few percent over a wide range for a given fluid. This reformulation does not involve any assumption concerning the nature of boiling; it follows simply from the nature of the saturation properties, ρ_f, ρ_g, h_{fg}, σ, k_f, μ_f, c_{pf}. It is not necessary to determine all those properties for each application.

It has long been known that fluid properties are not fully independent. Many workers have used reduced properties to express other properties, invoking the law of corresponding states. Some workers, such as Borishanskii [2], have used reduced properties in boiling. However, they generally produced multiterm algebraic expressions for properties or for $F(p_r)$ in Eq. (3). What is newly shown in Cooper [1] is that the interdependence of the relevant saturation properties can be expressed in the particularly simple and useful form of Eq. (4). That is particularly useful for boiling because it shows that, for a given fluid, the simple formulation of Eq. (5) is equivalent to the multiproperty formulation of Eqs. (1) or (2), with sufficient accuracy for boiling.

C. TWO FURTHER SIMPLIFICATIONS

The least squares program is easily modified to exclude terms or to include others, so alternative, simpler formulations were readily tested, initially for matching to properties and later for direct production of correlations. These simplifications lead to correlations dependent only on p_r, which has considerable practical advantages. Most experimenters measure and quote system pressures with more accuracy and confidence than corresponding saturation temperatures. To obtain T_r from p_r it is still necessary to consult property tables, with all their possibilities of introducing errors and inconsistencies.

1. Omitting T_r

Of the three quantities in Eqs. (4) and (5), T_r has the least effect, as its variation between triple point and the CP is from, say, 0.4 to 1.0, while p_r varies from 0.0001 to 1.0 and $(1 - T_r)$ varies from 0.6 to 0.0. If T_r is omitted and a, c, and e (for constant 10^e) are reevaluated to produce a shortened Eq. (4), the quality of fit to property data for any one fluid is not greatly impaired, as discussed in Cooper [1]. By substitution in Eq. (2), that leads to correlations in the form of a shortened Eq. (5), with reevaluated m, A, C, C_E, giving almost unimpaired agreement with the multiproperty formulation of Eq. (1) or (2). Correlations of this type are also produced directly by matching to boiling data, as described in Section IV. The fit is hardly affected by omitting T_r, as discussed in Cooper [1], and in more detail in Appendix A, Section C,1.

2. Alternatives to $(1 - T_r)$

In Eq. (4), or its shortened version omitting T_r, the effect of the term $(1 - T_r)$ is to provide a quantity that goes to zero at the CP. A term $(1 - p_r)$ would also go to zero, but it goes in a rather different way, and tests prove that it is not so satisfactory [1]. A term $[-\log_{10}(p_r)]$ does go to zero at the CP in much the same way as $(1 - T_r)$, and it can be used instead, giving an even simpler representation of properties, again of adequate accuracy, as discussed in Cooper [1]. Substitution in Eq. (2) gives the correspondingly simple form of correlation:

$$h/(q/A)^m = p_r^A \times (-\log_{10} p_r)^D \times C_E \qquad (6)$$

which again provides adequate accuracy for representing the original multiproperty formulation, as discussed in Cooper [1], and for correlating experimental data, as illustrated in Appendix A, Section C,1.

Either $(1 - T_r)$ or $(-\log_{10} p_r)$ is equally good for a single fluid, and, as discussed further in Section VI,D, it is not clear whether one or the other is in some sense more "fundamentally" involved in property variation and in boiling.

At present it appears that the "ideal set of tools" is p_r and $(-\log_{10} p_r)$, but we must keep in mind that further study may suggest a change. It is a simple matter to extend the set, by reintroducing T_r or otherwise.

III. Consequences

A. Consequences for Practical Applications

Correlations in the form of Eqs. (5) or (6) lead to great simplification for practical use, particularly for the less common fluids, where the many properties demanded for Eq. (1) can be obtained only by tedious searching for property tables, which sometimes differ, or by estimation. For Eq. (6) we need for each fluid only the critical pressure, together with the values of the indices m, A, D, the constant C_E, and the system pressure.

It would be inconvenient if every fluid required an individual set of values of m, A, D, C_E, but, as discussed in the rest of this paper, a common set of values may well suffice for a group of similar fluids. For a wider range of fluids, we see tentatively some dependence on molecular weight M or on (p_{cr}/T_{cr}) (Section VII,C).

This simple formulation also facilitates optimization by differentiation, since the right-hand side of Eq. (6) is an analytically differentiable function of p_r, and the right-hand side of Eq. (5) is also differentiable, given $(dT/dp)_{sat}$ from the Clausius–Clapeyron equation.

It is to be hoped that practical users will also gain from more accurate correlations to be developed as discussed later. Such correlations will be somewhat more complex than Eq. (6) but much less complex or demanding than Eqs. (1) or (2). When any correlation is applied in practice, its accuracy will be limited by uncertainty about conditions, since, for example, the detailed physicochemical condition of the heater surface may be unknown and varying with time.

B. Consequences for Research

There are also important consequences for research, partly explored in Cooper [1]. Here they are considered in two categories. The first is mentioned briefly and the second underlies the remainder of this paper.

1. Interdependence of Properties: "N-CD" Groups

If each of the relevant properties could be precisely represented as Eq. (4), or a shortened version, then so too could the five or six algebraically independent dimensionless groups that can be formed by combinations of those properties. Those dimensionless groups therefore could not be numerically independent, and it would be possible to combine them into compound dimensionless groups that were constant. In fact, the representation of properties as Eq. (4) is not precise, but it was shown in Cooper [1] that compound dimensionless groups can be derived that are constant within a few percent over a wide range for a given fluid. These nearly constant dimensionless groups (n-cd groups) have two major nuisance effects:

(a) Two correlations may look very different because they have widely different combinations of properties, but the correlations will in fact differ only by a (disposable) multiplying factor if their ratio is expressible as a product of powers of n-cd groups for the fluid considered.

(b) If we seek to correlate experimental data (usually subject to scatter of several percent) for any one fluid, by means of statistical regression or other analysis applied to all the properties or only to the dimensionless groups, then there cannot be a unique solution, because any n-cd groups could be present or absent without materially affecting the fit to the data.

These points are spelled out in Cooper [1] and will not be further elaborated here.

2. Accelerated Data Analysis

Analysis of experimental data is greatly speeded, as shown in Cooper [1]. The main aim of this paper is to continue the exploitation of that.

IV. Data Analysis: Aims and Methods

A. AIMS

Many data are available, and their scope will be discussed in Section V. The basic aim of the data analysis is to find any parameter that is having a discernible effect on heat flow in boiling. Many parameters have long been considered to be important in boiling or have been shown to be important in experiments directed specifically at them, as discussed in Section VII.

By a simple method, described later, correlations are readily produced from the experimental data, directly into the form of Eqs. (5) or (6), instead of using all properties, as in Eq. (1). That permits rapid processing of the very large amount of existing experimental data and permits further developments, producing correlations that are more detailed and subtle in several ways:

1. by using raw data points instead of representing a dozen or more points at given pressure by a few characteristic points or by a straight line of best fit;
2. by developing correlations in which the index m of Eqs. (2)–(6) is dependent on the system pressure (e.g., $m = m_1 + m_2 p_r$ or $m_1 + m_2 \log_{10} p_r$);
3. by incorporating terms involving roughness, possibly dependent on p_r, such as $(R_p)^{F \times (1 - p_r)}$, suggested by Nishikawa et al. [3], or more convenient variants;
4. by allowing for variation of $F(p_r)$ for different fluids—done chiefly by multiplying $F(p_r)$ by a number dependent on the fluid, though some data suggest that the form of $F(p_r)$ varies between fluids, as $F(p_r, M)$;

5. by linking to existing expressions for natural convection and to correlations for burnout heat flux.

B. METHODS

1. *Data Input*

Raw data were read in from original publications, where they had been presented as graphs or as tabulations, relating h and q/A, or q/A and ΔT or h and ΔT. Tabulations were used where available, since graphs can introduce errors of several percent from plotting, printing, and reading. Whatever the quantities used in the source document, they are correlated in the form of h and q/A, because most existing correlations do so.

2. *Correlation of Raw Heat Transfer Data*

Correlations in the form of Eqs. (5) or (6) are produced directly from raw data by a straightforward extension of the least squares program which has already been used to produce a fit among properties and p_r, $(1 - T_r)$, and so on. Here the fit is among $\log(h)$, $\log(q/A)$ (using raw data points), and a selection of $\log(p_r)$, $\log(T_r)$, $\log(1 - T_r)$, $\log[-\log(p_r)]$, together with 1.0, for the constant C_E, regarded as 10^E. It is also extended for the further developments just described, involving $m = m_1 + m_2 p_r$, a roughness term and dependence on M, by including in the selection $[p_r \log(q/A)]$, $[(1 - p_r) \log(R_p)]$, $[\log(M)]$, respectively, or many variants, to be described below. To give the greatest flexibility in operation while avoiding frequent modification of the main program, a short, easily edited data file was used to control many different functions. It selected the required terms and selected which data sets were to be used to produce the correlation by least squares fit and which sets were merely to be compared with that correlation. In addition, it controlled printout, diagnostics, plotting, and so on.

The data input to be matched by least squares is now values of h, q/A at given p_r (and T_r), numbering hundreds or thousands, in place of the 10 or 12 values of properties formerly matched. The computation remains brief, because, in least squares analysis, the data are rapidly reduced to a small matrix that is readily manipulated. Even when the process is iterated, as will be described, the computation rarely needs more than 10 sec of CPU time on a VAX 11/780, equivalent to a second or so on a fast current mainframe computer. Quality of fit to the experimental data is similar to that obtained by much slower methods using all properties or dimensionless groups, as discussed in Cooper [1].

3. *Variation of Fluid Properties*

Most existing correlations of the form given in Eqs. (1) or (2) predict strong dependence of heat flow rate on fluid properties, but that is arduous to check, requiring many properties of many fluids at many states. Section II refers to a convenient and accurate method of dealing with the variation of properties of a single fluid. Section VI describes a corresponding approach to the variation of properties between fluids, which is much less accurate, but is of assistance in seeking systematic variations in boiling heat flow, for the limited range of fluids chiefly used in boiling research.

4. *Advanced Statistical Techniques*

No advanced statistical techniques are used. The basic method is simple least squares fit, applied to a chosen set of experimental conditions and results. This requires an element of judgment concerning which parameters are or may be important. It is appropriate to leave that judgment "to the computer" in cases where there are many parameters of known value but unknown importance. That is not the situation here, where we have only a very limited number of parameters that are stated for all or nearly all data sets.

In any case, the more complex the statistical technique, the less easy it is for the reader (and sometimes the user) to see just what has been done.

C. DIMENSIONAL ANALYSIS

The analysis is necessarily subject to requirements of dimensional homogeneity. The true relation between h and (q/A) may be more complex than any of Eqs. (1), (2), (3), (5), and (6), each of which amounts to a single term [a product of powers of h, (q/A) and properties] equated to a constant. Whatever the true form of the relation, it must be dimensionally homogeneous, because all physical laws must be expressible in dimensionless form, as Eq. (1) already is. For Eqs. (5) and (6) to be dimensionally homogeneous, the constant C_E must have the dimensions of $h/(q/A)^m$. Equations developed below require in addition a length to nondimensionalize a roughness parameter. In principle, these dimensional requirements should help in determining what properties are present, for example, in the constant C_E, though the n-cd groups mentioned earlier will make that difficult in practice.

SI units are used throughout this work, so the units of C_E, being $h/(q/A)^m$, or $(q/A)^{1-m}/\Delta T$, are $W^{1-m} m^{-2(1-m)} K^{-1}$. The only case in which a multiple or submultiple is used is R_p, which is in micrometers.

V. Data Available: Scope, Accuracy, Reproducibility

A. SCOPE

There have been many experimental studies of boiling heat transfer by many experimenters under various conditions. They usually aimed either to determine heat flow for design with a particular heater–fluid combination or to determine the effect of varying some quantity, of operating conditions (pressure) or of heater (material, shape, roughness) or fluid. Some sets are decades old, and some lack information, particularly about the surface of the heater. Any data not included in the original publication are sometimes lost after comparatively few years.

Table I summarizes the main categories of data, showing the number of data points and (in parentheses) the number of sets of data included in each category. Appendix C lists, in the same categories, the sources for those sets. The same source is listed there more than once if it contains data in several categories of Table I. References in the text are listed in the usual way, independently of Appendix C. The author realizes that he must have missed many data sets, and he would be grateful for references to them.

1. Distribution

This distribution of data is not ideal for analysis, and it will be suggested that it would be useful to fill some gaps and to extend some experiments

TABLE I

DATA BANK, JUNE 1983, AND DISTRIBUTION OF DATA POINTS[a]

Heater		Fluid			
Shape	Material	Water	Refrigerants	Cryogens	Other
Plate	Copper	299 (8)	1247 (24)	169 (11)	330 (6)
	S/S	222 (4)	nil	nil	nil
	Other	46 (3)	nil	371 (8)	nil
Cylinder	Copper	nil	361 (12)	233 (4)	116 (2)
	S/S	573 (6)	nil	nil	525 (6)
	Other	30 (1)	nil	177 (2)	245 (4)
Wire	S/S	58 (2)	59 (1)	nil	nil
	Platinum	231 (5)	143 (1)	236 (3)	155 (2)
	Other	15 (1)	nil	nil	nil

[a] The numbers in parentheses correspond to the number of data sets listed under each subsection of Appendix C.

(e.g., as regards range of pressure or of fluid), which may be possible with modest effort.

Clearly, most heaters are copper plates or copper cylinders, with some data for water on stainless steel cylinders and a few data on wires, usually platinum, sometimes stainless steel. This bedevils any attempt to separate the effects of change of material from effects of plate or cylindrical geometry (Section VII,B).

The simplest way to fill gaps is to find more data sets in the literature, with the help of readers, as invited previously.

2. *Information Contained*

The ideal data set gives details of apparatus, including roughness, and contains many points over a wide range of pressure and several fluids. Ideally, each fluid would appear in many such sets.

A numerical measure of roughness is given for nearly all the many data sets using refrigerants, but there is only one major study that includes combined effects of varying pressure, roughness, and fluid [3, 4]. In Nishikawa *et al.* [3], three refrigerants were studied for a wide range of roughness ($0.02 < R_p$, $\mu m < 4.3$) and a modest range of pressure ($0.08 < p_r < 0.9$). There are other smaller studies of effects of roughness, some at a single pressure and some indicating that effects on boiling cannot be represented by any one numerical measure of roughness.

There are 30 data sets for water, some of them at a single pressure (usually about 1 atm), but several have a usefully wide range of pressure, a few having (max p_r/min p_r) > 100. They are largely on copper plates or stainless steel cylinders or wires, with none on copper cylinders.

For refrigerants as a class, there are more data, all on copper heaters, unevenly divided between cylinders (largely from Germany) and flat plates (largely from Japan). Most tests are at pressures above 1 atm ($p_r > 0.03$), though a few go lower. The pressure range covered by a single set is usually below 10; only for a few sets does it exceed 15, and it never exceeds 50.

Cryogens have also been studied, often with unusual heater material. Scatter tends to increase as experimental difficulties increase at lower temperature.

Among "other fluids," ethanol has the largest number of data sets, 11 in all, of which several have a wide range in pressure.

It would be useful to have extensive and comparable data sets on fluids with properties that differ radically from the general run. As will be discussed, there are few such fluids, and the data sets are very few and difficult to compare. Nobody has produced data for, say, neon on the same apparatus as that designed for refrigerants; the reason is obvious.

B. ACCURACY AND REPRODUCIBILITY

It is difficult to assess the reliability of general boiling data, including such old sets, so no attempt has been made here to do that for all data. For some recent data it is possible to assess accuracy (of individual readings taken in a single test run) and reproducibility (comparing readings taken in the same experimental program, in the same or closely copied apparatus, after considerable lapse of time). These assessments depend on some analysis, though the main discussion of analysis is in Section VII.

The experiments most suitable for assessing accuracy and reproducibility are those that give results from the same fluid under closely similar conditions, or under conditions (e.g., surface roughness) that were varied in a closely controlled way. Two major series of this kind have been reported in recent years from Germany, based on Karlsruhe [5–7], and from Japan, based on Kyushu [3, 4], and they are examined in detail. The intention is not to criticize that work, but to find the accuracy given by the best experimental techniques. For the other data sets, no such detailed examination is made here, and analysis in Section III,D suggests that some old sets are outliers for unknown reasons.

1. The German (Karlsruhe) Data

Originating from Karlsruhe, various workers in Germany [5–7] have obtained data for boiling of several refrigerants in a particular type of pool boiling apparatus, with horizontal cylindrical heater of copper, usually 8 mm in diameter, with roughness in the range $0.2 < R_p \ (\mu m) < 0.9$. They gave very careful attention to detail and planned the experiments to operate in regimes that minimized their sensitivity to factors that are difficult to control. Temperature differences were measured with accuracy claimed to be ± 0.005 K, as was vital for tests at very high p_r, reported in Bier et al. [5].

Certain points at low q/A and low p_r are omitted as they do not seem to be on a boiling curve; they presumably refer to single-phase convection. As seen in Fig. 4 of Bier et al. [5], the transition from single phase to boiling appears very sharp, so it can be clearly identified, providing a clear criterion for omitting the points below it. Certain points at high q/A and high p_r appear to be influenced by proximity to burnout, and some allowance should be made there for the transition from the lower, straight sections of the curves. However, those points do refer to boiling, so they are, for the present, analyzed as they stand. The possibility of making allowance for transition to natural convection and to burnout is considered in Section VII,E.

Various refrigerants were studied at very high p_r [5] and at low p_r [6]; R12 was also studied at moderate p_r [7]. The results for R12 in terms of

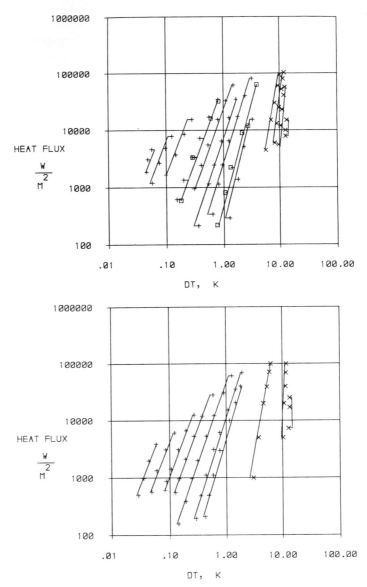

FIG. 1. Experimental results of boiling heat transfer compared with correlations from Table V, Appendix A,C,1: (a) Refrigerant R12, Bier *et al.* [5], +; Engelhorn [6], ×; Götz [7], □—Table V, line 5. (b) Refrigerant R13B1, Bier *et al.* [5], +; Engelhorn [6], ×—Table V, line 6.

$\log(q/A)$ and $\log(\Delta T)$ are repeated here in Fig. 1a. Different symbols refer to different workers, and the experimental points are linked to corresponding straight lines determined by a best fit correlation of the current type. Figure 1b is a counterpart for R13B1. The correlations include allowance for variation of roughness between the experiments, and details are given in Appendix A,C,1, with readily computed numerical accuracy of fit to the correlations. Here we seek to assess internal consistency, rather than fit to correlations of given assumed form. No assumption therefore can be made here, concerning the line or curve on which the data should lie; hence that numerical analysis of fit is not relevant here. The graphs show that the 10 or 15 individual points at a given pressure generally lie within a few percent of a common line or "reasonably smooth" curve. It is rare for any data point to lie away from the curve defined by its fellows by more than ±10%.

Such a low level of scatter in individual readings would normally imply that an average or a curve of best fit through a batch of some 10 points at given p_r should be reliable to a few percent, unless there is a problem of longer-term reproducibility. That reproducibility could not be verified as between Bier *et al.* [5] and Engelhorn [6], because they concerned different ranges of p_r, but Bier *et al.* [5] and Götz [7] studied R12 in overlapping ranges of p_r. Close examination of Fig. 1a and the associated computer printout shows that each set of points from Götz [7] defines a line lying, on average, some 10% further to the left than Bier *et al.* [5] would suggest. This implies further discrepancies, in addition to the slight scatter around the lines. In fact, there seems to be an error of reproducibility in ΔT or $h/(q/A)^m$, amounting to some ±10%, in these very careful experiments.

2. The Japanese (Kyushu) Data Sets

Professor Nishikawa and others at Kyushu in Japan have conducted various experiments exploring the effects on boiling resulting from variation in pressure and fluid [4], and more recently, variation in surface roughness as well [3]. Again, numerical accuracy of fit to specific correlations is discussed in Appendix A,C,1. Again, graphs are given here (Figs. 2a for Nishikawa *et al.* [3], 2b for Nishikawa *et al.* [4]), which show that each series of 10 or 15 points under given conditions and pressure could be represented by its own line or curve with very few points lying away from that curve by more than ±10%. These graphs also give an indication of reproducibility, since they both refer to data obtained for R114 on a flat copper plate, polished with emery of grade 0/4. The lines on the figures show a single correlation obtained by fitting these two data sets, together with similar sets for R21 and R113. On average, the points in Fig. 2a lie 7%

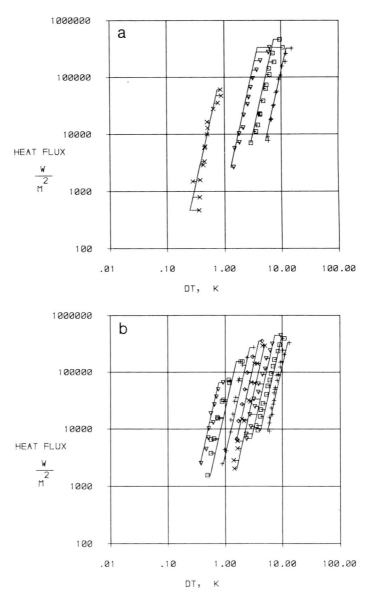

FIG. 2. Experimental results of boiling heat transfer of R114, with $R_p = 0.125 \,\mu$m (Nishi-kawa et al. [3, 4]), compared with correlation from Table VI, Appendix A,C,1: (a) Nishi-kawa et al. [3], $p_r = 0.09$, +; 0.30, □; 0.57, ▽; 0.90, ×. (b) Nishikawa et al. [4], $p_r = 0.09$, +; 0.18, □; 0.30, ▽; 0.45, ×; 0.81, +; 0.57, □; 0.90, ▽.

above the lines, and the points in Fig. 2b lie 12% below the lines, again suggesting reproducibility of order ±10%. Similar reproducibility is shown for R21 and R113 in Appendix A,C,1.

3. *Other Sets*

Among the many other data sets, many original papers include graphs of the data, from which it can be seen that accuracy does not in general improve on the two sets discussed previously. Few offer opportunities to check reproducibility.

On comparing data from different experimenters in different laboratories using nominally similar conditions, much larger discrepancies arise, which will be discussed in Section VII,D, below, on unexplained differences between experimenters.

4. *Comment: Accuracy and Reproducibility*

The intention in closely examining these recent experiments from Germany and Japan is not to denigrate the work. Instead it is to emphasize that, even with the greatest care in experimentation within a single laboratory, it is very difficult to produce results in which ΔT is accurate (consistent within a single test run) to ±10%, or reproducible (between experimenters using the same or closely copied apparatus) within ±10%. Comparability between different laboratories is hard to assess, since apparatus is rarely reproduced. An analysis disregarding detailed differences in apparatus is given in Section VII,D and suggests a scatter band of +40%, −30%.

This can act as a warning that any proposal for research that would depend on obtaining results with higher accuracy may not achieve its aims. It also affects the present work or any other attempt to extract information from the general boiling data, as will be discussed. If subsequent analysis of data implies that some change of fluid or conditions causes a difference of the order of ±10%, then it must, of course, be treated with reserve.

VI. Examination of Fluids and Existing Correlations

To determine the effects on boiling due to variation of properties among many different fluids, it would be desirable to have a further "set of tools," analogous to the set p_r, $-\log_{10} p_r$, and so on, which was developed in Section II to describe the variation of properties of a single fluid. Variation of properties between fluids is more complex, but some simple approximate patterns exist and will be of value here. In this section we therefore first consider briefly and in very general terms how properties tend to vary among those fluids that are used in boiling research; then we examine the existing boiling correlations with those tendencies in mind.

Much skilled effort has been devoted to accurate estimation of all properties of all fluids, using many numbers, factors, and parameters to characterize differences between molecules, as summarized in Reid *et al.* [8]. Our present concern is with a limited number of saturation properties that are considered relevant to boiling, and the limited number of fluids, of limited types, that have been widely used in boiling research. Our concern is at first to identify major changes or trends in variation of these properties, resulting from variation of fluid, to compare with general trends among existing correlations and among heat flow data. It is to be hoped that effects of minor changes in properties will be of interest in due course, but at present they would be lost in the scatter of general data on boiling heat flows.

This examination also recognizes and quantifies the fact that, at least among those fluids commonly used for boiling research, the properties are not independent. A change from one fluid to another will generally cause properties at a given reduced pressure to change en bloc; it is not possible to change only one property. This interdependence has positive and negative effects. Positively, it enables the present general comparison of fluids to be put on a simpler, though approximate, basis. Negatively, it effectively prevents the direct experimental determination of the influence of individual properties in boiling.

In the following subsections we consider molecular weight M in connection with properties (Section VI,A), then in connection with existing correlations (Section VI,B), then some alternatives to M (Section VI,C), and also parameters describing the differences in shape of the liquid vapor saturation line, as p_r against T_r (Section VI,D).

A. Thermodynamic Properties: Broad Variation with M

Among parameters that characterize molecules, M is readily known and is prominent in thermodynamics through the importance of molar quantities in general and the importance of more or less universal charts of compressibility $[pv/RT = f(p_r, T_r)$, the law of corresponding states] and molar enthalpy deficit $(H - H_0)/RT_{cr}$. Certain properties, notably ρ_g, h_{fg}, c_{pf}, vary widely between fluids, while their molar counterparts do not vary so much. For the fluids that concern us, several further properties vary broadly with M, albeit with considerable scatter and an occasional outlier, exemplified by k_f and μ_f in Figs. 3a and 3b, which are evaluated at $p_r = 0.1$. These and similar graphs suggest the following sweeping approximations for the common boiling fluids, at given p_r:

$$\rho_g \propto M, \qquad h_{fg} \propto M^{-1},$$

$$k_f \propto M^{-1/2}, \qquad \mu \propto M^{2/3}, \qquad c_{pf} \propto M^{-3/4}$$

(7)

ρ_f and σ show little systematic variation. Both ρ_f and σ are low for fluids (cryogens) with very low M.

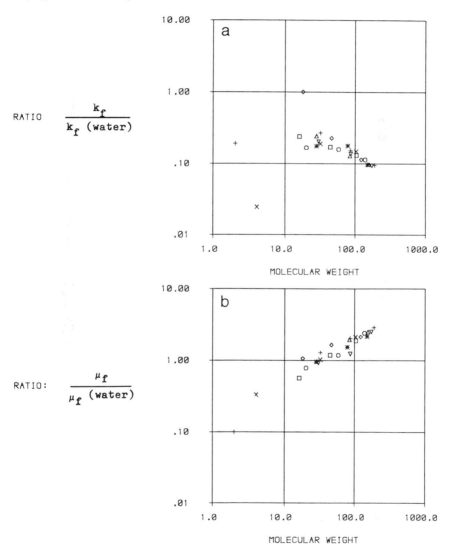

FIG. 3. Trends in property variation at $p_r = 0.1$, with molecular weight for fluids listed in Table VII, Appendix A,C,2: (a) k_f, thermal conductivity of saturated liquid; (b) μ_f, viscosity of saturated liquid; (c) k_f/σ; (d) p_{cr}/T_{cr}.

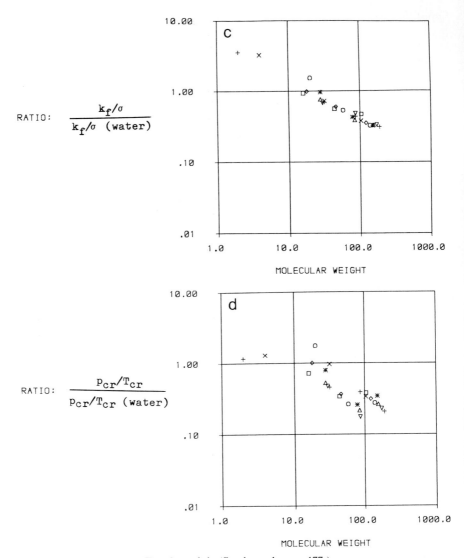

RATIO: $\dfrac{k_f/\sigma}{k_f/\sigma \text{ (water)}}$

RATIO: $\dfrac{p_{cr}/T_{cr}}{p_{cr}/T_{cr} \text{ (water)}}$

FIG. 3c and d. (See legend on p. 177.)

Several of these will be the despair of experts in property estimation. Several could be immediately improved by use of p_{cr} or T_{cr}, notably $\rho_g \propto M p_{cr}/T_{cr}$, $h_{fg} \propto T_{cr}/M$, at given p_r. These are left in abeyance at present, for clarity, since the variation of M is one hundredfold, whereas the variations of p_{cr} and T_{cr} are generally less than twofold, except for the cryogens, which receive some special treatment later, though the experimental data are few. Water is a major exception, as regards p_{cr}, T_{cr}, σ, and (Fig. 3a) k_f, which are all high, so it is considered separately in various contexts here.

A determined attempt to "buck this trend" was made by Mayinger and Holborn [9], seeking fluids that boiled at reasonable temperature and pressure with significantly differing values of μ_f. They found the physics of fluids was against them, and after a systematic search they decided to use isopropyl alcohol and isopropyl benzene, both at subatmospheric pressures corresponding to p_r about 0.0045. That gave boiling temperatures and viscosities of 50°C, 0.00101 kg/(m sec), 90°C, 0.00038 kg/(m sec), respectively. The ratio 2.7 is quite modest and principally attributable to differences in molecular weight using Eqs. (7), since M is 120 and 50, respectively, and $(120/50)^{2/3}$ is 1.8. The residual difference, 1.5, reflects the fact that, because of lower T_{cr}, the isopropyl alcohol was boiling at lower temperature. That could no doubt be allowed for, but it is not profitable to set up a patchwork of approximations.

There are modest systematic variations between types of fluid; e.g., on replacing a paraffin by a refrigerant, the effect is partly described by Eqs. (7), and there is also an increase in ρ_f and reduction in c_{pf}, causing the product ($\rho_f c_{pf}$, or thermal capacity per unit volume) to remain nearly unchanged.

These Eqs. (7) for properties are quite rough, but sufficiently accurate to guide our present rough search of the general heat flow data and to show or explain trends among existing correlations. This enables guidance to be given for optimizing searches through existing data for effects of fluid properties and for optimizing any future experimental programs.

B. EXISTING CORRELATIONS: BROAD VARIATION WITH M

1. Existing Correlations: Structure and Evaluation

Most correlations of multiproperty type [Eqs. (1) and (2)] predict strong dependence of heat transfer rate on certain properties of the fluid. Correlations differ widely with regard to selection of properties and their indices [α_i in Eq. (2)], because the authors have different views of the importance of heat flows caused by different mechanisms. Those may include the nucleation of bubbles or the flow of latent heat to maintain bubble growth, or the consequent fluid flow effects, such as pumping action or effects of forced convection on a small scale. No attempt is made here to assess or even review all the theories. Bubble dynamics is considered briefly in Appendix B, largely as an illustration of how one particular set of mechanisms may be expected to influence boiling heat transfer. In

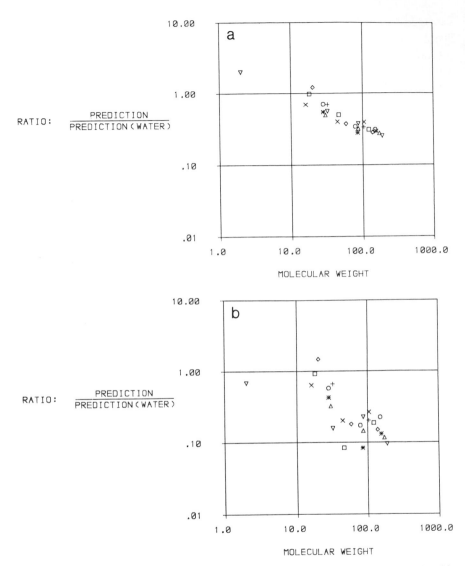

FIG. 4. Trends in prediction of boiling correlations at $p_r = 0.1$, with M for fluids listed in Table VII, Appendix A,C,2: (a) Chen correlation [10], typical of many; (b) Rohsenow correlation [18], omitting fluid-correction factor.

particular, that can lead to predictions that heat flow is reduced by high σ (hindering nucleation) and increased by high k_f (assisting conductive flow of heat). It also tends to suggest that the product $\rho_g \times h_{fg}$ is more important than either ρ_g or h_{fg} in its own.

If we evaluate correlations for a range of fluids at the same reduced pressure, then many of them predict much the same trend among the fluids. They predict that $h/(q/A)^m$ (with m about 0.6–0.8) will show a downward trend with increasing molecular weight M. This is shown in Fig. 4a for the correlation of Chen [10], at a typical value of p_r of 0.1. Similar graphs for the correlations in [11–15] and for p_r of 0.3 or 0.03 look very similar. Thus in many cases the differences between correlations are more apparent than real. The correlations in [16, 17] show a reduced fall for $M > 50$, and that in Rohsenow [18] receives special consideration below.

In this evaluation of correlations at given p_r, it is convenient to use the expressions for properties in the form of Eq. (4) or variant with $(-\log_{10} p_r)$, using different values of the indices for each property of each fluid. That assists interpolation and causes no significant loss of accuracy; indeed, it may even improve accuracy by smoothing out any misprints in property tables. Equations (7) would introduce significant inaccuracy, so they are not used for this evaluation.

2. Trends in Correlations with M

The reason for this variation of correlations with M can be given in a general approximate manner, by using the approximate relations of Eqs. (7): If we take a typical correlation of the form of Eq. (1), transform it to the form of Eq. (2), and then substitute Eqs. (7), to show its broad dependence on M, then most correlations predict that C_E will fall with M, generally as $M^{-0.4}$–$M^{-0.6}$. That arises because a typical correlation will give equal powers to the properties most strongly dependent on $M(\rho_g$ and $h_{fg})$; hence their effects largely cancel out. The powers given to σ, k_f, μ_f, and c_{pf} are typically -0.4, 0.4, -0.2, 0.2, so application of Eqs. (7) gives $M^{-0.48}$.

These are very broad generalizations. The occurrence of ρ_g and h_{fg} as the product $(\rho_g h_{fg})$ arises from the Clausius–Clapeyron equation, or from bubble dynamics, as discussed in Appendix B. In fact, using the better approximations $\rho_g \propto M p_{cr}/T_{cr}$, $h_{fg} \propto T_{cr}/M$, the T_{cr} term also cancels in the product $\rho_g h_{fg}$, leaving only p_{cr} uncanceled, and that is usually between 30 and 40 bar (water being a major exception). The occurrence of σ and k_f to opposite powers is plausible, as discussed earlier; high σ would hinder nucleation and high k_f would aid heat flow. It arises in correlations because they often have σ in a length d such as $\sigma/\Delta p$ where Δp is some form of pressure difference, or as $[\sigma/(\rho_f - \rho_g)g]^{0.5}$; that length and k_f often occur in $Nu = hd/k$, and also in groups on the right-hand side of Eq. (1), such as Re, Pr.

As noted earlier, water has several outlying properties, and most existing correlations originating as Eq. (1) have σ and k_f to nearly equal and opposite powers. Since σ and k_f for water both lie about a factor of 3 above the general trend, that factor 3 cancels in most correlations.

One well-known correlation that does not give equal powers to ρ_g and h_{fg} is that by Rohsenow [18], which omits ρ_g and has $h_{fg}^{-0.667}$. That would lead to strong dependence on M, but it is partly counteracted by inclusion of Pr to the unusually high power of 1.7. That brings dependence on M more into line with other correlations. The correlation remains unusual in another respect, that σ and k_f do not cancel. Different fluids may therefore be expected to need their own correction factor, and the Rohsenow correlation does have such a factor, varying with fluid and heater material. Without that correction factor, the predictions would be very scattered, as shown in Fig. 4b.

3. *Summary: Trends in Correlations with M*

Most correlations suggest a downward trend in boiling heat flow with increasing M, and most correlations suggest that the two chief outlying properties of water (σ and k_f) have little combined effect. As will be seen, that is in broad agreement with the trend of the experimental data (Fig. 8 and Section VII,C,2). The experimental data are so scattered that we cannot decide between correlations, which often look different but in fact give much the same answers, because of the various approximate interrelations between those properties.

As will be seen, experimental results show no great differences between water and other fluids, so it is no coincidence that, in any relations that are successful in correlating boiling data for water and for a wide range of other fluids, σ and k_f should either be absent or cancel. Also, if experimental data show a general trend as $M^{-0.5}$, then successful correlations should reflect that fact, either by containing combinations of properties that are broadly equivalent to $M^{-0.5}$ (Fig. 4a) or alternatively by having a disposable constant dependent on the fluid. But the question to be discussed later is whether any of the complicated correlations fit the data any better than a simple term such as $M^{-0.5}$. Is the complication justified?

To determine these trends, it is clearly desirable to use fluids with a wide range of M, including those with low M, the cryogens. That can cause confusion here because cryogens have both low M and low T_{cr}; hence quantities that correlate with M will tend also to correlate with T_{cr}. Water is even more of an outlier in this context, as it combines high T_{cr} with quite low M.

C. ALTERNATIVES TO M

Although a term like $M^{-0.5}$ will be seen to lie within the scatter of the boiling data, it is not clear whether M and the variation of properties with M are truly significant for boiling. Broad variations of the type being considered in terms of M can also be expressed in terms of other quantities, at least for the limited range of fluids considered. If, for many fluids, several properties $P_1, P_2, P_3, \ldots$ were each accurately proportional to certain powers of M, then of course $P_2, P_3, \ldots$ would each be accurately proportional to certain powers of P_1. In any relationship such as a boiling correlation involving M, no significant change would be caused by replacing M throughout by P_1 to some power. If the proportionality of P_1, P_2, P_3 to powers of M is not accurate, then the replacement would affect relative positions of some fluids. If this type of argument is pursued at length, it soon reaches increasing inaccuracy and decreasing value amid the scattered experimental data, but one case of possible value is mentioned here and applied to boiling data in Section VII,C,3.

For many fluids, the critical compressibility Z_{cr}, $p_{cr}v_{cr}/(RT_{cr})$ is nearly the same, and for those fluids, Mp_{cr}/T_{cr} does not vary greatly, or $T_{cr}/p_{cr} \propto M$. As shown by Fig. 3d, that is broadly true, though it is not accurate, and there is a significant difference between the refrigerants and most other fluids. This is similar to the better-known difference between ρ_f for refrigerants (usually >1400 kg/m³) and for most other fluids (usually <1000). Given sufficient data for refrigerants and the other fluids, it would be possible to see whether boiling data varied more closely with M or with (T_{cr}/p_{cr}), and that will be discussed in Section VII,C,3.

D. PARAMETERS RELATING TO $(p_r, T_r)_{sat}$

It is not clear whether boiling is more fundamentally related to temperature and T_r, as implied by use of $h = (q/A)/\Delta T$, or to pressure and p_r, which is more conveniently and accurately measured in experiments and is frequently used in correlations. This is briefly discussed in Appendix B. If the relation between T_r and p_r were universal for all relevant fluids, then boiling could be expressed in terms of either T_r or p_r, but the relation is not universal. It is shown in Fig. 5, where the reduced saturation line $(p_r, T_r)_{sat}$ is drawn on log $\times$ linear scales for a range of fluids. It is of broadly similar shape for all fluids, but there are significant residual differences between fluids, with cryogens on the right (steeper curves), alcohols on the left (less steep). The differences can be quantified by use of various parameters. One is the Pitzer acentric factor, ω [$= 1 - \log_{10}(p_r)$ (at $T_r = 0.7$)], which appears to a linear scale on Fig. 5. It has the advantage of referring to a temperature near the median for most liquids, since T_r at the

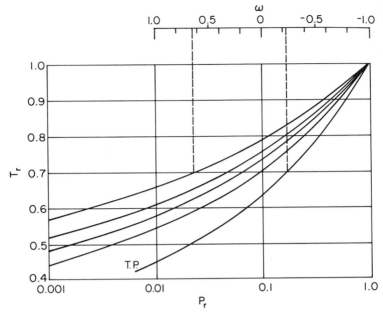

FIG. 5. Relation between reduced pressure and saturation-reduced temperature for several liquids, showing Pitzer acentric factor = $\omega = -1 - \log_{10}(p_r$ at $T_r = 0.7)$. Curves refer to ethanol (highest), water, R12, oxygen, hydrogen (lowest) (for ω, read from intercept on $T_r = 0.7$).

triple point is typically about 0.4 or 0.5, which is off Fig. 5 for most fluids. There is a general tendency for high ω to be associated with high M, though water is again among the major exceptions, as it has fairly high ω, comparable to refrigerants, but, of course, much lower M. Simple scaling of the abscissa in Fig. 5 suggests that the graph of T_{sat} against $(\log_{10} p_r)/(1 + \omega)$ would be nearly universal, as it is. In fact,

$$(-\log_{10} p_r)/(1 + \omega) \approx (1/T_r - 1)/(1/0.7 - 1)$$

which is related to the Clapeyron approximation, but many other approximations have been suggested [8]. In the widely used Antoine approximation, $\log(p_{sat}) = A_a + B_a/(T + C_a)$, the parameters A_a, B_a, C_a are usually chosen so that the expression matches the mid-range on Fig. 5; hence it does not match near the CP—a major snag for the present work. An alternative to ω is the Riedel parameter, evaluated at the critical point $(dp_r/dT_r)_{cr}$. One of these may prove useful in due course, leading to a clearer understanding of the fundamental importance of temperature and pressure in boiling.

VII. Data Analysis: Results

Using the computation, the data bank, and the fluid property summary described earlier, this section presents the results of an analysis of general saturated nucleate pool boiling data, attempting to find what parameters affect the heat transfer. The text would become unwieldy if all details were presented here. This section therefore amounts to an extended summary of results, while supporting details are given in Appendix A, subsections there being denoted A, B, and so on, to correspond to subsections VII,A, VII,B, and so on, here.

The parameters that are generally assumed to be important can be put into three identifiable categories and a fourth emerges, namely:

1. operating conditions—determined by pressure in saturation boiling;
2. heater—roughness, shape, material;
3. fluid—thermodynamic properties;
4. unidentified differences between experimenters.

These parameters are considered later, in the preceding sequence, since that generally places the clearest and most important first.

Although the data bank is extensive, its rather poor distribution and the well-known scatter among boiling data prevent full determination of the effects of these parameters.

Most existing analyses have implicitly assumed that effects of these parameters are independent, in the sense that they could be examined separately, and represented by separate terms that could then be multiplied together. The following analysis will proceed largely on that same assumption, though it proceeds differently in places, because some recent data suggest that the effect of roughness is different at different pressures.

Different data sets or combinations of sets are appropriate for different aspects of the present search; a single set from a single experimental program with unchanged or carefully copied apparatus is ideal for determining effects of changes in pressure and fluid, but naturally sheds no light on differences between heaters.

The work amounts to successive refinements of

$$h/(q/A)^m = f(p_r, R_p, \ldots) \tag{8}$$

Errors are quoted as errors in $h/(q/A)^m$ at given (q/A), which is the same as the error in h at given q/A, or (inverted) the error in ΔT at given q/A, if we prefer to consider the data as relating q/A and ΔT. In the latter graph, error in q/A at given ΔT would be greater, since $q/A \propto \Delta T^n$, $n = 3$ or 4, so 10% in ΔT corresponds to some 35% in q/A.

A. EFFECT OF OPERATING CONDITIONS (PRESSURE)

The general effect of pressure has long been clear from correlations presented as Eq. (3), in which the right-hand side is a function $F(p_r)$. This present method produces that function in the simple form of Eq. (6), or variants, direct from experimental data, and the result is in general agreement with existing correlations whether they were presented as in Eq. (3) or in the multiproperty form of Eq. (1).

This analysis produces different indices for Eq. (6), that is, different forms for $F(p_r)$, depending on the data sets used. Indices can be produced for any set or sets, but if the data are confined to $p_r > 0.1$, then the resulting $F(p_r)$ may go far astray when extrapolated to p_r far below 0.1.

1. $F(p_r)$ from Data Sets with Wide Range of Pressure

To obtain an expression for $F(p_r)$ that is to be trusted over a wide range, it is necessary to base it on data sets having a wide range of p_r. As discussed in Appendix A, Section A,1, there are 14 such sets, and taken together they yield the curve given by full lines in Fig. 6, which is the right-hand side of

$$h/(q/A)^{0.73} = p_r^{0.1}(-\log_{10} p_r)^{-2/3}C_E \qquad (9)$$

Taken one at a time, the individual wide-ranging sets produce different values for the index of (q/A), which are not our present concern, and also a cluster of curves for Fig. 6. When normalized to a common value at $p_r =$

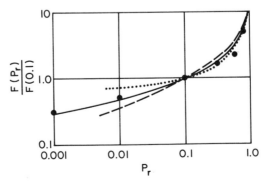

FIG. 6. $F(p_r)$, describing variation of $h/(q/A)^m$ with p_r, for data sets with wide range of p_r: average, solid curve; hydrogen, dashed curve; oxygen, dotted curve. Points from graph by Borishanskii [2].

0.1, they lie between the extremes shown on Fig. 6, the steepest (dashed line) for hydrogen and the flattest (dotted line) for one of the oxygen sets. As will be discussed, such differences in shape of $F(p_r)$ may be partly attributable to differences in roughness, which was not stated for those sets, and perhaps partly attributable to differences between fluids and experimenters.

The central line in Fig. 6 is in close agreement with a similar curve by Borishanskii [2], except around $p_r = 0.5$. A test was therefore run by including an additional term $(1 - p_r)^D$, which would have large effect at high p_r. The result is detailed in Appendix A,A,1 and suggests that the curve in Fig. 6 is a better fit to this data than Borishanskii's curve, which was based on a narrower range of data.

2. $F(p_r)$ for Refrigerants

Despite the large amount of good data for refrigerants, no data set for refrigerants is used in the preceding discussion, because none covered a wide range of pressure. As discussed in Appendix A, Section A,2, they correspond to $F(p_r)$ between the extremes shown in Fig. 6. Those data are of greater value in discussing roughness, since R_p is virtually always given for refrigerants.

There is clearly a need for extending existing experimental data, to produce sets for more fluids with a wide range of p_r, from, say, 0.9 to 0.01 or below, though that is subatmospheric for most fluids.

B. EFFECTS OF HEATER

It is well known that boiling heat transfer is strongly affected by varying the surface condition of the heater and that no one measure such as R_p is sufficient to describe the effect, since different behavior is observed when the same R_p is obtained by emery or by etching. Effects of shape and material are less well established, partly because of the unfavorable distribution of data discussed in Section V,A,1 and partly because of the problem that any change of heater shape or material involves change of heater and is liable to introduce an uncontrolled and unrecognized change in surface condition, which can cause changes exceeding those from shape or material.

1. Heater Roughness

A recent large group of data sets from Japan [3] suggests strongly that the effect of roughness is greater at low pressure; specifically, the originators suggest $h/(q/A)^m$ is proportional to $(8R_p)^{0.2(1-p_r)}$. As discussed in

Appendix A, Section B,1, the present analysis was readily extended to include such a term. That gave a very close confirmation of their finding, with 0.2 replaced by 0.203. An alternative term $(R_p)^{G(-\log_{10}p_r)}$ was also tried, using the new analysis. With the constant $G = 0.206$, that gives equally good fit, over the rather limited range of p_r reported. More data are needed, particularly at lower p_r, to decide whether the index of R_p should involve $(1 - p_r)$ or $(-\log_{10} p_r)$. The latter continues to change as p_r falls from 0.08 (the lowest limit of data in Nishikawa *et al.* [3]) to 0.01 and beyond, whereas $(1 - p_r)$ changes little. For most fluids $p_r < 0.03$ implies pressure below 1 atm.

If we adopt the form $(R_p)^{G(-\log_{10}p_r)}$, it can be transformed using $x^{\log y} = b^{\log x \log y} = y^{\log x}$ (where b is the base of the logarithm) to $p_r^{G(-\log_{10}R_p)}$, showing more clearly that it is equivalent to a change in slope of the left-hand part of Fig. 6. At this stage the equation is of the form

$$F(p_r, R_p) \propto p_r^{A+B\log R_p}(-\log_{10} p_r)^C \tag{10}$$

which can be represented graphically in the form of a product of two terms, shown in Fig. 7a and b. Each term has a disposable index, and that has the effect of stretching the vertical scales on Fig. 7a and b. The suggestion is that roughness operates by stretching the vertical scale of the straight line term (i.e., altering its slope). Further data at lower p_r would indicate whether that is to be preferred to the original suggestion, with $(8R_p)^{0.2(1-p_r)}$ (Fig. 7c) and whether the index of $(-\log_{10} p_r)$ should also be varied.

Earlier studies of effects of roughness may appear at first to be incompatible with the new work. However, as discussed in Appendix A, Section B,1, a reconsideration, in the light of the influence of p_r, shows that some at least are consistent.

There is a view that roughness has little effect, provided it exceeds a certain value. That view is not supported by Nishikawa *et al.* [3] or by the general data used here, and Appendix A, Section B,1 discusses how it might be tested and incorporated [e.g., with Eq. (10)], in the light of the influence of p_r.

Roughness is presumably having effect through nucleation, which has been much studied [19]. Appendix B briefly surveys bubble nucleation and growth, and Appendix A, Section B,1 discusses how that basic research might contribute here.

2. *Heater Shape*

The ideal data to show effects of shape would be for boiling of the same fluid on heaters of the same material but different shapes. As shown by

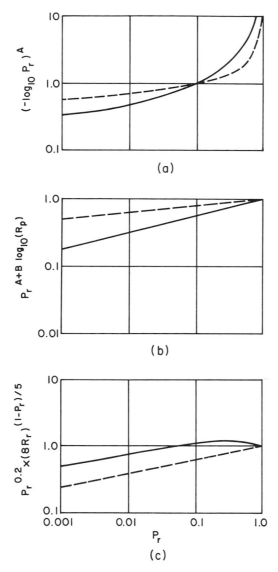

FIG. 7. Components of $F(p_r)$, showing dependence on roughness R_p: (a) $(-\log_{10}p_r)^{-1.0}$, solid curve; $(-\log_{10}p_r)^{-0.5}$, dashed curve; representing $(-\log_{10}p_r)^A$, not dependent on R_p (b) $p_r^{0.25}$, solid curve; $p_r^{0.1}$, dashed curve; representing $p_r^{A+B \log_{10} R_p}$ (c) $p_r^{0.2} \times (8R_p)^{(1-p_r)}$; with R_p = 3.0, solid curve; with R_p = 0.1, dashed curve; tested in Nishikawa et al. [3] for $0.022 < (R_p, \mu m) < 4.3$ and $0.08 < p_r < 0.9$.

Table I, such data exist for refrigerants on copper heaters in the shape of flat plates or cylinders. Conveniently, there are many such data with roughness stated, plate data largely from Japan, cylinder data largely from Germany, each using several refrigerants. There are no data on the same refrigerant in the two laboratories, but the two ranges of refrigerants are interleaved in respect of molecular weight and most thermodynamic properties.

Those data are therefore reasonably well suited to this problem, and the analysis, discussed in Appendix A, Section B,2, shows a dependence on shape, cylinders having C_E some 1.7 times that of flat plates. That is quite plausible, analogous to natural convection being greater for cylinders than for flat plates.

It might appear from Table I that some similar comparisons are possible: for water on stainless steel plates, cylinders, and wires; for cryogens on copper plates and cylinders. These are discussed in Appendix A, Section B,2, but the data can support only a tentative suggestion that C_E for stainless steel wires may be less than for other stainless steel heaters.

3. Heater Material

Here we require ideally data for the same fluid boiling on heaters of the same shape but different material. As shown by Table I, most flat plates are of copper; the few noncopper flat plate heaters are listed in Appendix A, Section B,3 and are insufficient to support a comparison, partly because they refer chiefly to cryogens, a very varied group, with T_{cr} ranging from 5.2 K for helium to 154 K for oxygen.

Among cylindrical heaters, most are copper or stainless steel. Unfortunately, these are for very different fluids, with the copper heaters being used largely for refrigerants (none for water) and the stainless steel ones largely for water (none for refrigerants). No doubt there are good experimental reasons for this division, but it effectively prevents the desired comparison.

As described in Appendix A, Section B,3, a less selective attempt was made to find an effect of k_{wall} among all data, by including a term (k_{wall}) to some power in the least squares analysis. The result indicated a very weak residual dependence on k_{wall}, provided the computation already made allowance for the factor 1.7 noted earlier for improved heat transfer from copper cylinders.

4. Heater Shape and Material

We can make a combined comparison and draw a tentative conclusion from data for water boiling on copper plates and on stainless steel cylin-

TABLE II

Processing Data from the "3 × 3" Set [20][a]

Fluid (M)	Heater			
	Cu plate	S/S cylinder	Pt wire	Average for fluid
Water (18.0)	2.6	2.5	2.7	2.6
Ethanol (46.1)	1.8	1.1	1.9	1.6
Benzene (78.1)	1.2	1.0	0.9	1.0
Average for heater	1.9	1.5	1.9	

[a] Value of C_E in the correlation: $h/(q/A)^{0.776} = p_r^{0.073}(-\log_{10} p_r)^{-0.747} \times C_E$

ders. The relevant sets are listed and analyzed in Appendix A, Section B,4, showing that there is little difference in C_E. That suggests a cancelation has occurred, as if the change from plate to cylinder were canceled by the change from copper to stainless steel. If we assume that the difference between boiling of refrigerants on cylinders and on flat plates is also applicable to water, then we can apply to water the numerical value 1.7 that was observed with refrigerants but that is risky with such a different fluid. Again, the suggestion is plausible, since the thermal conductivity of copper is some 20 times that of stainless steel, but the argument is tenuous.

Some support comes from a data set [20], which gives a "3 × 3" comparison, using three types of heater (copper plates, stainless steel cylinders, and platinum wires) for three fluids (water, ethanol, and benzene), with results given in Table II. Although most tables of results are relegated to appendices, this table is given here because it will be found very useful in several contexts here and later. Copper plate and stainless steel cylinder again yield similar results for water and for benzene but not for ethanol. For each fluid, copper plate and platinum wire yield similar results.

Among refrigerants, there is a tendency for platinum wires to give heat flow less than copper cylinders, as will be seen below in Fig. 9.

The best suggestion we can make at present is to accept that boiling heat flow on copper cylinders is about 1.7 times higher than that on other heaters. That has been determined only from comparisons with copper plates and stainless steel cylinders, so it is not strictly logical to apply it to all other heaters, but there are simply not enough data on them.

It would be preferable to have the data needed to avoid such combined comparisons, and that means data for refrigerants with stainless steel plates and cylinders and for water with stainless steel plates and copper cylinders. There may be great experimental obstacles to some of these; the fact that the data have not yet been produced itself suggests that there may be problems.

5. *Effects of Heater: Summary*

The effect of roughness has long been known and has recently been shown to be dependent on pressure. Earlier studies did not show this but are not necessarily inconsistent with it. The recent investigation could usefully be extended to lower pressures and to other geometries.

An effect of geometry emerges from comparison of boiling certain refrigerants on copper cylinders in Germany and other refrigerants on copper plates in Japan, the former having C_E some 1.7 times the latter. The evidence for that is not overwhelming and would be rejected in some fields as being too close to the scatter, or subject to differences between fluids and laboratories, but if we do not at this stage accept it as a hypothesis to be considered and tested, then we will get little out of boiling data.

The data do not support any direct estimate of the effect of changing heater material. However, for water and for benzene, a combined change from copper plate to stainless steel cylinder is reasonably clearly shown to have little effect, so there is a tenuous suggestion of a cancelation: perhaps the change from copper to stainless steel cancels the change from flat plate to cylinder. In the absence of sufficient data, we take it that boiling heat flow for cylindrical heaters with high conductivity (i.e., copper) is 1.7 times that for all other heaters.

Again, the experiments could usefully be extended, as will be discussed more comprehensively in Section VII,F. The chief needs arising here are:

1. for roughness, extension to lower p_r and high R_p;
2. for material, the use of plates other than copper, and the use of stainless steel cylinders for fluids other than water.

C. EFFECTS OF FLUID PROPERTIES

Most correlations predict that some particular selection of properties will affect C_E. Ideally, experiments would confirm or deny that selection of properties. To achieve that, the data bank would need data sets from identical apparatus for fluids with radically different values for some properties, or—best of all—for one property. As discussed in Section VI,A, for most changes from one fluid to another (except water), the individual properties will change en bloc. It was also shown in Section VI,B that, although correlations have widely differing selections of properties, most selections amount to a decrease with M, represented roughly by $M^{-0.5}$, even for the exceptional case of water. Experimentally, the changing of the fluid requires and generally receives much care, as it interrupts an experimental run and may have important unwanted effects on the physicochemical state of the heater, even apart from the possible effect of simple contamination by the previous fluid or other material.

In these circumstances, there are two main approaches:

1. To consider a particular set of data from the same apparatus with different fluids; any consistent difference between fluids exceeding the reproducibility limit (say 10%) should be significant. Various German and Japanese data sets are examined separately in this way. The method is to set up a correlation by least squares fit, confined to the particular data under examination. That correlation will look rather different from Eq. (9) and will be unsuitable for extrapolation, but it aids detailed examination of the data in question.

2. To use all data for all fluids and compare it with the best prediction so far obtained, making no specific allowance for different fluid, thus determining a different constant C_E for each fluid. Any trends in the value of C_E are then examined, bearing in mind the general variation of fluid properties with M and the two outlying properties k_f, σ, for water, discussed earlier.

1. Different Fluids under Similar Conditions

As detailed in Appendix A, Section C,1, a few groups of data cover widely different fluids, notably the "3 × 3" group [20], which used water, ethanol, and benzene. The results have already been shown in Table II and discussed in connection with change of shape and material of heaters. Here it is noted that these fluids do not behave very differently, though C_E does fall with M, possibly as $M^{-0.5}$, in accordance with many correlations.

Many groups of data use several fluids, which are different but of similar type, chiefly refrigerants [3–7], but also hydrocarbon gases (methane, ethane, propane, n-butane) [21]. Closer examination of the data for refrigerants on cylinders (mostly from Germany) and (separately) for refrigerants on plates (mostly from Japan) yields confusing results, detailed in Appendix A, Section C,1. Under nominally identical conditions, values of C_E for different refrigerants can differ by 20 or 30%, but they are "out of step" with M, that is, they are not related in any simple or obvious way to differences in M or in properties, since the highest C_E may occur for the refrigerant having intermediate values for M and most properties. At this level of abstraction, we have few pieces of information; hence it will always be possible, with a bit of ingenuity, to contrive some combination of properties to fit that information, but that would be overloading the data.

The conclusion to be drawn at this stage is that if there is any difference between C_E for refrigerants, then it is small, as are the differences between properties. Before we can use refrigerants to draw any further conclusion under the current heading, "Different Fluids under Similar Conditions," we seem to need additional accurate data on a wider variety of refrigerants. That is not necessarily of high priority.

The data on hydrocarbon gases are less numerous and are also seen in Appendix A, Section C,1 to have confusing results.

The main requirement seems to be for further data for boiling of widely differing fluids under carefully reproduced conditions. The "3 × 3" set points the way [20].

2. General Data

Since there are few "ideal" sets, covering wide ranges of fluids under closely similar conditions, we now examine more general data, to see how heat flow varies among fluids. Inevitably, data from varied experimental conditions and experimenters are scattered (Section VII,D), but the hope here is that the scatter may average out, leaving a pattern of variation among fluids.

As shown in Table I, most data fall into two large coherent groups, for water ($M = 18$) and for refrigerants ($86 < M < 200$, mostly $103 < M < 187$). There are also many data for cryogens, but cryogenic fluids are more varied than refrigerants, as they range from helium and hydrogen to oxygen, nitrogen, and methane ($2 < M < 32$). There are few points for fluids with $30 < M < 100$. There is a danger that any statistical analysis or regression will be unduly influenced by its fit to water and refrigerants, while other fluids are not reasonably represented. There is a further danger that a regression may be unduly influenced by data for fluids with very low M (H_2, D_2, He) obtained with great experimental difficulty and subject to considerable scatter. Instinctively, one feels that helium may be an outlier anyway, because of its peculiar properties.

We first produce a correlation combining all 116 data sets, giving a single, averaged, value of C_E to each fluid, as detailed in Appendix A, Section C,2. In view of the very broad dependence of most properties on M, the results are plotted in Fig. 8, as C_E against M. It shows much scatter, particularly at low M (cryogens), but also a trend, generally falling, but possibly rising slightly for refrigerants, $M > 100$. That rise creates

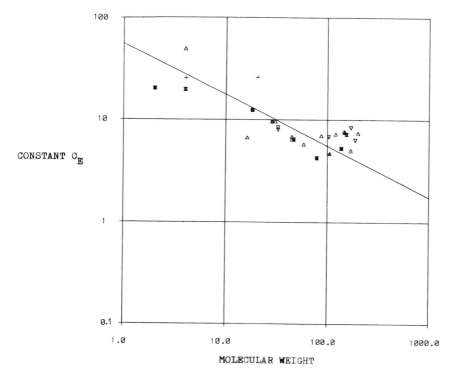

CONSTANT C_E

FIG. 8. Constant, C_E, describing variation of heat flow with molecular weight, one averaged point for each fluid, as listed in Table VII, Appendix A,C,2; h for horizontal copper cylinders reduced by factor 1.7. Horizontal heaters: flat plate, $\triangledown$; cylinder, $\triangle$; wire, $+$ (composite symbols if more than one type of heater).

a "trough" at about $M = 80$, and if it is to be regarded as significant, then it presents a problem. It is not represented by any of the existing correlations (e.g., Fig. 4a and 4b), nor can it be represented by a simple power of M (a straight line on Fig. 8), nor by a simple power of any quantity that itself varies as a power of M. For example, Fig. 3a shows that a power of k_f will not follow the trough. It might be possible to follow the trough by putting together an ingenious combination of all the properties, but that would again appear to be overloading the data. If the trough is ignored for the moment, then the data as a whole can be approximated as a power of M, either by drawing a line of best fit on Fig. 8 (slope about -0.5) or by introducing yet another term into the least squares analysis, to show dependence on a power of M, as M^H.

If we accept a term $M^{-0.5}$ and apply it to the refrigerants, its effect is small, since most refrigerants have $103 < M < 187$, with a few extending to 86 and 200, and $(103/187)^{-0.5}$ is 1.35.

It was already noted in Section VII,C,1 that there are small differences between refrigerants that appear significant but that are definitely "out of step" with M.

Figure 9 is based on the same correlation as Fig. 8, but shows individual values of C_E for each data set, again plotted against M. That shows graphically the dominance of data on water and refrigerants and illustrates the scatter among data for the same fluid from different experimenters, which will be considered in Section VII,D. That scatter casts more doubt on any precise analysis of the trough in Fig. 8, since it depends crucially on a few sets of data: one for butane ($M = 58$), with 57 points, and six for benzene ($M = 78$), with 365 points, which are now seen to be well scattered.

It is perhaps inadvisable to ignore the trough in Fig. 8, since the points to the right of it are some of the best data we have, all for refrigerants, including many accurately reproduced sets. Taken with the data for water, they really set the trend, and the term $M^{-0.5}$, since their numbers

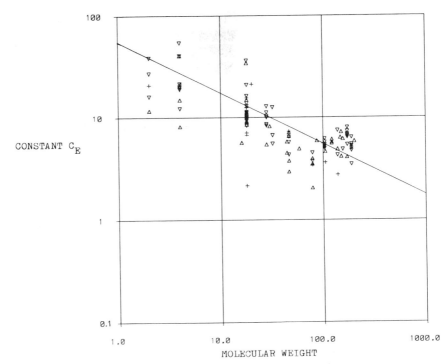

FIG. 9. Constant, C_E, describing variation of heat flow with molecular weight, one point for each data set (fluids as listed in Table VII, Appendix A,C,2); h for horizontal copper cylinders reduced by factor 1.7. Horizontal heaters: flat plate, ∇; cylinder, $\triangle$; wire, $+$.

dominate the data, as discussed earlier. More complex terms to follow the trough are considered tentatively in Section VII,C,3 and in Appendix A, Section C,3.

No precise explanation is attempted here. Instead, it is suggested that, at present, we may as well represent the difference between fluids by a small negative power of M. It is to be hoped that a better understanding and clearer representation will emerge, possibly involving a change of the shape of $F(p_r)$, analogous to that found for roughness. With that in mind, a term $p_r^{\text{const} \times M}$ is discussed in Section VII,C,4, but it also overloads the data.

3. Alternatives to M^H

It was pointed out in Section VI,C that the variation of C_E could be related to some parameter other than M. In particular, it was noted that the product Mp_{cr}/T_{cr} does not vary greatly among most boiling fluids, apart from refrigerants, for which it is high. Variation of properties (or boiling data) with M could be seen as a variation with T_{cr}/p_{cr}, in which case the relative position of refrigerants is altered. Little is gained by long pursuit of such arguments, but it is worth giving here one figure (Fig. 10), which uses T_{cr}/p_{cr} as abscissa in place of M, showing the altered position of the refrigerants relative to most other fluids. The low-M fluids, which are isolated on the left of Fig. 8 and may therefore unduly influence a regression as M^H, are not so isolated in terms of T_{cr}/p_{cr}, and in fact some of them fall better into line here. However, any impression that Fig. 10 improves on Fig. 8 is partly an illusion, resulting from compression; the vertical scatter about any line of best fit remains much the same in both figures. Also, detailed examination of the refrigerants on the right of Fig. 10 shows that it has unfortunately not resolved the two problems noted in Sections VII,C,1 and VII,C,2: data that were "out of step" with M in Fig. 8 remain "out of step" with T_{cr}/p_{cr} in Fig. 10; the slight rising trend among refrigerants is still present in Fig. 10, so the trough is still present.

The data are too few and too scattered to support any clear preference for Fig. 8 or Fig. 10. That is a useful reminder that an open mind is needed: we do not have the data to determine whether C_E is better regarded as proportional to $M^{-0.5}$ or to $(p_{cr}/T_{cr})^{2/3}$ or otherwise. Other properties might be tried, when more data are available, including parameters such as ω, which largely describes the variation in shape of the saturation line as p_r against T_r (Sections VI,C; VI,D; Fig. 5). It was noted in Section VI,D that ω tends to increase with increasing M, apart from water, which is again a major exception, having ω comparable with refrigerants. We can now see from Fig. 8 that the many data for water and for refrigerants

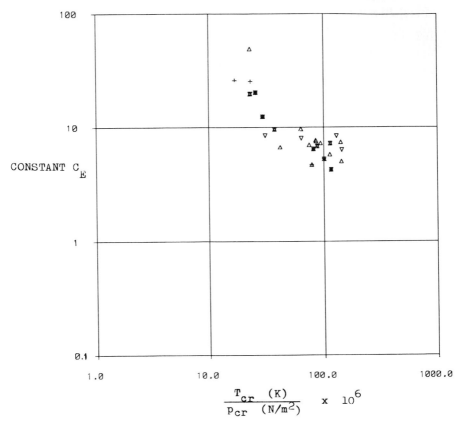

FIG. 10. Constant, C_E, describing variation of heat flow with T_{cr}/p_{cr}, one averaged point for each fluid, as listed in Table VII, Appendix A,C,2; h for horizontal copper cylinders reduced by factor 1.7. Horizontal heaters: flat plate, ▽; cylinder, △; wire, + (composite symbols if more than one type of heater).

are significantly different, so it would be detrimental to replot Fig. 8 with ω as abscissa. As argued in Section VI,B,2 when discussing k_f and σ, it is possible that ω is nevertheless involved, but canceled by one of the many other exceptional properties of water (k_f, σ, p_{cr}, T_{cr}).

It was noted in Section VI,A that some properties depend strongly on T_{cr} as well as on M, notably ρ_g and h_{fg}, but also μ_f. These matters were then left in abeyance, and we could now consider a term T_{cr} to some power as well as $M^{-0.5}$, but it is of doubtful value here, since T_{cr} varies much less than M. The only fluids with T_{cr} below 200 K are the cryogens, all with low M, giving a possibility of confusion between variation with M and with T_{cr}, as noted in Section VI,B,3.

4. $F(p_r)$ Dependent on Fluid

It is of course possible that a change of fluid is not represented by a simple multiplier applied to $h/(q/A)^m$, as assumed so far. The function $F(p_r)$, already altered to $F(p_r, R_p)$ in Section VII,B,1, might need further amendment to $F(p_r, R_p, M)$, or otherwise. The one data set for hydrogen that was sufficiently wide-ranging to be used in determining $F(p_r)$ showed relatively strong dependence of h on p_r (a steep line on Fig. 6). It is tempting to suggest it is related to the strong dependence of T_r on p_r for hydrogen (a steep line on Fig. 5). However, the suggestion is not supported by oxygen, which has a quite steep line on Fig. 5, and the flattest line on Fig. 6. Alternatively, we might suggest that it is related to the low value of σ/p_{cr} for hydrogen, which arises in nucleation (Appendix B, Section B,1), but water also has low σ/p_{cr}. Many similar suggestions can be made concerning such scattered data.

The computer program is easily adapted for establishing $F(p_r, R_p, M)$ of many forms, or indeed the index m in $h/(q/A)^m$ might be dependent on M, as $m = m_1 + m_2 M$, or even $m_1 + m_2 p_r + m_3 M p_r$. Again, we have a relatively small number of pieces of information at this level of abstraction, and considerable scatter, so there seems no justification for pursuing these details in this context or others. Answers obtained could be quite wrong. For example, an expression for $F(p_r)$ including the term

$$p_r^{A + B \times M \times \log_{10}(p_r)}$$

might be much affected by the fact that low values of p_r (<0.03) occur largely for water, with $M = 18$, while there are many data sets with exclusively higher p_r, largely for refrigerants, with M typically 150. The resulting expression with values of $A, B,$ might therefore be reflecting some cross relation between aspects of the distribution of the data, rather than aspects of the data values themselves.

5. Residual Scatter among Fluids: Proposals for Research

Any deduction from Figs. 8, 9, and 10 must be regarded as very tentative. It is even possible to draw a horizontal line across the figures. There is always a danger of "forcing" the data into a particular pattern, when in fact it would fit better into a quite different pattern. However, the figures do summarize the data used and suggest that experiments should be conducted to verify or contradict the patterns, by testing a range of fluids with different M in the same apparatus.

Data for fluids with intermediate M ($50 < M < 100$) would show whether or not the trough in Fig. 8 is significant. Scatter and uncertainty are particularly large at present for certain cryogens. The few data from hydrogen, helium, and methane do not fit well. Among these, the data for methane are particularly few; hydrogen and helium may well be outliers for other reasons.

D. UNEXPLAINED DIFFERENCES BETWEEN EXPERIMENTERS

In principle, differences between experimenters could be very useful. If two experimenters produced results that differed consistently and markedly, by using the same fluid in two sets of apparatus that differed in only one characteristic, then that would be a valuable clue to the importance of that one characteristic. Unfortunately, two sets of apparatus from different experimental programs never do differ in only one characteristic, but in a large number of ways, so it is impossible to determine which characteristic is important—the more so because we do not know which characteristics do determine, say, the nucleation behavior of a given fluid–surface combination.

The possibility of systematic error is always also present, in the other person's readings if not in your own, since precise measurements of thermal quantities, such as small temperature differences or heat flow rates, are notoriously difficult to obtain, even at experimentally convenient temperatures and pressures.

Differences between experimenters are examined here, largely to produce a measure of the scatter arising from unknown (unrecognized) differences in apparatus or technique. For this examination we do not seek wide-ranging results from a single experimenter, as used earlier, but results from different experimenters, using the same fluid under nominally similar conditions.

Different Experimenters Using Water

Water is the fluid most widely used in boiling research, and Table I shows there are 30 data sets, totaling 1474 points. Variation between experimenters could take many forms, but the simplest form is assumed here, in which different experimenters are assumed to differ in respect to a multiplying constant (different C_E), but not in respect to the manner of dependence of h on (q/A) and on p. Trials, using the sets that contained readings at more than one pressure, showed no consistent tendency for the index m to depend on pressure, so m was assumed constant. Roughness was rarely stated, so its effect was ignored, and the data were matched by least squares to a correlation of the simple form of Eq. (6), with individual values of C_E for each set, using five iterations. In addition, the data sets that referred to only one pressure (usually about 1 atm) were not used for the least squares fit, but their values of C_E were obtained by comparison with that correlation. The resulting correlation and values of C_E are given in Table VIII in Appendix A, Section D,1, where discussion includes reasons for omitting a few very high and low points. The conclusion is that variation between experimenters using water in nominally similar experimental conditions can be of order $+40\%$, -30%, that is, there can be a factor of 2 between lowest and highest. That is not attributable to different geometry or material of the heater, since, as mentioned in Section VII,B,3, there is no tendency for copper plates, stainless steel cylinders, or platinum wires to give higher heat flow. The points for water in Fig. 9 confirm these findings, though the scatter among water data is somewhat higher in Fig. 9, since it refers to a correlation based on all data for all fluids.

That indicates the degree of scatter to be contended with when stirring in all results from all experimenters. The scatter is considerable, but not large compared with the differences between predictions of various existing correlations when applied to a wide range of fluids. This must be borne in mind during all attempts, such as the present one, to form a correlation from all boiling data.

E. Burnout and Natural Convection

The full boiling curve of $\log(q/A)$ against $\log(\Delta T)$ is of inclined $\int$-shape, and the discussion so far has concerned correlations of the form $h/(q/A)^m = F(p_r, R_p, \ldots)$, which refer to the central, nearly straight section, often known as fully developed boiling $(q/A)_{FDB}$. At the upper end, the curve is limited by the phenomenon called burnout, or boiling crisis, or departure from nucleate boiling (DNB). At the lower end, heat transfer cannot fall below that for single-phase natural convection of the particular liquid and geometry, $(q/A)_{NC}$. Expressions for the maximum heat flow, $(q/A)_{max}$ and for $(q/A)_{NC}$ are well established and are generally much more accurate than those for $(q/A)_{FDB}$. Examples are quoted in Appendix A, Section E. These can be evaluated and taken in conjunction with $(q/A)_{FDB}$ to produce the full $\int$-shaped curve, using one of various theories for the transitions from section to section.

That involves three separate calculations, to be subsequently combined. The expressions for $(q/A)_{max}$ and $(q/A)_{NC}$ require various thermodynamic properties from tabulations, so it is worth considering whether they could be expressed, like $(q/A)_{FDB}$, in simpler terms, to produce a single combined expression for the whole boiling curve, either for a single fluid, using p_r, $(-\log_{10} p_r)$, and so on, or extended to all fluids using M or other simple means. In part, that can be done: with a flat-plate heater, $(q/A)_{max}$ can be expressed simply and accurately in those terms, for a single fluid, but its extension to a range of fluids using M is insufficiently accurate; $(q/A)_{NC}$ depends in important and agreed ways on parameters of the apparatus, which are, of course, not expressible in the terms used in $(q/A)_{FDB}$. These matters are discussed briefly in Appendix A, Section E, with some pointers for future progress.

The computer program incorporates the means to include both $(q/A)_{max}$ and $(q/A)_{NC}$, and produce an $\int$-shaped curve. Some success was reported in Cooper [1], for one fluid, but this matter is not pursued here, since the main result of this paper, a new survey of fully developed boiling, has not yet received critical examination. That examination and further data may call for substantial changes.

F. Further Data Needed: Summary

Several of the preceding subsections have ended with suggestions for more data, and they are brought together here. In some cases the suggestion may involve great experimental difficulty (e.g., because of problems of flammability or toxicity of fluid or because of greater temperature gradients in heaters when made of stainless steel instead of copper). In other cases, the suggestion may be relatively easy to implement on existing apparatus. Easiest of all, some data may already exist, and the author again stresses his wish to be informed of these.

Section VII,A: To establish $F(p_r)$—data covering a wide range of p_r, for fluids other than water (since there are already several sets for water), ideally several different fluids in the same apparatus.

Section VII,B,1: To establish $F(p_r, R_p)$—further data using a variety of roughnesses at a wide range of pressures, ideally for a wide range of fluids. Also, data to establish R_t (a roughness beyond which any further increase in R_p has little effect).

Section VII,B,2: To establish effects of heater shape—data for water on stainless steel plates.

Section VII,B,3: To establish effect of heater material—data for water on copper cylinders and stainless steel plates and for refrigerants on stainless steel cylinders and plates.

Section VII,C: To establish effects of different fluids—data for a wide range of pressures, using several widely differing fluids (widely differing M) in a single apparatus, with checks for aging, or using closely copied apparatus, with check calibrations between copies of apparatus. There is a shortage of data for fluids with M between its values for water and refrigerants (say $20 < M < 100$), which is surprising, since that includes hydrocarbons of practical interest. There is also a shortage of data for fluids with low M (usually cryogens), especially with a wide range of p_r or with different heater roughness, though that is less surprising in view of experimental difficulties. Data for fluids with $M > 200$ are lacking at present and would be of research value for extrapolating the trends if suitable fluid exists.

VIII. Resulting Correlation

A. EMPIRICAL CORRELATION

A correlation can now be built up empirically, by various steps, several of which have been known for some time:

1. For a given fluid on a surface of given roughness, at constant pressure

$$h \propto (q/A)^m$$

The average value of m, determined from all data sets, is 0.67. It is significantly higher at low p_r for some data sets, but not for others, so the average value is used here.

2. For a given fluid on a surface of given roughness, as pressure is varied, the ratio $h/(q/A)^m$ depends on p_r and is conveniently represented as:

$$\frac{h}{(q/A)^m} \propto F(p_r) \approx p_r^A \times (-\log_{10} p_r)^D$$

To determine the indices A, D, we need to use data sets covering a wide range of p_r. That has been done, yielding $A \approx 0.1$, $B \approx -2/3$, but these results are superseded by the next step:

3. For a given fluid as surface roughness and pressure are varied, the form of $F(p_r)$ varies. One possible formulation is:

$$\frac{h}{(q/A)^m} \propto F(p_r, R_p) \approx p_r^{A+B \times \log_{10} R_p} \times (-\log_{10} p_r)^D$$

but there are alternative formulations [e.g., setting $B = 0$ and using $R_p^{0.2(1-p_r)}$]. More data are needed to determine the best formulation.

Ideally, the indices A, B, D would be determined from data sets for many fluids, each having wide ranges of p_r and R_p, but there are no such sets. Instead, B is determined from the data sets in Nishikawa *et al.* [3], which refer to three refrigerants, each with a wide range of R_p. They give $B \approx -0.2$, but have only a narrow range of p_r.

4. Comparison of heater shapes suggests that, for horizontal copper cylinders at given (q/A) and p_r, h is some 1.7 times higher than for

horizontal copper flat plates or stainless steel cylinders, so a factor for geometry and material can be applied, giving:

$$\frac{h}{(q/A)^m} \propto F(p_r, R_p) \approx p_r^{A-0.2\log_{10}R_p} \times (-\log_{10} p_r)^D$$

$$\times 1.7 \text{ for horizontal copper cylinders}$$

This is not logically consistent, but it stands for the present, since there are few data on flat plates other than copper.

5. Comparison of different fluids suggests that h falls with increasing M, giving:

$$\frac{h}{(q/A)^m} \propto F(p_r, R_p) \approx p_r^{A-0.2\log_{10}R_p} \times (-\log_{10} p_r)^D \times M^{-0.5}$$

$$\times 1.7 \text{ for horizontal copper cylinders}$$

Many existing correlations that involve a product of properties are broadly consistent with this, because of broad variation of properties with M.

6. Indices A and D are now determined using all data sets, taking the $\log(R_p)$ term to be zero ($R_p = 1 \mu m$) for cases where it is unspecified. This gives $A = 0.12$, $D = -0.55$.

7. The correlation is now complete apart from the constant of proportionality, which is obtained from a correlation using those indices and averaging over all data points, to reach a suggested correlation:

$$\frac{h}{(q/A)^{0.67}} = 55 \times p_r^{0.12-0.2\log_{10}R_p} \times (-\log_{10} p_r)^{-0.55} \times M^{-0.5}$$

$$\times 1.7 \text{ for horizontal copper cylinders}$$

in which the units are W/(m² K) for h, W/m² for (q/A), and μm for R_p.

Modification may be needed as more data are obtained, but it is hoped that the general method and approach will remain of value.

There remains much scatter between different experimenters, which effectively conceals further variation between different geometries or fluids. No clear systematic variation with k_f or σ or any other fluid property is apparent, despite wide variation in some of those properties between fluids.

8. As an alternative to item 7, the comparison of fluids can be seen as dependence on $(p_{cr}/T_{cr})^{2/3}$ or on $(k_f/\sigma)^{0.8}$ instead of $M^{-0.5}$, and they may be attractive to some theoreticians, but the supporting data are sparse, the dependence may not be better, and the simple form with $M^{-0.5}$ is more attractive to the practical man.

These are modest results, and this author claims little novelty in them. Item 1 has long been known, and so has item 2. Item 3 derives from a single data set, and the experimenters had themselves observed such dependence on R_p. Items 4, 5, 6, 7, and 8 are new contributions, dependent on the present rapid data processing, but they are rather tentative results.

As regards accuracy, the new correlation amounts to drawing a straight line with slope -0.5 as shown on Fig. 8 (which perhaps does not look too bad) and across Fig. 9 (which reemphasizes that there is much scatter, even in that figure, where each set of data is reduced to a single, averaged point; there is even more scatter if we look at all individual raw data points).

Correlations of the new kind can easily be devised to fit any subset of the existing data (e.g., data for one particular fluid, or for all refrigerants, or for hydrocarbons with $40 < M < 90$). Such correlations of subsets give much better fit to the existing data, as indicated by Figs. 1a, 1b, 2a, 2b, but it is far from clear whether they would give a good fit for any new data from a different experimental program.

Existing correlations, whether for all fluids or for subsets, would apparently be no better, while demanding much effort to find property values and evaluate complex expressions.

As regards showing the structure and pattern of variation, the new correlation is clearer, so it is easier to see how experimental programs can be devised to check (or demolish) it.

As with all such findings, it must be asked whether the data have been examined sufficiently closely, since other effects have been reported in studies directed specifically at observing effects of specific changes, such as the diameter of a cylindrical heater. The answer is that any such effects are lost among general boiling data, because of its variability, $\pm 10\%$ in carefully copied experiments in the same laboratory; $+40\%$, -30% or more for nominally similar experiments in different laboratories. Either the statistical processing of existing data must be greatly refined, which would require more accurate specification of exact experimental conditions in different laboratories (not always possible for experiments performed decades ago), or further accurate, wide-ranging experimental studies must be aimed at the specific phenomena. Studies using a single fluid at one pressure are not so convincing as multifluid, multipressure studies, from which internal consistency can be estimated.

B. THEORETICAL CORRELATION

This is a short section. As stated in the first paragraph of Section I, there is no shortage of theories. The author does not wish to add another. This work shows why the many theories, each matched to some data by disposable parameters, are all giving much the same answers. This may lead to more crucial tests of existing data, or more experiments, to decide among the theories.

No theoretical reason is suggested here for the new findings, such as that $h/(q/A)^m$ should be dependent on $M^{-0.5}$ or $(p_{cr}/T_{cr})^{2/3}$ or $(k_f/\sigma)^{0.8}$. After all, there is still no agreed theoretical reason for the long-established facts that $(q/A) \propto \Delta T^n$ ($h \propto (q/A)^m$) or that the ratio $h/(q/A)^m$ rises with p, and sharply near p_{cr}.

IX. Conclusions

1. Analysis of boiling heat flow data is facilitated if we recognize that the saturation thermodynamic properties normally used in correlations of heat flow rates in boiling can be expressed in a particularly simple and convenient single-term form, Eq. (4). This expression is possible because there is a form of redundancy among those properties, for any given fluid. It is convenient for various reasons, including the fact that, by substitution, most typical correlations of boiling heat transfer data for any given fluid can be reformulated in a similar simple way such as

$$h/(q/A)^m = p_r^A \times T_r^B \times (1 - T_r)^C \times C_E \qquad (5)$$

or (with different indices $A, \ldots$ and constant, C_E)

$$h/(q/A)^m = p_r^A \times (-\log_{10} p_r)^D \times C_E \qquad (6)$$

The reformulation is purely algebraic and depends only on the nature of the saturation properties involved. It implies no further assumptions concerning the nature of boiling. Any discrepancy between values given by these formulations and by the original formulations using properties is small compared with the scatter in experimental data for boiling heat transfer.

The particular quantities p_r, T_r, $(1 - T_r)$, $(-\log_{10} p_r)$ constitute a "set of tools" to simplify direct analysis of the raw experimental data of boiling heat transfer. Using this "set of tools," or a selection from them, statistical analyses of the raw data for a given fluid can be simplified, avoiding the redundancies among its saturation properties. Also, more complex analyses can be readily tackled, giving significant improvement in fit to the data, such as analyses in which the left-hand side of Eq. (5) involves a power of (q/A) dependent on T_r, or the right-hand side involves roughness or conductivity of the heater, or molecular weight of the fluid.

2. No "set of tools" of comparable accuracy is available to represent the variation of those saturation properties among different fluids, but some less accurate generalizations are possible. The principal one is that, among the fluids commonly used for boiling research, many of the rele-

vant properties vary broadly as a power of M, the strongest variations being for ρ_g and h_{fg}, dependent on M and $1/M$. These can be refined by introducing p_{cr} and T_{cr}, since it is more accurate to say $\rho_g \propto Mp_{cr}/T_{cr}$ and $h_{fg} \propto T_{cr}/M$.

As an alternative to the use of M, we can see the properties as varying broadly with a power of (T_{cr}/p_{cr}), or of (k_f/σ), since each of those varies broadly with M. There is not enough data to decide whether one of these is of more fundamental significance for boiling than the variation with M, and it will be difficult to obtain enough, but they may trigger some useful theoretical arguments.

3a. Existing correlations typically predict boiling heat flow rates that decrease with M, as $M^{-0.5}$.

3b. Existing correlations typically predict heat flow for water lying in the same pattern as for other fluids, despite σ and k_f for water both being high by about a factor of 3.

The reasons for these can be seen in the light of conclusion 2. For 3a we note that most existing correlations contain the strongly M-dependent terms, ρ_g and h_{fg}, as the product $(\rho_g h_{fg})$, so their dependence on M (and T_{cr}) cancels, but they contain other M-dependent terms to moderate powers. For 3b we note that most correlations contain k_f and σ to low powers, of opposite sign, so their strong variation for water tends to cancel.

4. Using the set of tools described in conclusion 1, and the M-dependence noted in conclusions 2 and 3, with a rapid least squares analysis, many data have been examined (probably constituting an important fraction of the world stock of saturation pool boiling data) involving more than 5000 data points for more than 100 experiments and using fluids with widely differing values of the transport and other properties, alleged to have large effect on boiling heat transfer. Despite the large amount of data, there emerge only the following rather limited results (some already known) for fully developed saturation nucleate pool boiling:

(a) The heat transfer coefficient h is related to heat flux q/A as $h \propto (q/A)^m$, m about 0.7 (equivalent to $q/A \propto \Delta T^n$, $n = 1 - 1/m$, about 3 or 4).

(b) The effect of increasing system pressure is to increase $h/(q/A)^m$ roughly as $p_r^{0.2}$, but more strongly near CP, perhaps by a factor of 6 as p_r rises from 0.1 to 0.9, and it can be represented by

$$\frac{h}{(q/A)^m} = p_r^A \times (-\log_{10} p_r)^D \times C_E \tag{6}$$

(c) The effect of heater roughness is not expressible as a simple multiplier applied to $h/(q/A)^m$, since roughness has greater effect at low pressure. This has been shown by a single wide-ranging recent study, and its results can be represented in various ways, of which a convenient one is

$$\frac{h}{(q/A)^m} = p_r^{A+B(\log_{10}R_p)} \times (-\log_{10} p_r)^D \times C_E$$

but better alternatives may emerge when such studies are extended to lower pressure.

(d) The effect of heater shape is not well established by the data, but there is some evidence that, at given p_r, $h/(q/A)^m$ for horizontal cylinders is 1.7 times that for horizontal flat plates, at least for copper heaters with refrigerants at p_r above 0.07.

(e) The effect of heater material is not established directly by the data, because the data are very ill distributed for that purpose; most heaters are copper, except for water, which has some data on stainless steel cylinders. However, there is indirect evidence that, at given p_r, $h/(q/A)^m$ for copper cylinders is 1.5 or 2 times that for stainless steel cylinders. In the absence of sufficient data on plates other than copper, we combine this with preceding paragraph (d) in the statement that $h/(q/A)^m$ on horizontal copper cylinders is 1.7 times that on all other heaters. This is not logically consistent and may well be superseded when more data are available.

(f) The effect of changing fluid is not well established by the data, since the data are ill distributed for that purpose (few experimental programs have covered a wide range of fluids of different types), but there is evidence that $h/(q/A)^m$ varies approximately as $M^{-0.5}$, which (unsurprisingly) is in accordance with many existing correlations, noted in conclusion 3a.

These lead to:

$$\frac{h}{(q/A)^{0.67}} = 55 \times p_r^{0.12-0.2\log_{10}R_p} \times (-\log_{10} p_r)^{-0.55} \times M^{-0.5}$$
$$\times 1.7 \quad \text{for horizontal copper cylinders}$$

in which the units are W/m^2 K for h, W/m^2 for (q/A), and μm for R_p, but this is tentative in several respects.

It is not highly accurate, largely because the data are scattered. That scatter and the general accuracy of the correlation are indicated on Fig. 9, where each of the 116 points represents an average from a set of data, and the line represents the correlation.

Its great advantage is its (relative) simplicity, since it avoids the need for many fluid properties, requiring only the system pressure and the values of M and p_{cr} for the fluid. Most existing correlations are in the more complex multiproperty form of Eqs. (1) or (2), which would be preferred if they offered greater accuracy, but they do not seem to.

5. It is not possible to disentangle the effects of any individual fluid property from the general boiling data. It is to be hoped that such disentangling may be possible in the future, but the tendency of properties to vary together (e.g., with M) may prevent it.

6. It is all too easy to "force" scattered data into a pattern that is not really appropriate. Perhaps the data should really be fitted to T_r, not p_r. Perhaps the relationship is not, in fact, reducible to a single-term form as is assumed here (a product of powers of h, q/A, and a subset of p_r, T_r, $(-\log_{10} p_r)$, and so on), or any other single-term form, although that is implied by most existing correlations. The proposed relation does at least escape some of the faults of others, in which some property appears to a high power and leads to great errors. That can happen for fairly normal fluids if ρ_g or h_{fg} appears alone, instead of as the product $\rho_g \times h_{fg}$. It happens for hydrogen with ρ_f, which is very low.

7. Further work to improve understanding at modest cost can be specified in the light of the analysis and how the analysis is hampered by the distribution of the present stock of data. Requirements are detailed in Section VII,F and amount to extended work with a wider range of pressures and (simultaneously) of roughness, and certain particular combinations of heater and fluid—water on copper cylinders, refrigerants on stainless steel cylinders, more use of stainless steel plates, more use of fluids other than water and refrigerants, preferably in the same apparatus. Some of these may well present experimental difficulties, but others appear to be modest extensions of previous studies.

Ideally, such tests would compare heaters and fluids having more divergent characteristics, while (by some means) avoiding uncontrolled differences in geometrical and other factors. Neither of these requirements is easily met; small, unidentified differences in the physicochemical interaction of fluid and heater surface are known to have major effects on boiling, through nucleation characteristics; fluid properties tend to vary en bloc.

An experimenter planning a future study of any parameter should bear in mind the accuracy and extent of the work reported in Nishikawa et al. [3], since the effect of any parameter other than roughness will be smaller. This is not intended as a counsel of despair, but a counsel of caution.

8. There is nothing surprising in demands for more data. What is now claimed is that by use of the "set of tools" from reduced properties (conclusion 1), the analysis can be simplified so much that it is now feasible to subject the large mass of existing data to analysis of greater subtlety, without being intolerably and pointlessly complicated by all the saturation properties. This points to directions in which progress can be expected.

By suggesting these particular crucial experiments, the author invites the rapid destruction of the detailed forms of conclusion 4 but it is hoped that the method and approach will remain of value.

9. The reader(s) will consider this is not a final answer. The author agrees and hopes for comments, suggestions, and more data. Meanwhile he returns to more fundamental studies of bubble dynamics.

X. Appendix A

A. EFFECTS OF OPERATING CONDITIONS (PRESSURE)

It has long been known that heat flow rate (q/A) rises with wall temperature (T_w), initially as dictated by natural convection, but rising more sharply as nucleate boiling starts, leading to a regime of "fully developed nucleate boiling." In that regime, if T_w is varied at a given operating pressure, then $(q/A) \propto (T_w - T_{sat})^n$, $n \approx 3$ or 5, or $h \propto (q/A)^m$, $m = 1 - 1/n \approx 0.67$ or 0.8. As discussed in Section II, the constant of proportionality (and perhaps the index m) depends on the pressure. Most correlations take the form of $h/(q/A)^m = f$(properties), where f(properties) is a product of properties as in Eq. (2), or an explicit function of p_r, as $F(p_r)$ in Eq. (3). For correlations formulated as in Eqs. (1) or (2), the properties involved are very different for different fluids, but there is little variation in fact, because, for all fluids, that product of properties varies with p_r in much the same way, rising at about $p_r^{0.2}$, then rising more sharply as p_r approaches 1.0. There are differences in detail, arising partly because each correlation was based on different groups in Eq. (1), and based on data from a limited range of p_r, and also partly because each experimenter used different fluid and apparatus. For the correlations that are formulated as Eq. (3), $F(p_r)$ itself rises in a similar way, again with some variation between experimenters.

When the present method of correlating by least squares fit to simple terms in p_r, $(-\log_{10} p_r)$ and so on, is used to produce correlations, the results for a wide range of individual sets, or from combinations of sets, are equivalent to a value of m and an expression of $F(p_r)$, which, on evaluation, turns out to be in general agreement with existing correlations, whether they were formulated in the form of Eq. (1) or Eq. (3).

The general agreement is not surprising, since each correlation involved some matching to some data.

The simplicity of the present method of producing correlations enables many combinations of data for many fluids to be analyzed in new ways. One such way is to allow the index m to vary with p_r or T_r or $(-\log_{10} p_r)$.

Several data sets, including that of Fedders [22], show a strong tendency for that index to be larger for low p_r, and the fit to such data is improved by allowing for that, as discussed in Cooper [1] for several data sets, and illustrated in Section C,1 of this appendix for two sets. However, a few data sets show an opposite tendency, so m will be taken as constant for most of the following data processing.

1. $F(p_r)$ from Data Sets with Wide Range of Pressure

Different sets of data produce different value of the indices in these correlations [i.e., different expressions for $F(p_r)$]. It is clearly desirable to obtain the best estimate of $F(p_r)$ from the data available, and the present method is well suited to do so, when applied to the most appropriate data sets. For this purpose, sets obtained at a single pressure give no information, and the best sets are those that cover a wide range of p_r, say (max p_r/min p_r) exceeding 100. As noted earlier (Section V,A,2), that excludes refrigerants; one experimenter has produced a wide-range set for hydrogen; another has produced two for oxygen; another has produced five sets, two for benzene and three for ethanol. Out of 28 sets for water, there are only five such sets, but a sixth is included because it has (max p_r/min p_r) over 90. In order to establish the form of $F(p_r)$ over a wide range of pressures, correlations were produced by least squares fit, using only those 14 data sets, giving a different value of C_E to each data set and using five iterations. The resulting expression for $F(p_r)$ is:

$$p_r^{0.0948}(-\log_{10} p_r)^{-0.6598}C_E \qquad (A1)$$

or, with slightly rounded numbers,

$$p_r^{0.1}(-\log_{10} p_r)^{-2/3}C_E \qquad (9)$$

which is plotted in Fig. 6. The unrounded version lies within the thickness of that line ($\pm3\%$). The version of Borishanskii [2] is also within a few percent, as shown by the marked points on Fig. 6, except around $p_r = 0.5$. Data there in Borishanskii [2] are sparse and scattered anyway, but a simple alteration to the analysis enabled a further check of the form in that region: an additional term, $(1 - p_r)$ would have much effect there, so a least squares fit was obtained with such a term in addition to those in Eq. (A1). The least squares analysis showed that the inclusion of that term led to

$$p_r^{0.0974}(1 - p_r)^{-0.0136}(-\log_{10} p_r)^{-0.6479}C_E$$

which differs from Eq. (A1) by only a few percent, up to $p_r = 0.9$. This has two implications: (1) where Borishanskii's expression differs from Eq.

(A1), the latter is a better fit to this data; (2) of the two terms $(1 - p_r)$ and $(-\log_{10} p_r)$, which both go to zero at CP, the latter gives better agreement to this wide-ranging boiling data. That might be expected from its better agreement to property data, mentioned in Section II,C,2.

If each of these wide-ranging sets is treated separately, we obtain varying values for m and a cluster of graphs for $F(p_r)$. The extremes of the cluster are plotted on Fig. 6, normalized to a common value at $p_r = 0.1$, showing that the steepest curve is for hydrogen and the flattest is for one of the oxygen sets. As will be discussed in Section B,1 of this appendix, such differences can arise from different roughness, which is not given for either of those extreme sets. Section C of this appendix will discuss whether the differences can arise from differences between fluids. Section VII,C,4 discusses whether they can arise from differences between experimenters.

2. $F(p_r)$ from Data for Refrigerants

As shown in Table I, refrigerants as a class have the largest amount of boiling heat flow data, exceeding even that for water, and they claim attention, because of their weight of numbers and their internal consistency. However, in this context, their contribution is weakened by the narrowness of their range of p_r, as $(\max p_r / \min p_r)$ never exceeds 50 and rarely exceeds 10, so they provide a less reliable determination of $F(p_r)$ than the wide-ranging sets used earlier. An equation analogous to Eq. (9) can be produced from all refrigerant data and compared with Fig. 6. Within the limited range of p_r for refrigerants, that curve lies very slightly steeper than Eq. (9), but within the extremes shown on Fig. 6. This suggests that there is no case for using the refrigerant data to modify Eq. (9).

B. Effects of Heater

It is well known, from experiments directed specifically at them, that the microscopic topography of the heater surface, and, with less certainty, the general shape of the heater and its material, can affect boiling heat transfer, though such experiments have not yet produced agreement on the relevant parameters. We consider these effects in succeeding subsections, taking roughness first because it has the clearest and most important effect.

The roughness is generally stated for the experiments using refrigerants, but seldom for other fluids. As shown in Table I, the data are not favorably distributed for analysis of effects of other heater characteristics. Most boiling data have been obtained with a regrettably limited range of heater materials. Most flat-plate heaters are copper, and so are most cylindrical heaters, though a few used for water are stainless steel. Most wire heaters are platinum or stainless steel. These were presumably chosen for good reasons of experimental convenience in this difficult experimental field. There are exceptions, and interest focuses on them, but they generally involve greater experimental difficulty (e.g., use of a stainless steel heater in place of copper will increase temperature gradients, making it harder to deduce the precise temperature at the boiling surface from extrapolation of temperatures given by sensors buried in the heater). This must make it difficult to obtain the high accuracy that is needed, particularly near the CP, where $(T_w - T_{sat})$ is very small.

Many of the tests aimed at changes in heater produce only a small number of data points, under a limited range of conditions and fluids (e.g., one fluid at one pressure). They therefore do not contribute many points to the data bank, but are best examined individually. Some major sets of wide-ranging data including changes to roughness have been produced and will be examined.

Among changes that can be made to heaters, a change of orientation is apparently the only one that can be made during continuous running of a single test. Other changes, of heater roughness or material or shape, necessarily involve breaking that continuity, with much risk of introducing unintended and uncontrolled changes, which can have large effects on heat flow, through changes in nucleation characteristics. Attempts to avoid or minimize this problem have included tests with different heater material, each covered with a thin coating of (as far as possible) identical composition and surface topography and surface chemistry, hence identical nucleation characteristics. Such attempts are laudable, but they have their own problems. They introduce a further complication, since it is not clear whether a layer thick enough to establish standardized surface conditions for nucleation (including transient heat flow during the interval between bubbles) will be thin enough to have no effect on transient heat flow during subsequent bubble growth. Both nucleation and bubble growth are essentially involved with rapid transient heat flows, dependent on thermal properties on the wall very near the surface.

1. *Heater Roughness*

The detailed physicochemical state of the heater surface is a very important factor in boiling, but many experimenters give little or no information about roughness. For heaters in the form of thin wires, control and description of surface roughness are not as easy as those of flat plates. In any case, research has not yet shown that parameters are relevant. It has been shown that the various customary measures of physical roughness, such as R_p, CLA, and so on, are not enough on their own, because surfaces with the same R_p may have very different boiling characteristics, depending on whether that R_p was obtained by emery or by etching [23, 24].

Some individual sets of internally consistent results have been obtained in experiments where different surface finishes were obtained by carefully controlled treatment with emery or otherwise.

The Japanese laboratory has recently produced a wide-ranging study by Nishikawa et al. [3], discussed as regards accuracy in Section V,B,2. They systematically investigated effects of varying surface roughness of a flat-plate heater, using three refrigerants and reduced pressures from 0.08 to 0.90. Finish was obtained by using emery; the German Glättungstiefe, R_p, was used as a measure of surface roughness, and it varied from 4.3 to 0.022 μm (CLA approximately 2.2–0.024 μm). They obtained some 300 data points from which they noted that roughness has greater effect at lower p_r. That implies that the effect of roughness cannot be represented by a multiplier applied to $F(p_r)$: instead, the form of $F(p_r)$ is affected by roughness, so we really have $F(p_r, R_p)$. They correlated their data using graphical analysis, giving:

$$\frac{h}{(q/A)^{0.8}} = \text{alpha } p_r^{0.23}(1 - 0.99p_r)^{-0.9}(8R_p)^{0.2(1-p_r)} \qquad (A2)$$

where their alpha depends on the fluid and will be discussed in Section C,1 of this appendix, together with values of C_E arising shortly in Eqs. (A3)–(A5). Our present concern is with the R_p term.

When their data are subjected to the present method of analysis adapted as described earlier to include terms $(1 - 0.99p_r)$ and $\{(8R_p)^{(1-p_r)}\}^G$, we obtain:

$$\frac{h}{(q/A)^{0.748}} = p_r^{0.308}(1 - 0.99p_r)^{-0.732}(8R_p)^{0.203(1-p_r)}C_E \qquad (A3)$$

where the constant C_E is optimized separately for each fluid.

Comparing Eqs. (A2) and (A3), we see dependence on R_p in close agreement, and although the terms in p_r and $(1.0 - 0.99p_r)$ look rather different, there is little numerical difference between them in their experimental range $0.08 < p_r < 0.90$. Revised formulations have been tried, using $(1.0 - T_r)$ or $(-\log_{10} p_r)$ instead of $(1.0 - 0.99p_r)$, and $(1 - p_r)$, giving very similar fit to the data. An example is:

$$\frac{h}{(q/A)^{0.747}} = p_r^{-0.02}(-\log_{10} p_r)^{-0.612}R_p^{0.21(-\log_{10}p_r)}C_E \qquad (A4)$$

These revisions would cause differences in fit if the range of p_r were extended well below its present lower limit of $p_r = 0.08$, but the difference between $0.21(-\log_{10} p_r)$ and $0.203(1 - p_r)$ never exceeds 0.03 for the four particular pressures used, usually $p_r = 0.1, 0.3, 0.55,$ and 0.9. Equation (A4) can be transformed using $x^{\log y} = b^{\log x \log y} = y^{\log x}$ (b being the base of the logarithms) to give

$$\frac{h}{(q/A)^{0.747}} = p_r^{-0.02-0.21(\log_{10} R_p)}(-\log_{10} p_r)^{-0.612} C_E \qquad (A5)$$

This suggests that of the terms on the right-hand side, the term in p_r is most concerned with roughness, presumably through effects on nucleation.

These conclusions are different from those of previous workers, who did not find the dependence on p_r, but they are compatible in some cases, as will be shown.

The classical work of Berenson [25], which showed very marked effects, was done at 1 atm, with p_r about 0.03, so that is consistent with the new work. Unfortunately, it cannot be used to provide the desired quantitative extension to low p_r, since the roughness was specified by emery grades, instead of by a numerical measure. He also reported the anomaly of high heat flow for a lapped surface, which he tentatively attributed to embedded particles of lapping compound. His main concern was with transition boiling.

The recent German tests with refrigerants [5–7] used cylinders having carefully controlled and measured surface roughness, with R_p ranging from 0.2 to 0.9 μm. They found that such changes had little effect on heat flow, as shown concisely in Fig. 29 of Götz [7]. That may appear to contradict the Japanese finding, but in fact the data in that figure are confined to $0.3 < R_p < 0.7$ and to $p_r > 0.19$. The new Japanese work [3] would lead us to expect little effect, since the change in the $R_p^{0.2(1-p_r)}$ term is $(0.7/0.3)^{0.2(1-0.19)}$, which is 1.15. Closer examination of the data at lowest pressure in Fig. 29 in Götz [7] suggests that such an effect may be present, but it is practically lost in the normal scatter. At higher pressures the effect predicted by the Japanese work is smaller.

The German laboratory also produced data for boiling of refrigerants on a flat copper plate of various roughness, at p_r from 0.9 to 0.0043 [23], but the main conclusion was that R_p is not in itself a sufficient measure of surface condition, as described at the start of this section.

A term $R_p^{0.2}$ has also arisen in other work, referenced in Nishikawa et al. [3] (e.g., the brief report by Danilova [26] which gives no raw data but has graphs suggesting a term $R_p^{0.2}$, as above), but there is no suggestion that the index falls at higher p_r, even when p_r is 0.5.

In his extensive study of water on stainless steel cylinders, Fedders [22] used a moderate range of roughness, $0.18 < R_p$, μm < 3.6, and found the index of R_p was 0.1–0.13, with little or no difference between results at p_r of 0.08 and 0.012. That would be more consistent with Eq. (A2) than with Eq. (A4).

Vachon et al. [24] give data for water at 1 atm on plates of roughness R_p approximately 3.0, 1.8, 0.4, 0.09 μm. At such low p_r, a marked effect would be predicted by either Eq. (A2) or Eq. (A4), but their readings show no significant change as R_p is reduced from 3.0 to 1.8 and 0.4, and a fall of 20% on changing to R_p of 0.09.

It is suggested in some quarters (e.g., by Grigoriev et al. [27]) that roughness has no effect, provided it exceeds the Russian standard 7, sometimes described as roughness height of about 5 μm. This suggests an adaptation of Nishikawa's expression ($8R_p$), in which their number 8 corresponds to a quantity having the dimensions of a length and magnitude 0.125 μm. They perhaps chose that quantity in order that their previous work, in which R_p was 0.125 μm, should be unaffected by the new term. Since their new work does not suggest there is any fundamental significance in R_p of 0.125 μm, various roughness functions R_F could be set up to fit Nishikawa's requirement $R_F \propto R_p$ in his range, and also $R_F \propto 1.0$ when R_p is well in excess of transition value R_t, about 5 μm. Such an expression is

$$R_F = 1/(1 + R_t/R_p)$$

The effect of using this in Eq. (A5) would be to replace the term $p_r^{-0.02-0.21(\log_{10}R_p)}$ by $p_r^{-0.02+0.21[\log_{10}(1+R_t/R_p)]}$. For most industrial finishes, $R_p > R_t$, so $R_F \approx 1$, and the log term ≈ 0.

The value of R_t may also be related to the properties of the fluid, since the relevant basic study is nucleation, which has much detailed support [19]. As discussed in Appendix B, the theory suggests that nucleation occurs when the wall superheat corresponds to an excess of saturation pressure, $\Delta p = (p_{sat,T_w} - p_\infty)$ greater than $2\sigma/R_c$, where R_c is of the order of radii of cavities in the surface, of the order of R_p. Rewriting this as $R_p > 2\sigma/\Delta p = 2\sigma/(p_{cr} \Delta p_r)$ and noting the general similarity of saturation curves, as p_r against T_r (ignoring the variation shown in Fig. 5), we arrive at a length $2\sigma/p_{cr}$, which may be an appropriate measure of the effect of fluid properties on nucleation. If that is acceptable, then it suggests that R_t should be related to σ/p_{cr}, possibly a simple multiple of it. Values of σ/p_{cr} at $p_r = 0.1$ are 5 nm for refrigerants and many other fluids, 2 nm for water and hydrogen. That suggests that water and hydrogen may be more sensitive to a given value of R_p than other fluids. If so, it may emerge when more data are available, for water and hydrogen with varied R_p and for refrigerants at lower p_r, but at present it remains speculation.

At present, the effects of roughness are so scattered that an analysis like this cannot be conclusive, but it may suggest the most promising further experiments.

2. *Heater Shape*

As discussed in Section VII,B,2, the ideal comparison is between heaters of different shape but the same material, using the same fluid. The shape of the heater, whether flat plate, cylinder, or wire, and its orientation relative to gravity, are virtually always stated by experimenters. More details of size of flat plate or length and diameter of cylinder are usually given. Most experiments have been done on horizontal flat plates or horizontal cylinders or horizontal wires, so these three major categories are used here, with wires defined as cylinders of diameter less than 1 mm.

There are many data for refrigerants boiling on copper heaters in the form of flat plates or cylinders. Sources are listed in Sections A,2 and B,1 of Appendix C, and they give 34 data sets for which roughness is stated, 22 on plates and 12 on cylinders. They fit a common correlation of the type of Eq. (A4) with values of C_E averaging 2.40 for plates [standard deviation (SD), 0.14] and 4.06 for cylinders (SD, 1.2). The ratio is 1.69, or, rounded, 1.7. As discussed in Section VII,B,2, it is plausible that cylinders cause better heat flow.

No similar amount of data is available to compare other shapes. Table I suggests a comparison using water on stainless steel plates, cylinders, and wires. However, they are largely from different experimenters, so they show great scatter, as discussed in Section D of this appendix, on comparison of experimenters. In Table VIII, the values of C_E for stainless steel plate and cylinders are seen to be interleaved, so there is no support for the suggestion that cylinders produce higher C_E. The two data sets for stainless steel wires have the lowest C_E, so there is a very tentative suggestion that wires, like plates, may produce C_E lower than cylinders by perhaps 1.5 or 2. The distinction between cylinders and wires is, of course, simply a matter of size, so a heater of intermediate size (about 1 mm diameter) might act as a wire if bubbles are large, and as a cylinder if bubbles are small (e.g., at high pressure), but the present analysis does not show any such effect.

Table I also shows that there are data for cryogens on copper plates and cylinders, but they refer to several different cryogens, and the results in terms of C_E are interleaved.

There are isolated studies of change of shape, often using a single fluid at a single pressure, such as Cornwell *et al.* [28], but they are for individual study rather than for the present statistical analysis.

TABLE III

FLAT PLATE HEATERS, NOT COPPER

Heater material	Fluid	Data sets	First-named author
Stainless steel	Water	4	Vachon
Cu, plated with Ni	Water	1	Marcus
Nickel	Water	1	Madsen
Cu, plated with Ni	Water	1	Kotchaphakdee
Platinum	Nitrogen	1	Lyon
Platinum	Oxygen	1	Lyon
Cu, plated with Pt	Nitrogen	1	Kosky
Cu, plated with Pt	Oxygen	1	Kosky
Resistance alloy	Hydrogen	1	Class
Brass	Helium	1	Grigoriev
Sapphire	Nitrogen	1	Swanson
Aluminum	Helium	1	Jergel (1974)

3. Heater Material

As shown in Table I, there are 64 data sets with flat plate heaters, 49 of them being of copper. The others originate from sources listed in Sections A,5, A,6, and A,7 of Appendix C; their leading characteristics are given in Table III.

These are too few and too varied to support a comparison.

As discussed in Section VII,B,4, we cannot reach a fully consistent interpretation here, and for the present we suggest the simple compromise that h is greater by about a factor of 1.7 for copper cylinders than for any other heater. The computer program was readily modified to allow for that factor, and a search was then made for residual effect of k_{wall}, by seeking a term $(k_{wall})^L$. It emerged that the value of L was very small, of order -0.02, having very little effect, since the range of k_{wall} is from 18 for stainless steel to 380 for copper, and $(380/18)^{-0.02}$ is 0.94. The fact that it is less than 1.0 suggests that the factor 1.7 is slightly large, but the main point is that 0.94 is very close to 1.0, so there is no significant residual effect of k_{wall}.

4. Heater Shape and Material

As discussed in Section VII,B,4, combined comparison is possible between heaters that are copper plates and stainless steel cylinders, for several data sets using water, and for a few other fluids. For water, the relevant sources are listed in Sections A,1 and B,4 of Appendix C, and the leading characteristics and results are given in Table IV, showing that the

TABLE IV

Water Heaters of Copper Plate or Stainless Steel Cylinder

| Heater | No. of data points | Range of p_r | | C_E | First-named author |
		Min.	Max.		
Cu plate	61	0.0001	0.0046	5.537	Raben
Cu plate	42	0.0003	0.0088	3.178	Nishikawa
Cu plate	117	0.0003	0.3548	3.358	Fujita
Cu plate	40	0.0046	0.0046[a]	3.381	Gaertner
Cu plate	7	0.0046	0.0046[a]	4.775	Lorenz
Cu plate	15	0.0046	0.0046[a]	5.610	Shoukri
Cu plate	8	0.0046	0.0046[a]	3.427	Sultan
Cu plate	9	0.0046	0.0046[a]	3.594	Nakayama
S/S cylinder	86	0.0071	0.6646	3.035	Fedders
S/S cylinder	85	0.0045	0.8070	3.186	Borishanskii
S/S cylinder	38	0.0005	0.0046	2.890	Ponter
S/S cylinder	209	0.0046	0.8876	3.670	Borishanskii
S/S cylinder	37	0.0002	0.0090	5.643	Minchenko
S/S cylinder	118	0.0004	0.5322	3.318	Fujita

[a] Data sets at a single pressure are not used for least squares analysis.

average value of C_E is 4.16 for copper plates and 3.62 for stainless steel cylinders—a small difference as discussed in Section VII,B,4.

One individual data set from Kyushu [20] gives boiling heat transfer for water, benzene, and ethanol on copper plate, stainless steel cylinder, or platinum wire. They were first met in Section VII,B,4, where they were correlated and values of C_E were tabulated (Table II), which again shows that there is little difference between those three heaters, for any of the three fluids.

As discussed in Section VII,B,4, these results lead to a tenuous suggestion that the change from copper to stainless steel reduces C_E by the same amount as the change from plate to cylinder increases it, that is, by about a factor of 1.7.

5. *Effects of Heater: Summary*

The main text, Section VII,B,5, needs no further details here.

C. Effects of Fluid Properties

As explained in Section VII,C, the data are not ideal for observing whether a change of fluid has any discernible effect on boiling heat transfer. We first consider those few cases where different fluids can be clearly compared under nominally very similar conditions. Second, we consider a comparison including all data sets, looking for variation of C_E with fluid, despite scatter from different experimenters, assessed in Section VII,D.

1. Different Fluids under Similar Conditions

Widely differing fluids are seldom used in the same apparatus. The principal experiments that do so are those in the "3 × 3" group of sets [20], summarized in Table II of Section VII,B,4. In this context they are adequately discussed in Section VII,C,1 of the main text. Other groups of data cover a less wide range of fluids. The principal ones are refrigerants on plates [3, 4], refrigerants on cylinders [5–7], and hydrocarbon gases (alkanes) on cylinders [21], and these are now examined in detail. This examination also illustrates how closely individual sets of data can be fitted by correlations of the new simple form, and it also provides a basis for the assessment of accuracy and reproducibility in Section V,B.

Data for boiling R12 and R13B1 are given in Bier *et al.* [5], Engelhorn [6], and Götz [7] for apparatus that was closely copied except for surface roughness of the heater, $R_p = 0.9$ μm for Bier *et al.* [5], 0.2 for Engelhorn [6], 0.7 for Götz [7]. The results can be approximated by various expressions of the types described earlier. Although roughness is not varied widely, that variation must be allowed for; otherwise data at low R_p, low p_r do not fit other data. A term $R_p^{0.2(-\log_{10}p_r)}(= p_r^{-0.2\,\log_{10}R_p})$ is therefore used, in accordance with Section B,1 of this appendix. Five such expressions for R12 are given in Table V, lines 1–5, where the mean absolute error is also shown. This illustrates the statements in Section II,C, since lines 1–3 show that the mean absolute error is not much affected by omitting T_r or by changing $(1 - T_r)$ to $(-\log_{10} p_r)$, but lines 4 to 5 show improvement when m is allowed to vary. Line 5 is used to plot the straight lines on Fig. 1a. The percentage errors in h, quoted in Table V, correspond to the horizontal link lines in Figs. 1a and 1b as those show the errors in ΔT at given (q/A). Data from Bier *et al.* [5] and Götz [7] apply to overlapping ranges, providing evidence of reproducibility for Section V,B,1.

The boiling data for R13B1 in Bier *et al.* [5] and Engelhorn [6] can be treated similarly, with broadly similar results. One expression, for comparison with line 5, is given in line 6. It is used to plot the straight lines in Fig. 1b.

To compare the two refrigerants, a single expression was produced of the same type as lines 5, 6, but fitting data for both refrigerants. Line 7 shows the result, though without the constant C_E, since that differs for the two refrigerants, $C_E = 16.6$ for R12 and 16.1 for R13B1. That difference is small, of the same order as the reproducibility for a single fluid, as discussed in Section V,B,1. That is to be expected, since there is not

TABLE V

REFRIGERANTS ON HORIZONTAL CYLINDERS[a]

Expressions for h	Mean abs. error (%)
Refrigerant R12	
$(q/A)^{0.70}\, p_r^{-0.52-0.2\log_{10}R_p}\, T_r^{6.04}\,(1-T_r)^{-0.69}\,10^{0.59}$	11.3
$(q/A)^{0.70}\, p_r^{0.13-0.2\log_{10}R_p}\,(1-T_r)^{-0.77}\,10^{0.42}$	10.9
$(q/A)^{0.70}\, p_r^{0.04-0.2\log_{10}R_p}\,(-\log_{10}p_r)^{-0.76}\,10^{0.78}$	11.6
$(q/A)^{0.61+0.15(-\log_{10}p_r)}\, p_r^{0.78-0.2\log_{10}R_p}\,(1-T_r)^{-0.73}\,10^{0.84}$	8.3
$(q/A)^{0.61+0.15(-\log_{10}p_r)}\, p_r^{0.70-0.2\log_{10}R_p}\,(-\log_{10}p_r)^{-0.72}\,10^{1.18}$	8.0
Refrigerant R13B1	
$(q/A)^{0.61+0.24(-\log_{10}p_r)}\, p_r^{1.23-0.2\log_{10}R_p}\,(-\log_{10}p_r)^{-0.70}\,10^{1.28}$	5.8
Refrigerants 12 and R13B1	
$(q/A)^{0.60+0.18(-\log_{10}p_r)}\, p_r^{0.83-0.2\log_{10}R_p}\,(-\log_{10}p_r)^{-0.73}$	10.1

[a] From Refs. 5–7.

much difference between the properties of the two fluids, which are of similar type, with similar molecular weights, 120.9 for R12 and 148.9 for R13B1.

Another close comparison of data for refrigerants is obtained from the Japanese data on three refrigerants, R12, R114, R113, at varying roughness [3] (already examined in that context), together with earlier data for those same refrigerants in the same apparatus but with a single roughness (emery 0/4, $R_p = 0.125\ \mu$m) [4]. If we take the data from Nishikawa *et al.* [4] with those data from Nishikawa *et al.* [3] that were obtained at that same roughness, we have, for each of the three refrigerants, two data sets at different times. A common correlation for these six data sets is shown in Table VI. It was the basis for the lines in Fig. 2a and b, which were used to discuss reproducibility in Section V,B,2. Similar reproducibility can also be seen for R21 and R113 by comparing columns in Table VI for C_E for Nishikawa *et al.* [3, 4].

Looking now for differences between the refrigerants, comparison of rows in Table VI shows that the difference is small. For R114, the average of C_E from Nishikawa *et al.* [3, 4] is some 35% higher than that for the other refrigerants, despite the fact that M for R114 is intermediate between the others, and so are virtually all its properties at any given p_r. For their values of alpha in Eq. (A2), Nishikawa *et al.* [3] applied dimensional analysis (though not to the R_p term) and deduced that alpha $= p_{cr}^{0.2}T_r^{-0.9}M^{-0.1} \times$ (constant) after some controversial assumptions. These values are also shown in Table VI, where it is seen that they too give R114 a value intermediate between R21 and R113. Thus the experimental data indicate a significantly higher value of C_E for R114, and again it is difficult

TABLE VI

REFRIGERANTS ON FLAT PLATES, $R_p = 0.125 \ \mu m^a$

$$\frac{h}{(q/A)^{0.764}} = p_r^{0.102} \ (-\log_{10} p_r)^{-0.697} \times C_E$$

Fluid	C_E			M	Alpha [3]
	[3]	[4]	Average for fluid		
Refrigerant R21	1.42	1.24	1.33	103	1.75
Refrigerant R114	1.95	1.61	1.78	171	1.63
Refrigerant R113	1.44	1.18	1.31	187	1.42

a From Refs. 3 and 4.

to attribute that to differences in the usual thermodynamic properties, or any other function of this, such as alpha [3].

Another set of data for several fluids in the same apparatus is that of Sciance *et al.* [21], which used methane, ethane, propane, and *n*-butane, boiling on a gold-plated horizontal iron cylinder. As discussed in Cooper [1], the data for ethane show dependence on p_r quite different from the other fluids, despite ethane being an intermediate member of the group as regards M and properties. When the data for each fluid are averaged to obtain C_E, the value for ethane appears high, as shown by the lines marked [21] in Table VII below. However, when seen in conjunction with other fluids in that table, it seems that methane is the outlier, since it has low C_E despite low M.

2. *General Data*

Table II in Section VII,B,4 suggests that C_E for water is substantially higher than that for ethanol and benzene. This outlying behavior could therefore be correlated statistically with any of the outlying characteristics of water—high p_{cr}, high T_{cr}, high σ, high k_f, and (among the fluids in Table II), low M. A glance at Table II suggests that high C_E may be linked to low M, as many correlations would predict, but it provides no firm basis for that. If we seek further data, to decide these points, then we are forced to examine data from widely different apparatus. As discussed in Section VII,D differences between experimenters can be very substantial. If we accept that it is worth applying a "broad brush" approach, and assuming that such differences will average out, then we can stir all data into a single correlation, irrespective of heater material or geometry, obtaining a different value of constant C_E for each data set. Some sets show substantial values of average absolute error, implying that they do not really fit this single correlation. Either they require a different slope [different value of m in $h/(q/A)^m$] or they require a different "spread" with p_r [different $F(p_r)$].

Alternatively, a single value of C_E can be obtained for each fluid, and such results are given in Table VII with values of M, p_{cr}, T_{cr}, listed in order of molecular weight of fluid. There appears to be a general downward drift in the value of the constant C_E, as we go down the table, at

TABLE VII

ALL DATA SETS[a]

Fluid	M	p_{cr} (MN/m^2)	T_{cr} (K)	C_E	Ref.
Hydrogen	2.0	1.3	33.3	20.16	
Deuterium	4.0	1.66	38.3	25.29	
Helium II	4.0			48.94	
Helium	4.0	0.23	5.2	19.70	
Methane	16.0	4.6	190.6	6.71	[21]
Water	18.0	22.1	647.3	12.43	
Neon	20.2	2.6	44.4	25.72	
Nitrogen	28.0	3.4	126.2	9.60	
Ethane	30.1	4.9	305.3	9.67	[21]
Methanol	32.0	8.1	512.6	8.02	
Oxygen	32.0	5.0	154.6	8.51	
Propane	44.1	4.3	369.8	6.82	[21]
Ethanol	46.1	6.4	516.2	6.45	
n-Butane	58.1	3.8	425.1	5.79	[21]
Benzene	78.1	4.9	562.1	4.27	
Refrigerant R22	86.5	5.0	369.1	7.01	
Refrigerant R21	102.9	5.2	451.6	6.88	
Refrigerant R13	104.5	3.9	301.9	4.71	
Refrigerant R12	120.9	4.1	385.0	7.28	
Refrigerant R11	137.4	4.4	471.1	5.28	
Sulfur hex SF$_6$	146.0	3.8	318.8	7.72	
Refrigerant R13B1	148.9	4.0	340.1	7.59	
Refrigerant R115	154.5	3.1	353.2	7.24	
Refrigerant R114	170.9	3.2	418.9	8.41	
Refrigerant R226	170.0	3.1	432.8	5.04	
Refrigerant R113	187.4	3.4	487.3	6.36	
Refrigerant RC318	200.0	2.8	388.5	7.42	

[a] h for copper cylinders decreased by 1.7. Roughness term $R_p^{0.2(-\log_{10} p_r)}$ from previous analyses: $h/(q/A)^{0.666} = p_r^{0.12}(-\log_{10} p_r)^{-0.55} R_p^{0.2(-\log_{10} p_r)} \times C_E$.

least as far as the first refrigerants. That is confirmed by Fig. 8, which shows C_E against M.

3. *Alternatives to M*

Section VII,C,3 discusses chiefly (T_{cr}/p_{cr}) as one alternative to M in the correlation, by plotting C_E against (T_{cr}/p_{cr}) in Fig. 10. The falling and rising trend in C_E noted in Fig. 8 persists in Fig. 10 and cannot be matched by a simple power M^H or a simple power of (T_{cr}/p_{cr}).

To match the trough, we require a term such as $(c_1 M^{a_1} + c_2 M^{a_2})$. That is readily incorporated in the correlation as terms M^H and $(1 + M_E/M)^J$. The parameter M_E cannot readily be determined by least squares but is quickly optimized by trial and error, controlled from the easily edited control file. However, in view of the scatter among data sets, illustrated by Fig. 9, there seems no justification for these extra complications.

Many further alternatives could be considered, and no doubt given theoretical explanation. One such arises from the fact demonstrated in Fig. 3c that, for the 26 fluids tested, including hydrogen and helium, the ratio of k_f/σ is very nearly proportional to $M^{-0.6}$. Thus we could replace $M^{-0.5}$ in the correlation by $(k_f/\sigma)^{0.83}$. That would accord with some theoretical expectations. Also, the eye is caught by one point lying high both in the graph of C_E against M (Fig. 8) and in the graph of (k_f/σ) against M (Fig. 3c). Those points imply that the boiling data for that fluid, which is an outlier when plotted against M, would be more in line when plotted against (k_f/σ). We must be alert for such cases, comparing Fig. 8 with either Fig. 3 (properties) or Fig. 4 (correlations). However, there are reservations in the present case:

1. The fluid in question is neon, for which there is one data set, using a wire heater. That is not enough to support the use of (k_f/σ)

2. This is a chance discovery, and variation with $(k_f/\sigma)^{0.83}$ is very difficult to distinguish from variation with $M^{-0.5}$.

3. The practical man would surely prefer using the easily available molecular weight, rather than trying to find accurate values of k_f and σ for each fluid.

4. The true form of C_E may be far more complicated than either $M^{-0.5}$ or $(k_f/\sigma)^{0.83}$, and it could be misleading to suggest at this stage that any particular properties are positively known to be involved.

4. *F(p_r) Dependent on Fluid*

The main text, Section VII,C,4, needs no further details here.

5. *Residual Scatter among Fluids: Proposals for Research*

The main text, Section VII,C,5 needs no further details here.

D. UNEXPLAINED DIFFERENCES BETWEEN EXPERIMENTERS

As discussed in Section VII,D, the data most likely to show the extent of the differences between experimenters are those for water.

Different Experimenters Using Water

The 30 data sets, 1474 points, were used to produce a correlation as in Eq. (6), by least squares fit, excluding those sets that referred to only one pressure, and excluding the single set at $p_r > 0.979$ [29]. Those excluded sets were compared with the correlation and the results are given in Table VIII below for all sets using water.

For one data set C_E is outstandingly high, but the data were published in 1932 [30] and refer to a single pressure, 1 atm. One set has outstandingly low C_E, but that was the one obtained at extremely high p_r, from 0.979 to 0.990 [29], with much difficulty, since ΔT was in the range 0.14–1.67 K. It may be that the present simple correlation [Eq. (6)], cannot match data at that extremely high p_r. Further terms could be added to Eq. (6), to improve match at such p_r, but the extra complication is unwelcome, particularly as the values of ΔT are so low as to have little practical importance. Studies at high p_r may be of great value to theoretical research, indicating the correct theory to apply, but that is not the present aim. The next lowest two sets are the only ones on stainless steel wires [31, 32], and that may be why they are low, as discussed in Appendix A, Section B,2. Discarding those four sets, the remainder have values of C_E ranging from 3.2 to 6.0. That is a factor 1.9 between the lowest and the highest, or $+40\%$, -30% about their geometric mean. These discrepancies are in $h/(q/A)^m$, that is, in h or ΔT at given (q/A). The corresponding discrepancy in (q/A) at given ΔT is about 1.9^3 or 1.9^4 between lowest and highest, which are approximately 7 and 12, since $(q/A) \propto \Delta T^n$, $n \approx 3$ or 4. These are very large ranges, but fortunately the practical requirement is generally to obtain h or ΔT at given (q/A).

E. BURNOUT AND NATURAL CONVECTION

1. *Burnout or DNB*

Prediction of the maximum heat flux rate at burnout or DNB $(q/A)_{max}$ is usually based on a theory by Zuber and Tribus [33], with various subse-

TABLE VIII

All Data Sets Using Water

$$\frac{h}{(q/A)^{0.760}} = p_r^{0.12}(-\log_{10} p_r)^{-0.547} \times C_E$$

Heater			C_E	No. of data points	Range of p_r		First-named author
Geometry	Diameter (mm)	Material			Min.	Max.	
Horizontal cylinder	10	S/S	3.171	86	0.0071	0.6646	Fedders
Horizontal cylinder	6	S/S	3.304	85	0.0045	0.8070	Borishanskii
Horizontal cylinder	6	S/S	3.079	38	0.0005	0.0046	Ponter
Horizontal cylinder	5,6,7	S/S	3.811	209	0.0046	0.8876	Borishanskii
Horizontal cylinder	12	S/S	6.000	37	0.0002	0.0090	Minchenko
Horizontal cylinder	4	S/S	3.485	118	0.0004	0.5322	Fujita
Horizontal flat plate		Cu	5.925	61	0.0001	0.0046	Raben
Horizontal flat plate		Cu	3.389	42	0.0003	0.0088	Nishikawa
Horizontal flat plate		Cu	3.530	117	0.0003	0.3548	Fujita
Horizontal wire	0.9	S/S	2.178	49	0.0078	0.1178	Turton [31]
Horizontal wire	1.25	Pt	4.679	20	0.0266	0.0266	Sakurai
Horizontal wire	0.3	Pt	3.716	128	0.0002	0.5322	Fujita

Data sets below are not used for least squares analysis—data sets at single pressure

Horizontal cylinder	26.4	Brass	14.894	30	0.0046	0.0046	Cryder [30]
Horizontal flat plate		Cu	3.596	40	0.0046	0.0046	Gaertner
Horizontal flat plate		S/S	5.406	47	0.0045	0.0045	Vachon
Horizontal flat plate		S/S	5.817	46	0.0045	0.0045	Vachon
Horizontal flat plate		S/S	5.346	73	0.0045	0.0045	Vachon
Horizontal flat plate		S/S	4.537	56	0.0045	0.0045	Vachon
Horizontal flat plate		Cu,Ni	4.408	16	0.0046	0.0046	Marcus
Horizontal flat plate		Ni	3.803	17	0.0046	0.0046	Madsen
Horizontal flat plate		Cu,Ni	3.362	13	0.0046	0.0046	Kotchaphakdee
Horizontal flat plate		Cu	5.052	7	0.0046	0.0046	Lorenz
Horizontal flat plate		Cu	5.946	15	0.0046	0.0046	Shoukri
Horizontal flat plate		Cu	3.632	8	0.0046	0.0046	Sultan
Horizontal flat plate		Cu	3.818	9	0.0046	0.0046	Nakayama
Horizontal wire	0.3	Ni,Al	4.447	15	0.0046	0.0046	Calus
Horizontal wire	0.12	Pt	4.415	30	0.0046	0.0046	Haigh
Horizontal wire	0.12	Pt	4.334	12	0.0046	0.0046	Peterson
Horizontal tube	1.6	S/S	2.335	9	0.0046	0.0046	Lee [32]

Data set at very high p_r

Horizontal wire	0.1	Pt	1.371	41	0.9788	0.9946	Reimann [29]

quent contributions, and it is reasonably successful. For flat plates, the form is:

$$(q/A)_{\max} = \rho_g^{0.5}(\rho_f - \rho_g)^{0.25} h_{fg} \sigma^{0.25} g^{0.25} \times 0.149 \qquad \text{(A6)}$$

and its fit to the data is generally claimed to be a few percent, with much less scatter than among data for fully developed boiling. This function has been formulated as a graph against reduced pressure (see, e.g., Lienhard and Schrock [34]).

For any given fluid, the present method of formulating properties [Eq. (4)] can be applied to formulate an algebraic function equivalent to Eq. (A6) with high accuracy. For example, for water, the equivalent is:

$$p_r^{0.544}(-\log_{10} p_r)^{0.662} \times 13.3 \times 10^6 \quad \text{in SI units}$$

which agrees with Eq. (A6) within $\pm 1.0\%$ throughout the range $100 < T_{sat}°C < 365$ $(0.0046 < p_r < 0.90)$. An alternative, with slightly different indices, agrees within $\pm 1.2\%$ throughout the range $75 < T_{sat}°C < 365$.

The present method [Section VI,A, Eqs. (7)] of expressing approximate variation of properties between fluids can also be applied, but its accuracy is rather below the accuracy with which Eq. (A6) fits the data for $(q/A)_{\max}$. For what it is worth, Eqs. (7), applied to Eq. (A6) would suggest that $(q/A)_{\max}$ is proportional to $M^{-0.5}$. That looks usefully similar to variation of C_E with M, but in fact it is very inaccurate, largely because the terms strongly dependent on T_{cr} are ρ_g and h_{fg} which are not here present as a product, so their dependence on T_{cr} does not cancel here. Some improvement can be obtained by introducing p_{cr} and T_{cr}, leading to

$$(q/A)_{\max} \propto p_{cr}^{0.5} T_{cr}^{0.5} M^{-0.25}$$

which applies generally within $\pm 20\%$, except for water and the two alcohols methanol and ethanol, which are high by factors 3, 2, 2, respectively. That may be of some interest for future application when better understood, but there seems little point in pursuing this sort of approximation for burnout calculations, for all fluids, since Eq. (A6) is quite accurate, and the properties required for it are few and fairly readily available.

There may be some point in pursuing it for a single fluid, since it could be combined with the expression for developed boiling, $(q/A)_{FDB}$ to yield an expression that lies close to:

$$q/A = f(p_r, R_p) \, \Delta T^n = (q/A)_{FDB}$$

at low ΔT, but has an upper limit of $(q/A)_{\max}$. One algebraic device to achieve this is

$$q/A = [(q/A_{FDB})^\gamma + (q/A_{\max})^\gamma]^{1/\gamma} \qquad \text{(A7)}$$

When $\gamma < 0$, q/A will lie below the lower of $(q/A)_{FDB}$ and $(q/A)_{max}$. For example,

$$\text{If } \gamma = -1, \quad q/A = 1/[1/(q/A_{FDB}) + 1/(q/A_{max})] \tag{A8}$$

2. Natural Convection

Expressions for natural convective heat flow, although basically empirical, are even better founded and more accurate than those for DNB or $(q/A)_{max}$. They are typically of the form:

$$Nu = a \, Ra^b$$

where a and b are about 0.5 and 0.25, though dependent on Pr. The Rayleigh number Ra is Gr $\times$ Pr, so the equation involves a length and gravity, in ways which are crucial and well verified. That is unlike $(q/A)_{FDB}$, for which different theories and correlations predict quite different dependences on gravity and a length, none of which is well verified. The remaining properties no doubt could be expressed quite accurately in terms of p_r and $(-\log_{10} p_r)$ or similarly, for any one fluid, but there seems little point in doing so. Simple attempts to extend to other fluids by use of M would again forfeit too much accuracy.

If $(q/A)_{NC}$ is evaluated by some means, then it can be combined with $(q/A)_{FDB}$ by use of Eq. (A7), but now with $\gamma > 0$, since q/A will then lie above the larger of $(q/A)_{NC}$ and $(q/A)_{FDB}$, as required. The particular case of $\gamma = +1$ has often been used, giving:

$$q/A = (q/A)_{NC} + (q/A)_{FDB} \tag{A9}$$

though other values for γ have been proposed, and some theoretical arguments have been given for preferring one value of γ, or a different algebraic device to run from $(q/A)_{NC}$ to $(q/A)_{FDB}$.

F. FURTHER DATA NEEDED: SUMMARY

The main text, Section VII,F, needs no further details here.

XI. Appendix B: Pointers from Bubble Dynamics

A. SIMPLEST CASE: STAGES OF BUBBLE GROWTH

In the simplest case, a vapor bubble grows in liquid that has no initial field of gravity, temperature, or motion. In that case, the bubble can pass through three major stages in its life, linked by transitions, which may be lengthy.

1. The incipient bubble at a nucleation site remains small until its temperature (averaged in some sense) is sufficiently high for the corresponding saturation pressure to exceed system pressure p_∞ by more than the pressure from surface tension ($2\sigma/R$ for a sphere of radius R). Nucleation is therefore a matter of the available supersaturation pressure and the surface tension.

2. Once growth has started, $2\sigma/R$ rapidly diminishes, and early growth may approach the asymptotic case of "inertia control," in which the pressure in the bubble is equal to $p_{\mathrm{sat}\infty}$ (= the saturation pressure at a bulk temperature of T_∞), so the supersaturation pressure is at its maximum, ($p_{\mathrm{sat}\infty} - p_\infty$). That drives the bubble growth, by causing inertia stress, so

$$\rho_f \left[R \frac{d^2R}{dt^2} + \frac{3}{2}\left(\frac{dR}{dt}\right)^2 \right] = p_{\mathrm{sat}\infty} - p_\infty$$

giving, after a transient,

$$\frac{dR}{dt} = \left[\frac{2}{3} \frac{(p_{\mathrm{sat}\infty} - p_\infty)}{\rho_f} \right]^{0.5}$$

That asymptotic stage may never be closely approached, and if it is, that cannot continue indefinitely, since a temperature gradient is required to cause heat to flow from the bulk liquid to cause evaporation at the bubble surface. The temperature in the bubble is therefore never strictly T_∞, and it falls further below that, causing pressure to fall below $p_{\mathrm{sat}\infty}$.

3. Finally, another asymptotic case can be reached with "diffusion-controlled" growth, when the pressure in the bubble is close to p_∞, and the temperature is close to the corresponding saturation temperature, $T_{\mathrm{sat}\infty}$, with growth rate controlled by conduction of heat. That asymptotic case has long been analyzed for a spherical bubble, giving

$$\frac{dR}{dt} = \left(\frac{3}{\pi} k_f \rho_f c_{pf} \right)^{0.5} \frac{(T_\infty - T_{\mathrm{sat}\infty})}{\rho_g h_{fg}} t^{-0.5}$$

The spherical shape occurs if the bubble is growing in the bulk of the liquid, remote from any wall. A bubble growing at a wall can initially be nearly hemispherical if growth is fast, and it generally rounds off later, toward spherical shape.

B. ENERGY FLOWS

Results are required in terms of energy flows, not bubble dynamics, and are generally presented as heat flows against ($T - T_{\mathrm{sat}}$), rather than against $(p_{\text{sat}\infty} - p_\infty)$. These differences are approximately related by the finite difference form of the Clausius–Clapeyron relation $\Delta T \approx \Delta p(\rho_g h_{fg}/T)$, where ρ_g has been written for $1/(v_g - v_f)$, and the approximation is acceptable, provided the changes are not large.

Precise flows of work and heat are complicated, but the major energy flow in bubble growth is a rate of increase of enthalpy by vaporization in the bubble, which is

$$\frac{d}{dt}(\rho_g V h_{fg})$$

where V is the volume of the bubble.

Total energy flow in boiling will also include conduction–convection in the liquid, augmented by fluid motion, driven by bubble growth. It is arguable that to describe such effects we should not need additional parameters, since transport properties are already included; the relations become more complicated but involve no more properties.

C. BUBBLES IN BOILING

These simplified arguments do not apply directly to the complicated case of actual boiling, with initial fields of gravity, temperature, and motion in the liquid, so they can be used only as pointers.

It is noteworthy that in the preceding discussion (and in many correlations) the quantities ρ_g and h_{fg} appear only in the product ($\rho_g \times h_{fg}$). As mentioned earlier, ρ_g is broadly proportional to $M p_{cr}/T_{cr}$ and h_{fg} to T_{cr}/M, so they vary widely because of the very wide variation of M and T_{cr}, while their product varies only as p_{cr}, that is, it varies between fluids much less than either varies on its own. Applied to the Clausius–Clapeyron equation, $(dp/dT)_{\text{sat}} = \rho_g h_{fg}/T$, it shows that at a given p_r, $(dp/dT)_{\text{sat}}$ varies little among fluids, reflecting the broad similarity in the relation between p_r and T_r for common fluids (Fig. 5).

If ρ_g and h_{fg} are therefore not included as individual properties, but only as their product, then there are few quantities indeed that vary greatly among common nonmetallic fluids at the same p_r. We are left with substantial variation of c_{pf}, broadly as T_{cr}/M, some variation of ($\rho_g h_{fg}$), broadly as p_{cr} and the two outlying properties of water, namely, σ and k_f, both of which exceed other common fluids by a factor 3 or 4.

This suggests (though it does not prove) that it may be useful to pursue certain comparisons of boiling heat flow rates in different fluids:

1. Comparing different fluids at the same p_r, seeking systematic variation of heat flow rate with M and with p_{cr} and T_{cr}.

2. If phenomena of nucleation and incipient bubbles are important, then heat flow rates with water should be different from other fluids.

3. If phenomena of conduction of heat through the liquid are important, then, again, heat flow rates with water should be different from other fluids.

4. The boiling data for water suggest that items 2 and 3 might act in opposite directions and partly cancel each other.

XII. Appendix C: Sources of Data

A. PLATES

1. Plates, Copper, Water

Fujita, Y., and Nishikawa, K. On the pressure factor in nucleate boiling heat transfer. *Mem., Fac. Eng., Kyushu Univ.* **36**, 303–341 (1977).

Gaertner, R. F. Photographic study of nucleate pool boiling on a horizontal surface. *Gen. Electr. Rep.* No. 63-RL-3357C (1963).

Lorenz, J. J., Mikic, B. B., and Rohsenow, W. N. The effect of surface conditions on boiling characteristics. *Heat Transfer, Proc. Int. Heat Transfer Conf., 5th, Tokyo,* **4**, 35–39 (1974).

Nakayama, W., Daikoku, T., Kuwahara, H., and Nakajima, T. Dynamic model of enhanced boiling heat transfer on porous surfaces. *Natl. Heat Transfer Conf., 18th, San Diego, Calif., ASME/AIChE* 31–44 (1979).

Nishikawa, K., Fujita, Y., Nawate, Y., and Nishijima, T. Studies on nucleate pool boiling at low pressures. *Heat Transfer—Jpn. Res.* **5**(2), 66 (1976).

Raben, I. A. Beaubouef, R. T., and Commerford, G. A study of nucleate pool boiling of water at low pressure. *AIChE–ASME Nat. Heat Transfer Conf., 6th, Boston AIChE Prepr.* No. 28 (1963).

Shoukri, M., and Judd, R. L. Nucleation site activation in saturated boiling. *J. Heat Transfer,* **97**, 93–98 (1975).

Sultan, M., and Judd, R. L. Spatial distribution of active sites and bubble flux density. *J. Heat Transfer* **100**, 56–62 (1978).

2. Plates, Copper, Refrigerants

Fujii, M., Nishiyama, E., and Yamanaka, G. Nucleate pool boiling heat transfer from microporous heating surface. *Natl. Heat Transfer Conf., 18th, San Diego, Calif., ASME/AIChE* (1979).

Nishikawa, K., Fujita, Y., Ohta, H., and Hidaka, S. Heat transfer in nucleate boiling of Freon. *Heat Transfer—Jpn. Res.* **8**(3), 16–36 (1979).

Nishikawa, K., Fujita, Y., Ohta, H., and Hidaka, S. Effects of system pressure and surface roughness on nucleate boiling heat transfer. *Mem. Fac. Eng., Kyushu Univ.* **42**, 95–123 (1982).

Owens, F. L., and Florschuetz, L. W. Transient versus steady-state nucleate boiling. *J. Heat Transfer* **94**, 331–332 (1972).

Tanes, M. Y. "Influence of Heating-Surface Condition on Heat Transfer in Nucleate Boiling," Ph.D. Thesis. Technical Univ., Karlsruhe, 1976. (In German.)

3. Plates, Copper, Cryogens

Bewilogua, L., Knoner, R., and Vinzelberg, H. Heat transfer in cryogenic liquids under pressure. *Cryogenics* **15**, 121–125 (1975).

Bland, M. E., Bailey, C. A., and Davey, G. Boiling from metal surfaces immersed in liquid nitrogen and liquid hydrogen. *Cryogenics* **13**, 651–657 (1973).

Deev, V. I., Keilin, V. E., Kovalev, I. A., Kondratenko, A. K., and Petrovichev, V. I. Nucleate and film pool boiling heat transfer to saturated liquid helium. *Cryogenics* **17**, 557–562 (1977).

Grigoriev, V. A., Pavlov, Y. M., and Ametistov, Y. V. An investigation of nucleate boiling heat transfer of helium. *Heat Transfer Proc. Int. Heat Transfer. Conf., 5th, Tokyo* **4**, 45–49 (1974).

Ibrahim, E. A., Boom, R. W., and McIntosh, G. E. Heat transfer to subcooled liquid helium. *Adv. Cryog. Eng.* **23**, 333–339 (1977).

Jergel, M., and Stevenson, R. Static heat transfer to liquid helium in open pools and narrow channels. *Int. J. Heat Mass Transfer* **14**, 2099–2107 (1971).

Merte, H., Oker, E., and Littles, J. W. Boiling heat transfer to LN2 and LH2: Influence of surface orientation and reduced body forces. *Prog. Refrig. Sci. Technol., Proc. Int. Congr. Refrig., 13th, Washington, D.C., 1971* pp. 191–196 (1973).

Nakayama, W., Daikoku, T., Kuwahara, H., and Nakajima, T. Dynamic model of enhanced boiling heat transfer on porous surfaces. *Natl. Heat Transfer Conf., 18th, San Diego, Calif., ASME/AIChE* 31–44 (1979).

Ogata, H., and Nakayama, W. Heat transfer to subcritical and supercritical helium in centrifugal acceleration fields. 1. Free convection regime and boiling regime. *Cryogenics* **17**, 461–470 (1977).

4. Plates, Copper, Other Fluids

Chichelli, M. T., and Bonilla, C. F. Heat transfer to liquids boiling under pressure. *Trans. AIChE* **41**, pp. 755–787 (1944).

Fujita, Y., and Nishikawa, K. On the pressure factor in nucleate boiling heat transfer. *Mem. Fac. Eng., Kyushu Univ.* **36**, 303–341 (1977).

Lorenz, J. J., Mikic, B. B., and Rohsenow, W. N. The effect of surface conditions on boiling characteristics. *Heat Transfer, Proc. Int. Heat Transfer Conf., 5th, Tokyo,* **4**, 35–39 (1974).

Nishikawa, K., Fujita, Y., Nawate, Y., and Nishijima, T. Studies on nucleate pool boiling at low pressures. *Heat Transfer—Jpn. Res.* **5**(2), 66 (1976).

5. Plates, Stainless Steel, Water

Vachon, R. I., Tanger, G. E., Davis, D. L., and Nix, G. H. Pool boiling on polished and chemically etched stainless-steel surfaces. *J. Heat Transfer* **90**, 231–238 (1968).

6. Plates, Noncopper, Non-Stainless Steel, Water

Kotchaphakdee, P., and Williams, M. C. Enhancement of nucleate pool boiling with polymeric additives. *Int. J. Heat Mass Transfer* **13**, 835–848 (1970).

Madsen, N., and Bonilla, C. F. Heat transfer to sodium–potassium alloy in pool boiling. *Chem. Eng. Prog., Symp. Ser.* **56**, 251–259 (1960).

Marcus, B. D., and Dropkin, D. Measured temperature profiles within the superheated boundary layer above a horizontal surface in saturated nucleate pool boiling of water. *J. Heat Transfer* **87**, 333–341 (1965).

7. Plates, Noncopper, Non-Stainless Steel, Cryogens

Class, C. R., Descaan, J. R., Piccone, M., and Cost, R. B. Boiling heat transfer to liquid hydrogen from flat surfaces. *Adv. Cryog. Eng.* **5**, 254–261 (1959).

Grigoriev, V. A., Pavlov, Y. M., and Ametistov, Y. V. An investigation of nucleate boiling heat transfer of helium. *Heat Transfer, Proc. Int. Heat Transfer Conf., 5th, Tokyo* **4**, 45–49 (1974).

Jergel, M., and Stevenson, R. Contribution to the static heat transfer to boiling liquid helium. *Cryogenics* **14**, 431–433 (1974).

Kosky, P. G., and Lyon, D. N. Pool boiling heat transfer to cryogenic liquids. 1. Nucleate regime data and a test of some nucleate boiling correlations. *AICHE J.* **14**, 372–387 (1968).

Lyon, D. N., Kosky, P. G., and Harman, B. N. Nucleate boiling heat transfer coefficients and peak nucleate boiling fluxes for pure liquid N2 and O2 on horizontal platinum surfaces from below 0.5 atmosphere to the critical pressures. *Adv. Cryog. Eng.* **9**, 77–87 (1964).

Swanson, J. L., Bowman, H. F. Transient surface temperature behavior in nucleate pool-boiling nitrogen. *Heat Transfer, Proc. Int. Heat Transfer Conf., 5th, Tokyo* **4**, 60–64 (1974).

B. CYLINDERS

1. Cylinders, Copper, Refrigerants

Bier, K., Gorenflo, D., and Wickenhäuser, G. Pool boiling heat transfer at saturation pressures up to critical. *In* "Heat Transfer in Boiling" (E. Hahne and U. Grigull, eds.), Chap. 7. Academic Press, New York, 1977.

Engelhorn, H. R. "Heat Transfer during Nucleate Boiling in the Region of Low Boiling Pressures," Ph.D. Thesis. Tech. Univ., Karlsruhe, 1977. (In German.)

Götz, J. "Entwicklung und Erprobung einer Normapparatur zur Messung des Wärmeübergangs beim Blasenseiden," Ph.D. Thesis. Technical Univ., Karlsruhe, 1980.

2. Cylinders, Copper, Cryogens

Flynn, T. M., Draper, J. W., and Roos, J. J. The nucleate and film boiling curve of liquid nitrogen at one atmosphere. *Adv. Cryog. Eng.* **8**, 539–545 (1962).

Goodling, J. S., and Irey, R. K. Non-boiling and film boiling heat transfer to a saturated bath of liquid helium. *Adv. Cryog. Eng.* **14**, 159–169 (1969).

Tsuruga, H., and Endoh, K. Heat transfer from horizontal cylinder to subcooled liquid helium. *Proc. Int. Cryog. Eng. Conf.* **5**, 262–264 (1974).

3. Cylinders, Copper, Other Fluids

Engelhorn, H. R. "Heat Transfer during Nucleate Boiling in the Region of Low Boiling Pressures," Ph.D. Thesis. Univ., Karlsruhe, 1977. (In German.)

Labuntsov, D. A., Jagov, V. V., and Gorodov, A. K. Critical heat fluxes in boiling at low pressure region. *Heat Transfer, Int. Heat Transfer Conf., 6th, Toronto* **1**, 221–225 (1978).

4. Cylinders, Stainless Steel, Water

Borishanskii, V. M., Kozyrev, A., and Svetlova, L. Heat transfer in the boiling of water in a wide range of saturation pressure. *High Temp. (Engl. Transl.)* **2**, 119–121 (1964).
Borishanskii, V. M., Bobrovich, G. I., and Minchenko, E. P. Heat transfer from a tube to water and to ethanol in nucleate pool boiling. *In* "Symposium on Problems of Heat Transfer and Hydraulics of Two Phase Media" (S. S. Kutateladze, ed.), pp. 85–106. Pergamon, Oxford, 1966.
Fedders, H. Measurement of heat transfer in the nucleate boiling of water against metal pipes (in German). *Ber. Kernforschungsanlage Juelich* **Juel-740-RB** (1971).
Fujita, Y., and Nishikawa, K. On the pressure factor in nucleate boiling heat transfer. *Mem. Fac. Eng., Kyushu Univ.* **36**, 303–341 (1977).
Minchenko, F. P., and Firsova, E. V. Heat transfer to water and water–lithium salt solutions in nucleate pool boiling. *In* "Symposium on Problems of Heat Transfer and Hydraulics of Two Phase Media," pp. 137–151. Pergamon, Oxford, 1966.
Ponter, A. B., and Haigh, C. P. Sound emission and heat transfer in low pressure pool boiling. *Int. J. Heat Mass Transfer* **12**, 413–428 (1969).

5. Cylinders, Stainless Steel, Other Fluids

Borishanskii, V. M., Bobrovich, G. I., and Minchenko, E. P. Heat transfer from a tube to water and to ethanol in nucleate pool boiling. *In* "Symposium on Problems of Heat Transfer and Hydraulics of Two Phase Media" (S. S. Kutateladze, ed.), pp. 85–106. Pergamon, Oxford, 1966.
Fujita, Y., and Nishikawa, K. On the pressure factor in nucleate boiling heat transfer. *Mem. Fac. Eng., Kyushu Univ.* **36**, 303–341 (1977).
Huber, D. A., and Hoehne, J. C. Pool boiling of benzene, diphenyl and benzene diphenyl mixture under pressure. *J. Heat Transfer* **85**, 215–220 (1963).
Kutateladze, S. S., Moskvicheva, V. N., Bobrovich, G. I., Mamontova, N. N., and Avksentyuk, B. P. Some peculiarities of heat transfer crisis in alkali metals boiling under free convection. *Int. J. Heat Mass Transfer* **16**, 705–712 (1973).

6. Cylinders, Noncopper, Non-Stainless Steel, Water

Cryder, D. S., and Gilliland, E. R. Heat transmission from metal surfaces to boiling liquids. I. Effect of physical properties of boiling liquid on liquid film coefficient. *Ind. Eng. Chem.* **24**, 1382–1387 (1932).

7. Cylinders, Noncopper, Non-Stainless Steel, Cryogens

Kirichenko, Y. A., Levchenko, N. M., and Kozlov, S. M. Heat transfer with pool boiling of hydrogen. *Therm. Eng. (Engl. Transl.)* **24**(4), 58–61 (1977).
Sciance, C. T., Colver, C. P., and Sliepcevich, C. M. Nucleate pool boiling and burnout of liquefied hydrocarbon gases. *Chem. Eng. Prog., Symp. Ser.* **63**, 109–114 (1967).

8. *Cylinders, Noncopper, Non-Stainless Steel, Other Fluids*

Labuntsov, D. A., Jagov, V. V., and Gorodov, A. K. Critical heat fluxes in boiling at low pressure region. *Heat Transfer, Int. Heat Transfer Conf., 6th, Toronto* **1**, 221–225 (1978).

Sciance, C. T., Colver, C. P., and Sliepcevich, C. M. Nucleate pool boiling and burnout of liquefied hydrocarbon gases. *Chem. Eng. Prog., Symp. Ser.* **63**, 109–114 (1967).

C. WIRES

1. *Wires, Stainless Steel, Water*

Lee, L., and Singh, B. N. The influence of sub-cooling on nucleate pool boiling heat transfer. *Lett. Heat Mass Transfer* **2**, 315–324 (1975).

Turton, J. S. The effects of pressure and acceleration on the pool boiling of water and arcton 11. *Aeronaut. Res. Counc., U.K., Rep.* **ARC-29085** (1967).

2. *Wires, Stainless Steel, Refrigerants*

Turton, J. S. The effects of pressure and acceleration on the pool boiling of water and arcton 11. *Aeronaut. Res. Counc., U.K., Rep.* **ARC-29085** (1967).

3. *Wires, Platinum, Water*

Fujita, Y., and Nishikawa, K. On the pressure factor in nucleate boiling heat transfer. *Mem. Fac. Eng., Kyushu Univ.* **36**, 303–341 (1977).

Haigh, C. P., and Ponter, A. B. Sound emission from boiling on a submerged wire. *Can. J. Chem. Eng.* **49**, 309–313 (1971).

Peterson, W. C., and Zaalouk, M. G. Boiling-curve measurements from a controlled heat-transfer process. *J. Heat Transfer* **93**, 408–412 (1971).

Reimann, M., and Grigull, U. Free convection and film boiling heat transfer in the critical region of water and carbon dioxide (in German; English abstract). *Waerme- Stoffuebertrag.* **8**, 229–239 (1975).

Sakurai, A., and Shiotsu, M. Transient pool boiling heat transfer. Part 2: Boiling heat transfer and burnout. *J. Heat Transfer* **99**, 554–560 (1977).

4. *Wires, Platinum, Refrigerants*

Hahne, E., and Feuerstein, G. Heat transfer in pool boiling in the thermodynamic critical region: Effect of pressure and geometry. *In* "Heat Transfer in Boiling" (E. Hahne and U. Grigull, eds.), Chap. 8. Academic Press, New York, 1977.

5. *Wires, Platinum, Cryogens*

Astruc, J. M. Heat transfer in pool boiling liquid neon, deuterium, and hydrogen, and critical heat flux in forced convection of liquid neon (in French). *Rapp. CEA-R—Fr., Commis. Energ. At.* **CEA-R-3484** (1968).

6. *Wires, Platinum, Other Fluids*

Fujita, Y., and Nishikawa, K. On the pressure factor in nucleate boiling heat transfer. *Mem. Fac. Eng., Kyushu Univ.* **36**, 303–341 (1977).

7. Wires, Other Material, Water

Calus, W. F., and Rice, P. Pool boiling heat transfer from a vibrating surface. *Proc. Int. Heat Transfer Conf., 4th, Paris* **5**, Pap. B1.1 (1970).

NOMENCLATURE

a, b, c, d, e	indices in Eq. (4)	M	molecular weight
A, B, C, D, E	indices in Eqs. (5), (6)	Nu	Nusselt number
c, c', c''	indices in Eq. (1)	p	pressure, N/m²
c_p	specific heat capacity, J/(kg K)	Pr	Prandtl number
		(q/A)	heat flux rate, W/m²
C_E	constant in Eqs. (5), (6)	R	gas constant, J/(kg K)
h	heat transfer coefficient, W/(m² K)	R	universal gas constant, J/(kmol K)
H	molar enthalpy, J/kmol	Re	Reynolds number
h_{fg}	latent heat of vaporization, J/kg	R_p	roughness parameter (Glättungstiefe DIN 4762), μm
k	thermal conductivity, W/(m K)	T	temperature, K
m, m', m''	indices in Eqs. (2), (3), (5), (6)	v	specific volume, m³/kg
		V	molar volume, m³/kmol

Greek Symbols

α_{cr}	Riedel parameter at CP	μ	dynamic viscosity, kg/(m sec)
α_i	index in Eq. (2)		
ρ	density, kg/m³	σ	surface tension, N/m
		ω	Pitzer acentric factor

Subscripts

cr	relating to thermodynamic critical point (CP)	g	saturated vapor
		r	reduced (e.g., $p_r = p/p_{cr}$)
f	saturated liquid		

ACKNOWLEDGMENTS

This work forms part of a study of the fundamentals of nucleate boiling, financed by the Science and Engineering Research Council and by the Engineering Sciences Division, A.E.R.E., Harwell, where the information service of the Heat Transfer and Fluid Flow Service has been very valuable.

REFERENCES

1. M. G. Cooper, Correlations for nucleate boiling—formulation using reduced properties. *Physicochem. Hydrodyn.* **3**(2), 89–111 (1982).
2. V. M. Borishanskii, Correlation of the effects of pressure on the critical heat flux and

heat transfer rates, using the theory of thermodynamic similarity. *In* "Symposium on Problems of Heat Transfer and Hydraulics of Two Phase Media," (S. Kutateladze, ed.), pp. 16–37. Pergamon, Oxford, 1966.

3. K. Nishikawa, Y. Fujita, H. Ohta, and S. Hidaka, Effects of system pressure and surface roughness on nucleate boiling heat transfer. *Mem. Fac. Eng., Kyushu Univ.* **42,** 95–123 (1982).
4. K. Nishikawa, Y. Fujita, H. Ohta, and S. Hidaka. Heat transfer in nucleate boiling of freon. *Heat Transfer—Jpn. Res.* **8**(3), 16–36 (1979).
5. K. Bier, D. Gorenflo, and G. Wickenhäuser, Pool boiling heat transfer at pressures up to critical. *In* "Heat Transfer in Boiling" (E. Hahne and U. Grigull, eds.), Chap. 7. Academic Press, New York, 1977.
6. H. R. Engelhorn, "Wärmeübergang beim Blasensieden in Bereich niedriger Seidedrücke," Ph.D. Thesis. Technical Univ., Karlsruhe, 1977.
7. J. Götz, "Entwicklung und Erprobung einer Normapparatur zur Messung des Wärmeübergangs beim Blasenseiden," Ph.D. Thesis. Technical Univ., Karlsruhe, 1980.
8. R. C. Reid, J. M. Prausnitz, and T. K. Sherwood, "The Properties of Gases and Liquids," 3rd Ed. McGraw-Hill, New York, 1977.
9. F. Mayinger and E. Holborn, The effect of liquid viscosity on bubble formation and heat transfer in boiling. *In* "Heat Transfer in Boiling" (E. Hahne and U. Grigull, eds.), Chap. 17. Academic Press, New York, 1977.
10. J. C. Chen, Correlation for boiling heat transfer to saturation fluids in convective flow. *ASME–AIChE Heat Transfer Conf., Boston* Pap. 63-HT-34 (1963).
11. K. E. Forster and R. Grief, Heat transfer to a boiling liquid, mechanisms and correlations. *J. Heat Transfer* **81,** 43 (1959).
12. H. K. Forster and N. Zuber, Dynamics of vapour bubbles and boiling heat transfer. *AIChE J.* **1,** 531–535 (1955).
13. S. S. Kutateladze, "Fundamentals of Heat Transfer" (Engl. transl.), p. 362. Academic Press, 1963.
14. V. M. Borishanskii and F. P. Minchenko, cited in S. S. Kutateladze, "Fundamentals of Heat Transfer" (Engl. transl.), p. 362. Academic Press, New York, 1963.
15. C. H. Gilmore, Nucleate boiling—a correlation. *Chem. Eng. Prog.* **54,** 77–79 (1958).
16. M. J. McNelly, "A Study of Heat Transfer in a Long Tube Evaporator," Ph.D. Thesis. Imperial Coll., London, 1955.
17. K. Stephan and M. Abdelsalam, Heat transfer correlations for natural-convective boiling. *Int. J. Heat Mass Transfer* **23,** 73–87 (1980).
18. W. M. Rohsenow, A method of correlating heat transfer for surface boiling of liquids. *Trans. ASME* **74,** 969 (1952).
19. R. Cole, Boiling nucleation. *Adv. Heat Transfer* **10,** 86–166 (1974).
20. Y. Fujita and K. Nishikawa, On the pressure factor in nucleate boiling heat transfer. *Mem. Fac. Eng., Kyushu Univ.* **36,** 303–341 (1977).
21. C. T. Sciance, C. P. Colver, and C. H. Sliepcevich, Nucleate pool boiling and burnout of liquefied hydrocarbon gases. *Chem. Eng. Prog., Symp. Ser.* **63,** 109–114 (1967).
22. H. Fedders, Measurement of heat transfer in the nucleate boiling of water against metal pipes (in German). *Ber. Kernforschungsanlage Juelich* **Juel-740-RB** (1971).
23. M. Y. Tanes, "Influence of Heating-Surface Condition on Heat Transfer in Nucleate Boiling," Ph.D. Thesis. Tech. Univ., Karlsruhe, 1976. (In German.)
24. R. I. Vachon, G. E. Tanger, D. L. Davis, and G. H. Nix, Pool boiling on polished and chemically etched stainless-steel surfaces. *J. Heat Transfer* **90,** 231–238 (1968).
25. P. P. Berenson, "Transition Boiling Heat Transfer from a Horizontal Surface," Rep. no. 17. Heat Transfer Lab. MIT, Cambridge, Mass., 1960.

26. G. N. Danilova and A. V. Kupriyanova, Boiling heat transfer to Freons C318 and 21. *Heat Transfer—Sov. Res.* **2**(2), 79–83 (1970).
27. V. A. Grigoriev, Y. M. Pavlov, and Y. V. Ametistov, An investigation of nucleate boiling heat transfer of helium. *Heat Transfer, Proc. Int. Heat Transfer. Conf. 5th, Tokyo* **4**, 45–49 (1974).
28. K. Cornwell, R. B. Schüller, and J. G. Einarsson, The influence of diameter on nucleate boiling outside tubes. *Proc. Int. Heat Transfer Conf., 7th, Munich* **4**, 47–53. (1982).
29. M. Reimann and U. Grigull, Free convection and film boiling heat transfer in the critical region of water and carbon dioxide (in German; English abstract). *Waerme- Stoffuebertrag.* **8**, 229–239 (1975).
30. D. S. Cryder and E. R. Gilliland, Heat transmission from metal surfaces to boiling liquids. I. Effect of physical properties of boiling liquid on liquid film coefficient. *Ind. Eng. Chem.* **24**, 1382–1387 (1932).
31. J. S. Turton, The effects of pressure and acceleration on the pool boiling of water and arcton 11. *Aeronaut. Res. Counc., U.K., Rep.* **ARC-29085** (1967).
32. L. Lee and B. N. Singh, The influence of sub-cooling on nucleate pool boiling heat transfer. *Lett. Heat Mass Transfer* **2**, 315–324 (1975).
33. N. Zuber and M. Tribus, "Further Remarks on the Stability of Boiling Heat Transfer," UCLA Rep. No. 58-5. Univ. California, Los Angeles, 1958.
34. J. H. Lienhard and V. E. Schrock, The effect of pressure, geometry and the equation of state upon the peak and minimum boiling heat flux. *J. Heat Transfer* **85**, 261–272 (1963).

A Review of Turbulent-Boundary-Layer Heat Transfer Research at Stanford, 1958–1983

R. J. MOFFAT AND W. M. KAYS

Department of Mechanical Engineering, Thermosciences Division, Stanford University, Stanford, California

I. Introduction

Almost continuously for the past 25 years, there has existed in the Thermosciences Laboratory of the Mechanical Engineering Department at Stanford University a research program, primarily experimental, concerned with heat transfer through turbulent boundary layers. This program has consisted of a series of projects, with varying sponsors but with largely the same cast of faculty investigators, and with certain common themes that have carried through from one project to another.

The original faculty supervisors were W. M. Kays and S. J. Kline. W. M. Kays supervised the second project, and then W. M. Kays and R. J. Moffat supervised a long series of projects. In more recent years, J. P. Johnston joined the supervisory group.

The various results of the program have been published in numerous limited-distribution reports and various journals and symposia transactions. But many of the reports have long been out of publication, and nowhere is there a single source covering all the many phases of the program. Because of the continuity that has characterized the program, it was felt worthwhile to attempt to put together under a single cover a comprehensive summary of the more important contributions of these 25 years of effort. The objective of this paper is to do just that.

The reader will doubtless still find it desirable to return to the original source reports and papers for a detailed study of most of the topics covered, especially if tabular summaries of data are needed. All the results presented here are given in graphical form and have been chosen from a large number of examples to illustrate particular phenomena, rather than to provide extensive design data.

The paper first summarizes the early phases of the program, in which the topics considered were the simple zero-pressure-gradient turbulent boundary layer with constant and with varying surface temperature, and then the accelerated (negative-pressure) boundary layer.

Following the early phases, the next section of the paper introduces the concept of the equilibrium boundary layer and some of the consequences of this concept, because the program thereafter was very much centered around equilibrium boundary layers and a study of particular deliberate departures from equilibrium.

The next major topic is the transpired boundary layer. A series of projects covered the entire spectrum of blowing and suction with constant free-stream velocity and with strong acceleration. Finally, equilibrium adverse-pressure-gradient boundary layers with and without transpiration were considered.

All the preceding topics were concerned with aerodynamically smooth surfaces. There then followed a series of projects for a "rough" surface with and without transpiration.

Full-coverage film cooling using discrete circular holes in the surface provides the subject matter for the next section. Various hole angles and arrangements are considered.

All the preceding work was concerned with a flat plate. The effect of wall curvature, first convex and then concave, is the next topic. This work includes both constant free-stream pressure and a favorable pressure gradient, and then convex curvature with full-coverage film cooling.

The next topic considers the effect of buoyancy forces transverse to the main flow direction. The boundary layer is again turbulent. This project covered the full range of states between pure forced convection and pure free convection.

Finally, the paper concludes with a summary of a computation model that has been developed over most of the life of the program. This is essentially a single, continuous mixing-length model that will satisfactorily reproduce all the experimental data prior to the film-cooling and curvature projects. Models have been developed for the latter, but they are still in an evolutionary stage.

The Early Phases of the Program

The Stanford turbulent boundary layer program started in the mid-1950s, and the basic philosophy that characterizes the program has been the same since the very first project. The program has always been primarily experimental, but not exclusively so; the central idea has been that experiments come first, with analysis and models driven by the experiments, and not the other way around.

The first project sponsored by the (then) NACA was carried out because there were virtually no experimental data for even the most elementary case of turbulent boundary layer heat transfer—flow along a constant-temperature flat plate with no axial pressure gradient. Numerous analytic models were available in the literature, and there were some pretty good experimental data for high-velocity flow, including fairly high supersonic Mach numbers. On the other hand, the only low-velocity data were for rather low Reynolds numbers for which transition from laminar to turbulent flow confused the results.

The reason why low-velocity, high Reynolds number data were not available was that a large low-speed wind tunnel was necessary, and most such facilities were being used for aerodynamic studies. W. M. Kays and

S. J. Kline felt that there was a need for such data, which should also involve small temperature differences in order to avoid varying fluid properties. The 7.5-ft-diameter Guggenheim wind tunnel at Stanford University, with air speeds up to 130 ft/sec, was available, and its characteristics appeared to make feasible an experiment with Reynolds numbers (based on x distance) up to about 3.6 million.

An electrically heated test plate was constructed with a flow length of 60.5 in., a width of 60 in., and an active (heated) width of 34 in. The active surface consisted of 24 individually heated copper plates, each approximately 2.5 in. in the flow direction. Electric power to each plate could be separately controlled and metered and, with all pertinent temperatures measured by thermocouples, the heat transfer coefficient averaged over each plate could be determined. Heat leak through the back of the plate and out the ends of each strip was measured in special tests, and radiation heat transfer from the surface was calculated.

Temperature difference, plate to air, was generally held less than 25°F. Thus any effects of the temperature dependence of fluid properties were so small that the results could be described as "constant-property." The final results for Stanton number were corrected by the factor $(T_w/T_\infty)^{0.4}$, which was (and still is) believed to be a good approximation to the temperature-dependent property effect. The resulting correction was about 2%.

The use of small temperature differences to minimize uncertainty about the effects of temperature-dependent properties has since been a characteristic of all the Stanford projects. Similarly, low velocities eliminate compressibility effects. The central idea has consistently been to obtain a set of baseline data for equilibrium boundary layers with well-defined thermal boundary conditions.

Free-stream turbulence, in these early tests, was between 2 and 3%.

Although velocity and temperature traverses were made, the primary results were Stanton number and Reynolds number based on distance from the leading edge of the plate. There was a boundary layer trip on the nose section, so the boundary layer was already turbulent at the first active plate.

As soon as the original plate was designed, it was obvious that it could be used for obtaining experimental data with varying surface temperature, or at least for stepwise variations with 2.5 in. resolution. Experiments were carried out with a single step in surface temperature (the unheated starting-length problem) and for several cases with various other surface-temperature variations. These latter experiments provided a basis for testing theories of heat transfer when the surface temperature varied in the flow direction. Again, this idea has been pursued in most of the later

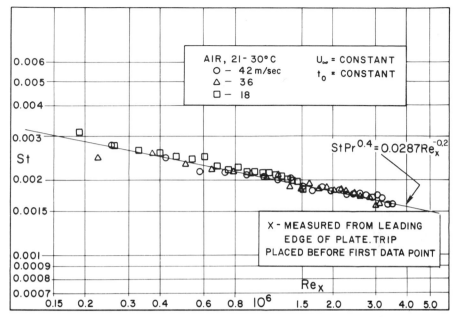

FIG. 1. Turbulent-boundary-layer heat transfer on a smooth, flat plate of uniform temperature.

Stanford projects and is consistent with the philosophy of data first, theory afterward.

The experiments were carried out by W. C. Reynolds, and the results were published in four NASA Memoranda in 1958 [34–37]. Figure 1 shows the test results for the constant-surface-temperature case for Reynolds numbers from 2×10^5 to 3.6×10^6. These are consistent with what has been obtained in every subsequent experiment in the Stanford program from entirely different test rigs. The equation for a straight line through the data is indicated.

The results for an unheated starting length are shown in Fig. 2. These data provided the basis for several theories of heat transfer for variable wall temperature. The data in Fig. 2 are correlated very well by:

$$St\ Pr^{0.4} = 0.0287\ Re_x^{-0.2}[1 - (\xi/x)^{9/10}]^{-1/9} \qquad (1)$$

This equation provides a convenient basis for solving variable-surface-temperature problems employing superposition theory.

The next phase of the program was started in the early 1960s. There was at the time considerable interest in heat transfer in supersonic nozzles, with particular interest in the boundary layer behavior at the throat.

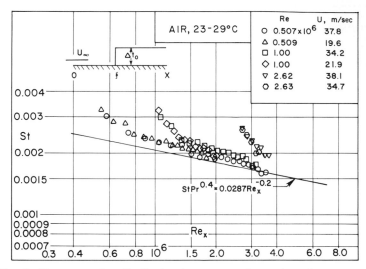

Fig. 2. Stanton number distribution downstream of steps in wall temperature.

Obviously, the primary application of interest was the rocket nozzle. The literature was full of proposed analytical procedures, and the experimental data available generally involved large temperature differences and nonequilibrium boundary layers, so it was difficult to draw any general conclusions about the influence of strong favorable axial pressure gradients.

At the time there was available in the Stanford laboratory a test plate consisting of 48 segments, each 1 in. wide (in the flow direction) by 20 in. long (normal to the flow direction). Each segment consisted of a $\frac{1}{8}$ in. copper plate bonded to a rectangular water tube (actually a wave guide). Between the plate and the water tube was bonded a heat meter and thermocouples to measure plate temperature. A rectangular air duct was built, using this plate as one wall, and controlled the flow acceleration by appropriately shaping the opposite wall.

With this apparatus, free-stream velocity could be varied in an arbitrary manner and local heat transfer coefficients measured over 1 in. flow distances. Constant surface temperature was used for most tests, but some with steps in surface temperature were obtained by closing off water tubes. The temperature difference, surface to air, was held to about 20°F, and, with air velocities less than 200 ft/sec, the effects of both variable properties and viscous dissipation were minimal.

Without much of an idea of what to expect, the experimenters began the work with a rather strong acceleration induced by shaping the oppo-

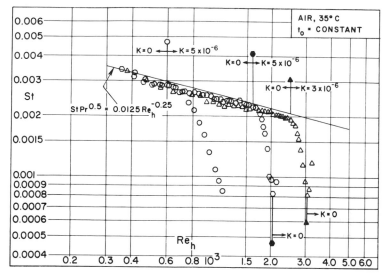

FIG. 3. Stanton number variations during strong acceleration.

site wall so that the flow cross section was linearly convergent. Some of the initial results with these strong accelerations are shown in Fig. 3, where Stanton number is plotted as a function of enthalpy-thickness Reynolds number. The results of three test runs are shown, and in each case the onset of strong acceleration was accompanied by a sharp reduction of Stanton number. In fact, it appeared that the turbulent boundary layer might be undergoing a reverse transition to a laminar boundary layer [29].

The straight-line correlation for St as a function of Re_h is derived directly from the equation for St as a function of Re_x on Fig. 1, assuming two-dimensional flow.

$$St\ Pr^{0.5} = 0.0125\ Re_h^{-0.25} \qquad (2)$$

When free-stream or wall conditions vary in the streamwise direction, x-Reynolds number is of little value in organizing heat transfer results. At a given x location, the local state (i.e., thickness and turbulence properties) depends on the upstream history and so will the heat transfer rate. Before one can judge whether the local heat transfer rate is "normal" or not, one must know the local boundary layer state. Enthalpy-thickness Reynolds number is at least a step in that direction. Upstream history is implicitly contained in the enthalpy thickness itself.

Moretti's experiments included only heat transfer measurements; velocity profiles were not measured, so it was not possible to determine the effect of strong acceleration on the hydrodynamics of the boundary layer. However, other results were becoming available concerning the momentum boundary layer under strong acceleration, in particular Launder and Jones [20] and Schraub and Kline [41] in the same laboratory as Moretti. Launder's experiments did not reveal the striking effects of acceleration that were evident in Moretti's heat transfer results, but he did use the same parameter to describe the strength of the acceleration. This nondimensional parameter, called K by both Launder and Moretti, is defined as follows:

$$K = (\nu/U_\infty^2) \, dU_\infty/dx \tag{3}$$

Flow between a pair of convergent planes will yield constant K, and in Moretti's experiments, K was maintained nearly constant in the accelerated region. Values of K for each of three tests are shown in Fig. 3.

Schraub and Kline [41], using a water table, injected dye into the viscous sublayer to visualize the turbulent "bursts" from the wall region that are characteristic of a turbulent boundary layer. His most significant observation was that acceleration reduced the frequency of bursts and that, at about $K = 3 \times 10^{-6}$, all bursting ceased and the boundary layer turbulence quickly decayed. These results were entirely consistent with Moretti's heat transfer data and went a long way toward explaining what was happening.

Moretti carried on further experiments at lower values of K, including experiments covering a constant-velocity recovery region following a strong acceleration. Some of these are shown in Fig. 4. Of significance here is the fact that even for K substantially below 3×10^{-6}, a marked reduction in Stanton number occurs. It has subsequently become apparent that a negative pressure gradient enhances the stability of the viscous sublayer, causing it to increase in thickness (in a nondimensional sense); at about $K = 3 \times 10^{-6}$, the sublayer simply engulfs the entire boundary layer, and all turbulence disappears. This phenomenon of "laminarization," or "retransition," involves a continuous series of boundary layer states and is not an abrupt event. These facts have been incorporated into the mixing-length model to be described later, and everything in Figs. 3 and 4 is now predictable using a mixing-length model in a finite-difference calculating procedure.

Moretti's work pointed out the importance of choosing the right parameter to hold constant during tests of boundary layers with complex boundary conditions. In the following programs, the integral equations provided guidance in parameter identification, always seeking "equilibrium" boundary layers of one kind or another. The integral equations can be written in several forms, and each form displays a different combination of terms that can be used. Equilibrium boundary layers and how to use

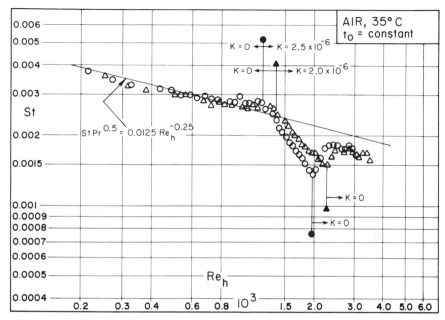

FIG. 4. Stanton number versus enthalpy thickness Reynolds number for two cases with moderate acceleration.

them in choosing the test conditions to define the most informative experiments are of great importance. The next section is entirely devoted to this issue.

II. Equilibrium Boundary Layers

The notion of equilibrium turbulent boundary layers has its roots in similarity analysis employed in laminar boundary layer theory. In laminar boundary layers, fixing the boundary conditions so as to hold constant a particular ratio reduces the partial differential equation to an ordinary one and permits a simple mathematical solution. The resulting velocity and temperature profiles are exactly self-similar. No such drastic benefit is available in turbulent studies. Combinations of boundary conditions can be proposed as being likely to lead to nearly self-similar behavior of the boundary layer, but the profiles must be found experimentally and are not usually completely self-similar, only approximately so. Turbulent boundary layers driven by those combinations of boundary conditions are known as "equilibrium" boundary layers—not truly similar, but nearly so. In the sections that follow, we shall describe the equilibrium parameters used in the Stanford study, classified by flow type.

A. TRANSPIRED FLOWS

The equilibrium parameter B_m comes from the similarity parameter used in the Falkner–Skan family:

$$U_\infty \propto x^m \tag{4}$$

and

$$v_0 \propto U_\infty(c_f/2) \tag{5}$$

It is particularly important to note that similarity is not obtained in general when v_0 is a constant, independent of x. The special case of $v_0 = 0.0$ and U_∞ a constant does yield similarity in velocity profiles, but, in general, a constant value of v_0 is an "arbitrary" variation of blowing as far as similarity is concerned. Similarity is achieved only if $v_0/(U_\infty c_f/2)$ is a constant with respect to x. This dimensionless group is usually called the *blowing parameter* and will be termed B_m. Thus:

$$B_m \triangleq \frac{\rho_0 v_0}{\rho_\infty U_\infty(c_f/2)} = \frac{\dot{m}''/G_\infty}{c_f/2} \tag{6}$$

The physical significance of holding B_m constant can be appreciated by rearranging it to display:

$$B_m = (\rho_0 v_0)U_\infty/\tau_0 \tag{7}$$

Thus B_m is the ratio of the transpired momentum deficit to the surface shear force. When these are kept in a fixed ratio along a surface, a laminar boundary layer develops with self-similar velocity profiles. It seems likely, then, that this ratio would also be important in turbulent boundary layers, and such is the case.

The energy equation can be approached in a similar way. Similar temperature profiles result when, in a laminar boundary layer having hydrodynamic similarity, the wall and free-stream temperatures are constant and a "heat transfer blowing parameter" B_h is held fixed. This parameter reflects the ratio of the transpired energy deficit to the surface heat transfer and is defined by:

$$B_h \triangleq \frac{\rho_0 v_0}{\rho_\infty U_\infty St} = \frac{\dot{m}''/G_\infty}{St} = \frac{\dot{m}''c(t_0 - t_\infty)}{\dot{q}_0''} \tag{8}$$

The blowing parameter B_m and the heat transfer blowing parameter B_h both arise in the reduction of the partial differential equations of the laminar boundary layer to the ordinary differential equation of the similarity situation. Both, however, are also visible in the integral equations of the boundary layer: a form that applies also to turbulent boundary layers.

B. FLOWS WITH AXIAL PRESSURE GRADIENTS

The velocity distribution specified by Eq. (4) results in a pressure gradient parameter β, which remains constant when m is constant. The parameter β is defined as follows:

$$\beta \triangleq \frac{\delta_2}{\tau_0} \left(\frac{dP}{dx}\right) \tag{9}$$

β may be interpreted as the ratio of the axial pressure force acting on the boundary layer to the shear force at the wall. Thus B_m and β should have similar influences on the development of the momentum boundary layer, and indeed if one examines the following form of the momentum integral equation of the boundary layer, this is seen to be the case:

$$\frac{d(U_\infty^2 \delta_2)}{dx} = \frac{\tau_0}{\rho_\infty} (1 + B_m + \beta) \tag{10}$$

Equation (10) expresses the rate of growth of the momentum deficit of the boundary layer. If B_m and β are held constant along a surface, the boundary layer maintains a similarity of structure as it develops.

The energy integral equation can be manipulated in such a way as to show the importance of B_h:

$$\frac{d[c\Delta_2 U_\infty(t_0 - t_\infty)]}{dx} = \frac{\dot{q}_0''}{\rho_\infty} (1 + B_h) \tag{11}$$

Equation (11) expresses the rate of growth of the axially flowing energy flux in the thermal boundary layer, and B_h is seen to have the same influence on the thermal boundary layer as B_m has on the momentum boundary layer. However, note that the pressure gradient parameter β has no direct effect on the thermal boundary layer.

Turbulent boundary layers have two distinct regions to consider: the inner and the outer. It is possible to have inner region similarity independent of the outer part of the boundary layer, though this latter may comprise most of the boundary layer thickness. The "law of the wall," which describes inner region similarity, is a good example. It seems to hold under many conditions regardless of upstream history and regardless of the distortion in the outer region. In short, it is possible for the inner region of a turbulent boundary layer to be in equilibrium even though the outer region is not.

Clauser [8] proposed that boundary layers having outer region similarity be called "equilibrium" boundary layers, defined as one for which the outer region velocity profile, plotted in velocity defect coordinates, was universal. This condition can be expressed by:

$$\frac{U - U_\infty}{U_\tau} = f\left(\frac{y}{\delta_3}\right) \quad \text{only} \tag{12}$$

where

$$\delta_3 = \int_0^\infty \frac{(U - U_\infty)}{U_\tau}\, dy \tag{13}$$

Clauser proposed a shape factor G that would be independent of x under these conditions:

$$G = \frac{1}{\delta_3} \int_0^\infty \frac{(U - U_\infty)^2}{U_\tau}\, dy \tag{14}$$

Experimentally, if either β or B_m acts alone and is constant, or both act together with $(\beta + B_m)$ held constant, then G remains constant. Thus it appears that the same relationship among the forces acting on the boundary layer yields similarity solutions for laminar boundary layers and results in equilibrium in the Clauser sense, for turbulent boundary layers.

A special case of equilibrium turbulent boundary layers occurs in flows where the acceleration parameter K is maintained constant.

$$K \triangleq (\nu/U_\infty^2)(dU_\infty/dx) \tag{15}$$

The significance of constant K can be appreciated if the momentum integral equation of the boundary layer is written in the following form:

$$\frac{d\mathrm{Re}_m}{U_\infty\, dx/\nu} = c_f/2 + v_0/U_\infty - K(1 + H)\,\mathrm{Re}_m \tag{16}$$

If K and v_0/U_∞ are independent of x, and K is finite and *positive*, the flow must inevitably approach a state of equilibrium for which Re_m is constant. This is often spoken of as an asymptotic-accelerating flow; a special case of an equilibrium boundary layer in which there is not only outer region similarity (constant G), but also inner region similarity. The velocity profiles are similar all the way to the wall, with the result that not only is Re_m constant, but so also are $c_f/2$ and the shape factor H. Thus, constant K and constant v_0/U_∞ together yield a family of similarity solutions for laminar boundary layers and a family of asymptotic-accelerating layers for the turbulent case. An interesting feature of Eq. (16) is the fact that for each positive value of K, and each value of v_0/U_∞, there exists a definite value of Re_m: as K increases Re_m decreases. Experiments indicate that it is impossible to maintain a turbulent boundary layer if Re_m is below about 300. The corresponding value for K is about 3×10^{-6}. In other words, if K is of the order of 3×10^{-6}, or greater, the turbulent boundary layer will tend to revert to a laminar boundary layer. Evidence of this trend was seen in the previous section.

If K is negative (i.e., a decelerating flow), no such asymptotic equilibrium can exist (except as discussed later). Note also that for a given value of K the rate of transpiration, whether positive or negative, will have a substantial influence on the asymptotic value of $\mathrm{Re_m}$.

Another related type of asymptotic flow can be recognized in Eq. (16). If v_0/U_∞ is negative, $\mathrm{Re_m}$ will approach a constant when K is zero or negative, so long as the v_0/U_∞ term is larger in the absolute sense than the last term. This type of boundary layer is frequently referred to as the *asymptotic suction layer* and may be either laminar or turbulent, depending on the magnitude of $\mathrm{Re_m}$ at the asymptotic condition. Note that for $K = 0.0$, $c_f/2$ approaches an asymptote, $-v_0/U_\infty$. Physically, the surface shear force is then precisely equal to the loss of momentum of the fluid that is brought from the free-stream to zero velocity at the surface. This represents a lower limit on $c_f/2$.

The energy integral equation of the boundary layer can be put in a form similar to that of Eq. (16), for the case of constant surface and free-stream temperatures.

$$\frac{d\mathrm{Re_h}}{U_\infty \, dx/\nu} = \mathrm{St} + v_0/U_\infty \qquad (17)$$

The important difference from Eq. (16) is that there is no term corresponding to the acceleration term (or pressure gradient term). Thus if K is a positive constant, the momentum boundary layer will come to equilibrium, with $\mathrm{Re_m}$ constant, but the thermal boundary layer will continue to grow. If K is maintained constant for a sufficient distance, the thermal boundary layer will grow outside of the momentum boundary layer as $\mathrm{Re_h}$ increases indefinitely.

Equation (17) suggests that an asymptotic thermal boundary layer exists only in the case of negative v_0/U_∞: an "asymptotic layer" similar to the momentum boundary layer case.

An equilibrium thermal boundary layer can be defined in an analogous manner to the equilibrium momentum boundary layer, that is, as a thermal boundary layer having outer region temperature profile similarity. Following the form of Eqs. (12) and (13), temperature-defect coordinates are defined,

$$t_d^+ = \frac{(t_\infty - t)}{(t_\infty - t_0)} \frac{\sqrt{c_f/2}}{\mathrm{St}} = f\left(\frac{y}{\Delta_3}\right) \qquad (18)$$

where

$$\Delta_3 = \frac{\sqrt{c_f/2}}{\mathrm{St}} \int_0^\infty \frac{(t_\infty - t)}{(t_\infty - t_0)} \, dy \qquad (19)$$

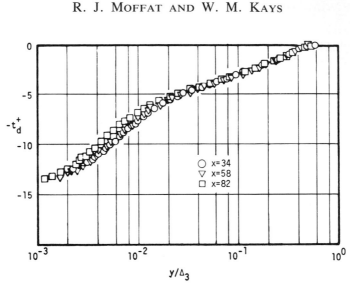

FIG. 5. Temperature defect profiles for a decelerating flow of $m = -0.2$.

Figure 5 shows three temperature profiles plotted in these coordinates for an adverse pressure gradient momentum boundary layer having velocity-defect similarity, that is, an equilibrium momentum boundary layer. The temperature profiles are universal everywhere outside of the sublayer. Thus a constant G, β, m equilibrium momentum boundary layer also yields equilibrium thermal boundary layers when the surface temperature is constant. The same is true if there is transpiration with B_m constant: B_h will become constant and similar temperature profiles will result.

III. The Apparatus and Techniques Used

The Moretti work showed the importance of being able to control the integral properties of the boundary layer when investigating complex turbulent boundary layer situations and led to the development, over the years, of a set of four heat transfer tunnels designed to provide careful control of the hydrodynamics, measurements within the boundary layer, and accurate measurement of the heat transfer. Planning for sustained investigations over long periods of time, each tunnel was built to provide one special feature, around which a program of broad interest could be built: smooth walls, rough walls, and curved walls with and without discrete hole injection. Each project, on each apparatus, was used to improve the capabilities of the apparatus, either by broadening its range, or by extending the flow range, or by upgrading the instrumentation. By adhering to this organized approach over the years, each apparatus has improved with age.

An apparatus used in convective heat transfer research must satisfy all of the requirements of a good tunnel for hydrodynamic studies and also meet the stringent requirements of a good heat transfer tunnel. It does no long-term good to carefully document the heat transfer for a flow situation that is not representative of its nominal description. In recognition of this fact, each tunnel used in the heat transfer program was put through a careful qualification program, beginning with the mainstream flow characteristics and ending with overall energy balances. This resulted in long development periods (typically 18 months to 2 years) before each new tunnel was put "on-line" for data production, and led to research projects that followed an evolutionary course, to take full advantage of the existing capabilities before starting a new "rig building" program.

In all, seven different research tunnels have been used in the course of the research reported here, covering the 25-year span of this program from Reynolds *et al.* [34–37] to Furuhama and Moffat [12].

The first two tunnels have already been discussed, though briefly, in the opening remarks. The present section begins with a description of HMT-1, used in studies of transpiration, acceleration, and deceleration in air, and ends with HMT-5, used for concave curvature studies in water.

A. The Smooth Plate Rig: HMT-1

Construction of HMT-1 was begun in 1964, and the tunnel was commissioned 27 months later, qualified both as to hydrodynamics and heat transfer. It has been used in the studies of transpiration, acceleration, deceleration, and free-stream turbulence effects; as of this writing (1983), it is now being used for laser anemometer studies of the fluid mechanics of backward-facing steps and heat transfer measurements in the reattaching region.

The working surface is 18 in. wide and 96 in. long and consists of 24 porous plates, each 4.0 in. in the streamwise direction. The plates, made of sintered bronze with the individual particles ranging in size from 0.002 to 0.005 in., were formed in polished steel molds. The surface feels smooth to the touch, even over the 0.020 in. wide insulating joints between plates, and displays an RMS roughness of 250 μin., using a standard Surfagage test probe. Experiments have shown that the working surface is aerodynamically smooth up to 38.5 m/sec, the tunnel limit. The plates are individually supported on linen-reinforced phenolic strips that reduce heat loss and form individual cavities under each test plate for the measurement of the transpiration air condition, when that feature is used.

A schematic of the test-plate installation is shown in Fig. 6. Each test plate is a separate heat transfer experimental apparatus, instrumented with thermocouples for temperature measurement and provided with metered transpiration air and electrical power. Corrections are made for heat losses from each individual plate to its support structure, based on the measured temperature difference between the plate and its support structure and using calibration constants for conduction heat loss determined from special energy loss tests. During data taking, the support structure is heated to the average plate temperature using a hot-water system controlled by the operator. The calibration constants for conduction heat loss are determined with the structure unheated, and applied with the structure heated, thus improving the absolute accuracy of the final data. Under most conditions, the correction amounts to 3% or less. By similar means, corrections are made for radiation (from the top and bottom faces of the test plates) and axial conduction (i.e., conduction to adjacent plates in the streamwise direction). The total of all corrections made, including those for the effects of wattmeter insertion losses as well as heat losses by all modes, amounts to about 6% for typical research conditions. The uncertainty in the reported heat transfer coefficients is believed to be ± 0.0001 Stanton unit, which includes all the stochastic components of experimental uncertainty plus the uncertainty in calibration of the instruments used. Overall energy balances (i.e., comparing the measured boundary layer energy flows with the surface heat transfer rates integrated up to the profile measuring station) closed within 3%. Base-line tests were run during the initial qualification runs to demonstrate that the data from HMT-1 agreed with the earlier data of Reynolds *et al.* [34–37], taken in the Guggenheim wind tunnel. Agreement was demonstrated within 1.3%.

Skin friction coefficient is more difficult to measure than heat transfer, and the values tend to be less certain. In addition, the skin friction data tend to be sparse (e.g., the data presented in this report provide only four or five values of skin friction per test run, whereas 24 values of Stanton number were obtained), and therefore admit of larger differences of opinion in interpretation. Thus, although the stochastic component of uncertainty in each measurement was $\pm 5\%$, there may be differences in interpretation of the same data by different investigators that are as large as 10–15%.

During the course of the many projects conducted on HMT-1, a pattern of recertification was followed in which each new research team would repeat the energy balance and flow field qualifications upon first taking custody of the rig. This provided a double benefit: the new certification "closed out" the preceding project, guarding against the possibility that

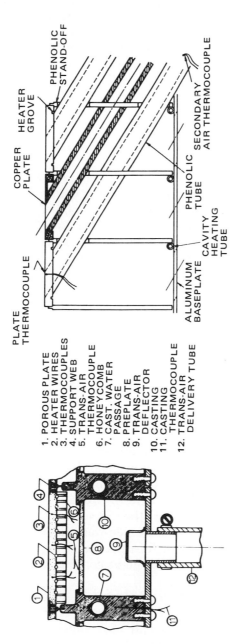

PLATE THERMOCOUPLE

COPPER PLATE

HEATER GROVE

PHENOLIC STAND-OFF

PHENOLIC TUBE

SECONDARY AIR THERMOCOUPLE

CAVITY HEATING TUBE

ALUMINUM BASEPLATE

1. POROUS PLATE
2. HEATER WIRES
3. THERMOCOUPLES
4. SUPPORT WEB
5. TRANS-AIR THERMOCOUPLE
6. HONEYCOMB
7. CAST. WATER PASSAGE
8. PREPLATE
9. TRANS-AIR DEFLECTOR
10. CASTING
11. CASTING THERMOCOUPLE
12. TRANS-AIR DELIVERY TUBE

FIG. 6. Schematic of the test plate installation of HMT-1 and the injection tube arrangement of HMT-3.

the rig had changed its characteristics during the preceding research, and also ensured that the new team understood the apparatus well enough to assure high-quality data. Each time a significant change was made in the rig structure or instrumentation, the energy balance and flow quality studies were rerun. Over the years, a considerable body of knowledge was assembled about the characteristics of HMT-1, all of which contribute to the assurability of the data it produces.

The test section is 15.24 cm high and 50.8 cm wide, 5.08 cm wider than the porous plates, the margins being smooth and flush with the porous surface, but unheated. This avoids some of the corner-flow effects on heat transfer.

The mainstream flow entering the test section was surveyed to document the distribution of mean velocity, the uniformity of the boundary layer at entrance to the test section, the location of the "virtual origin," and the turbulence properties of the entering flow. The mean velocity is uniform within 0.40% across the entrance plane, outside the boundary layers, and the temperature is uniform within 0.5°F. The momentum thickness of the boundary layer is uniform within ±2% across the center 6 in. portion of the test section. Momentum thickness distributions have been measured at the midlength and near the discharge end of the test section with similar results.

Electrical power for the tunnel is stabilized by a 10 kVA voltage-stabilizing transformer, to reduce the effects of line voltage fluctuations on the accuracy of the data.

Figure 7 shows a schematic of HMT-1. Air enters through a 5 μm filter box, passes through a centrifugal blower (2000 cfm, 15 hp), and enters a five-pass cross-counterflow heat exchanger through a tapered header and honeycomb flow straightener. The heat exchanger water is recirculated through a 10,000 gal reservoir under the building, which provides for stable operating temperature over long periods of time. On the exit side of the heat exchanger there are six layers of screen material, spaced apart and of different mesh sizes so as to most effectively reduce the turbulence scale and intensity. The air enters the test section through a 4 : 1 contraction, two-dimensional nozzle with side-wall suction slots. Discharge is directly into the room, with an unused exhaust length of about 30 cm to minimize the effects of the discharge unsteadiness on the measured data from the last test plate.

Each of the 24 test plates has a separately metered and instrumented transpiration flow delivery system, also shown on Fig. 7. Transpiration air enters the secondary blower through a 5 μm filter, passes through a cross-flow heat exchanger and a mixing box, and then enters a distribution header to which are attached the 24 individually controlled and flowme-

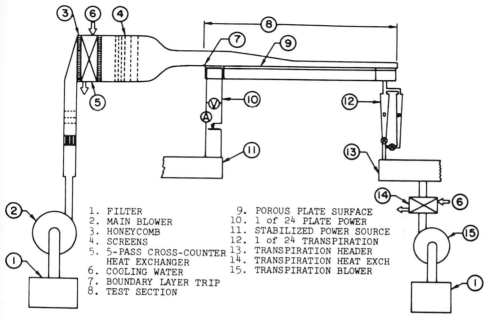

1. FILTER
2. MAIN BLOWER
3. HONEYCOMB
4. SCREENS
5. 5-PASS CROSS-COUNTER HEAT EXCHANGER
6. COOLING WATER
7. BOUNDARY LAYER TRIP
8. TEST SECTION
9. POROUS PLATE SURFACE
10. 1 of 24 PLATE POWER
11. STABILIZED POWER SOURCE
12. 1 of 24 TRANSPIRATION
13. TRANSPIRATION HEADER
14. TRANSPIRATION HEAT EXCH
15. TRANSPIRATION BLOWER

FIG. 7. Schematic of HMT-1.

tered delivery lines. The system is arranged so that either suction or blowing can be achieved.

The basic configuration of HMT-1 uses a one-piece top wall that can be adjusted to maintain a uniform static pressure in the streamwise direction, even with strong blowing or suction. The adjustment is iterative and is considered satisfactory when the static pressure is uniform within 0.002 in. of water from end to end. For studies of acceleration effects at constant K, a hinged top wall was built that allowed a constant K region to be started at any desired location along the test section and held until the maximum allowable velocity was reached. Deceleration is achieved by using a third top wall having 24 adjustable slots and running the tunnel at a slight positive pressure. Bleed flow through the slots causes deceleration of the mainstream. A potential flow-solving computer program is used in conjunction with the slotted top and predicts the in-slot opening distribution. This reduces the number of iterations required to achieve a desired deceleration profile.

In terms of general operating procedures to help achieve good data, the tunnel is always operated at a slight positive pressure (0.01 in. of water), so that any leakage will be out of the test section, rather than in. This prevents interference with the boundary layer structure by leakage fluid, which might otherwise appear as tiny jets at any leakage points. In addition, the mainstream air temperature is controlled to be equal to the room temperature as closely as possible, to avoid heat transfer within the approach nozzle.

B. The Rough Plate Rig: HMT-2

In 1970, work began on a study of the effects of roughness combined with transpiration. The objectives of the program included measurement of the turbulence properties of the transpired boundary layer over a rough surface. This required a surface with a uniform permeability that could also be used as a heat transfer surface without introducing ambiguity in surface temperature caused by low-conductivity roughness elements.

With some trepidation, it was decided to manufacture a porous surface by furnace-brazing copper spheres together in a mold cavity similar to that used with success in the earlier work with sintered bronze for HMT-1. Eight and one-half million copper spheres were plated with 0.0005 in. of nickel and brazed into 4.0 × 18.0 in. plates, 0.50 in. thick, using an atmosphere furnace. The resulting surface is shown in Fig. 8.

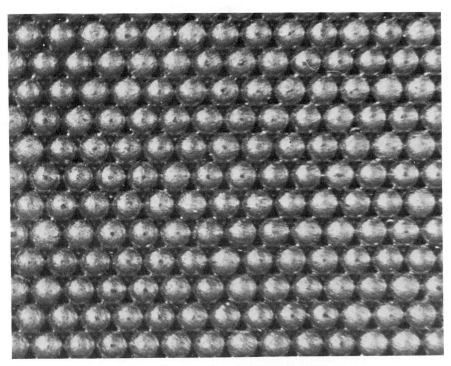

FIG. 8. Close-up of the deterministic rough surface.

Within the accuracy of modern hot-wire anemometry, the permeability of the surface was entirely uniform. The surfaces of the spheres lie all on the same plane, forming a deterministic model of sand-grain roughness that can be tested with either blowing or suction. The equivalent sand-grain roughness is 0.78 mm, using an accepted correlation for densely packed spheres (Schlichting, [39]). The bulk material thermal conductivity was about 80 Btu/(hr ft² °F), a value controlled by the plating thickness and the firing temperature (together these control the thickness of the "pipes" that join each sphere to its neighbors).

HMT-2 is in many respects similar to HMT-1, differing mainly in that it is a closed-loop tunnel instead of being open return. This decision was made in recognition of the growing importance of hot-wire anemometry in studying the fluid mechanics of boundary layers. The schematic is shown in Fig. 9, which also shows the slanted injection tubes used on its sister tunnel, HMT-3. There are no such tubes on HMT-2. The porous, rough plates are mounted in HMT-2 just as were the smooth plates in HMT-1, and are also electrically heated by wires embedded in grooves on the underside of the working plates. One difference in form, but not in function, is that this apparatus is entirely powered by dc rather than ac, which is used on HMT-1 and HMT-3. The intention here is to pave the way for

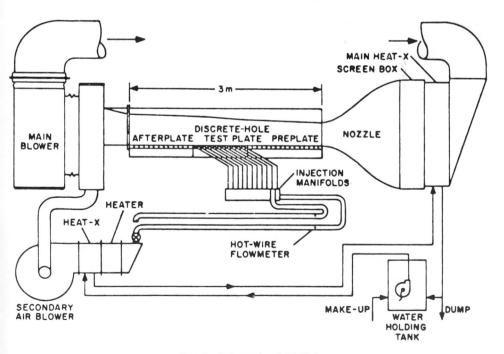

FIG. 9. Schematic of HMT-3.

computerized control, since dc voltages are easier to read with rapid scanning voltmeters.

One other feature worth mentioning is the temperature control system for the mainstream flow and the transpiration flow. Since there is no large water reservoir in the new laboratory, a recirculating system was designed that circulates 50 gpm through the main heat exchanger and uses a 20 gal mixing tank to blend in a controlled flow of make-up water so as to maintain a steady temperature. The 60 gpm rate assures a very low value of C_{min}/C_{max}; hence the heat exchanger surface remains at uniform temperature from top to bottom and from side to side, while the actual consumption of water (the make-up flow) is held to a minimum.

The heat exchanger, screen array, and nozzle were developed based on experience with HMT-1. The nozzle was made as a two-part Fiberglas-reinforced resin lay-up, using a polyurethane male mold contoured by hand to the desired coordinates.

A supercharging blower was added to the system so the tunnel could be maintained slightly above ambient pressure, for leakage control.

The temperature of the transpiration air is controlled independent of the mainstream, using a separate heat exchanger and electrical heater.

This apparatus was subject to the same qualification tests described already for HMT-1, and developed until the same level of performance was reached. The mainstream is uniform within ±0.25°F and ±0.15% in velocity. The momentum thickness of the boundary layer is uniform within ±3% at the end of the test section, and the free-stream turbulence intensity is less than 0.5%.

The tunnel was commissioned in 1972, after 2 years of development, and was used in the series reported here concerning roughness effects with and without transpiration, acceleration on a rough wall, and the effects of artificial thickening on boundary layer behavior. As of this writing, HMT-2 is involved in studies of the effects of high free-stream turbulence on boundary layer heat transfer.

C. The Full-Coverage Film-Cooling Rig: HMT-3

During the same time that the roughness rig was under construction, another apparatus of the same general form was being built to study the case of discrete hole injection with full-coverage film cooling. Figure 9 serves as the schematic for both. The details of installation of the injection tubes are shown in Fig. 6, which also shows the typical plate support structure of HMT-2. Three different injection tube geometries were tested in HMT-3, each requiring a separate test section with a new buildup of tubes. Figure 6 illustrates the 30° slant-angle injection. Also tested, but not shown, were perpendicular injection and compound-angle injection (30° downstream and 45° cross-stream).

Staggered arrays of holes were used for all geometries, with a basic pitch-to-diameter ratio of 5 : 1 between holes and rows. Eleven rows were used, alternately having eight or nine holes in each row. A preplate preceded the working section, providing a smooth approach section 1.25 m in length. On this section, the thermal and hydrodynamic boundary layer could be developed to the desired thickness before reaching the injection station. Initial profiles of velocity and temperature were recorded using hot-wire anemometers and boundary layer thermocouple probes. For a portion of the test every other hole, in both directions, was plugged and finished over to provide a pitch-to-diameter ratio of 10 : 1.

With discrete hole injection, the injected air is not necessarily at wall temperature, and, in fact, its temperature is an important variable in the problem and must be very accurately measured. This measurement poses some interesting problems, since the injected air will pick up heat from the tube walls, conducted down from the heated copper plate. This is accounted for by means of a special calibration, in which the injected air is trapped, mixed, and then measured to establish the correlation between the actual mixed mean injectant temperature and the measured injectant temperature. These calibrations have been conducted over a range of injectant flow rates and stored within the data reduction program.

There are either eight or nine holes in each of the eleven rows. To ensure uniform velocity in each injected jet, a hot-wire probe was positioned over each hole and the flow through that hole trimmed by adjusting a ball valve in the individual injection line until the reference value was attained. The trim valves are used only for that purpose, and their settings are adjusted only during calibration runs.

The injection tubes were kept as long as possible, considering the constraints of the geometry, to aid in obtaining fully developed flow within them. The average tube length is about 30 diameters. The copper plates were counterbored so that only copper is visible from the working face, and the diameter of the hole in the copper reamed to match the inner diameter of the delivery tube.

When compound-angle injection was being studied, the injected flow was cast to the right, looking downstream, which would have induced a significant large-scale vortex within the tunnel had precautions not been taken to prevent this. A yoke was added to the tunnel, which provided suction on the right sidewall of the tunnel and injection on the left (of the same amount). The amount of suction was adjusted experimentally until the near-wall streamlines of the boundary layer entered from the left sidewall at the same angle as they exited on the right. Under these conditions, the plate was assumed to behave as though part of an infinitely wide array. The suction required was a function of the injection rate and was adjusted appropriately for each new test condition.

It was in the study of the recovery region, downstream of full-coverage film cooling, that extensive use was first made of a triple-hot-wire ane- mometry system. Considering the vast amount of turbulence data needed to develop usable models of the hydrodynamics of full-coverage film cool- ing, a subproject was established that developed a high-speed analog processing system for the signals from a triple-hot-wire array. The system developed was capable of measuring the Reynolds stress terms within ±5% over a ±20° range of incidence angles of the flow. This system greatly speeded up the data acquisition process and made possible a de- tailed mapping of the recovery region hydrodynamics that could not have been done with conventional techniques. The system was subsequently expanded to a four-wire system by a later program and is now in use on the continuation of the full-coverage film cooling work (full coverage on a convex wall).

D. The Curvature Rig: HMT-4

Upon completion of the program on full-coverage film cooling on a flat wall, the HMT-3 apparatus was taken down and reformed into HMT-4, incorporating a 90° bend in the test section, as shown in Fig. 10. A curved working wall was assembled out of copper segments, each equipped with an electric heater and instrumented with thermocouples, following the typical construction used on previous equipment. These plates were then set into a 45 cm radius of curvature, using phenolic standoffs, as in HMT- 3. The plate assembly was installed in aluminum side plates equipped with guard heaters and instrumentation, as in the earlier equipment. Once the buildup was complete, the entire assembly was lathe-turned to a 45 cm radius. This ensured a smooth, constant-radius turn. Looking ahead to the next portion of the program, the curved section was equipped with injection tubes in the same pattern as for HMT-3 ($P/D = 5.0$, 30° slant angle), but the injection holes were plugged and finished over for the first series of experiments—heat transfer on a smooth convex wall.

For the curvature rig, the preplate was lengthened to 2 m so the ap- proaching boundary layer would be approximately 4.5 cm thick at entry into the curve. The preplate was equipped with two rows of holes, which could be used for suction or injection to control the thickness of the approaching boundary layer. A 2 m recovery plate was also installed, with the first half being equipped with injection holes in the same pattern as those in the curve: $P/D = 5:1$, 30° slant-angled injection, staggered.

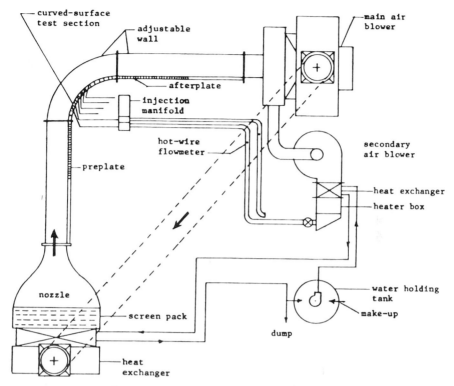

FIG. 10. Schematic of the curvature rig: HMT-4.

The tunnel was provided with a specially shaped wall opposite the working wall whose contour provided for a uniform static pressure on the working wall at the beginning and end of the curve. Without this specially shaped wall, there would have been unavoidable pressure disturbances at the inlet and exit of the curve, as the flow adjusted to the curvature. With the special control wall, however, the variation of static pressure was kept down to about 3% of the velocity head and limited in extent to a region only a few boundary layer thicknesses long at the inlet and the outlet of the curve. The design of this control wall is described by Gillis *et al.* [13] and Simon *et al.* [44], as are the details of the control-wall suction slots and the sidewall slots used to reduce secondary flow in the curved region. Boundary layer fences were installed in the curve and in the recovery region, to prevent sidewall boundary layer migration, which would have caused convergence of the flow on the test wall. With the combination of the control wall, sidewall suction, control-wall suction, and boundary layer fences, the secondary flows were kept under control.

The largest convergence angle measured in the recovery region (where the problem was most severe) was 3°, and this was only near the edges of the test plate, where it is believed to have had little or no effect on the measured heat transfer. Within ±6.5 cm of the centerline, the maximum convergence angle was 1.5°.

Projects on HMT-4 were among the first to employ large-scale computer-controlled traversing and data acquisition. Boundary layer probes were positioned both laterally and normal to the wall by stepper-motor-controlled actuators, with the wall position being checked manually before each traverse.

HMT-4 is currently being used in a detailed study of the streamwise and spanwise evolution of the boundary layer on a convex wall with full-coverage injection.

E. THE CONCAVE WALL RIG

The concave wall study was done in a specially constructed open-surface water channel, shown schematically in Fig. 11. Water enters the tunnel through a distribution header, calming tank, and screen section and

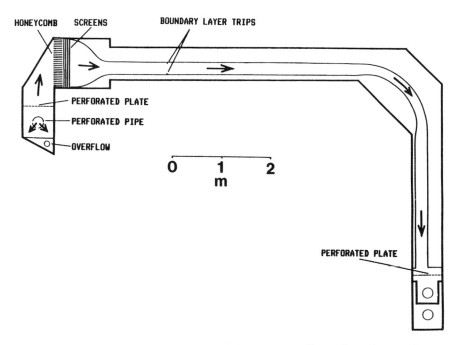

FIG. 11. Schematic diagram of the concave wall test channel.

is accelerated into the main channel through a 4:1 two-dimensional nozzle. The test section consists of a passage defined by two, parallel Plexiglas walls standing inside an outer container, also of Plexiglas but reinforced with steel braces to bear the load of the water. There is no load on the test section walls since there is water, to the same depth, both inside and outside of them. After a straight run of 4.9 m, the flow enters a 90° bend to the right (looking downstream). The working wall is the outer wall of the turn—the concave wall. The inner wall is a control surface that is shaped in much the same way as was the control wall on the convex rig, to provide a uniform static pressure on the working wall throughout the curve. The curved section is followed by a straight recovery section of 2.4 m, after which the flow dumps into a receiver and falls into the reservoir below. The reservoir is kept deaerated, free of algae and particles (using a recirculating flow filter system), and is temperature controlled. The temperature control is necessary to ensure accurate measurement of the heat transfer coefficient using the liquid-crystal-wall technique.

Water flow-rate control allows the tunnel velocity to be set at any value up to 18 cm/sec. Mean velocity can be measured with good accuracy by timing the travel of a dye spot injected into the core.

Velocity profiles made with hot-film probes and flow visualization studies made with dye injection provided evidence that the flow was spanwise uniform and was without significant vorticity, upstream of the curved test section.

The top edges of the external sidewalls of the tunnel support a pair of aluminum rails on which rides a motorized sled carrying probe-traversing gear and a remote-recording TV camera and strobe light system. The position and speed of the sled are controllable from the lab computer used with this experiment. At the present writing, a two-color laser system has been installed on the sled for hydrodynamic studies of the details of the flow near the concave wall.

The liquid-crystal system developed for the concave wall studies gives the sharpest resolution with monochromatic light. A battery of sodium lamps, filtered to remove the 404.7 nm line, was used for many of the photographs taken, although much was learned using white light and ordinary color film.

Two working walls were built, one for fluid mechanic visualization and one for heat transfer visualization. The fluid mechanic wall was equipped with carefully machined dye slots at streamwise locations, through which could be injected neutrally buoyant dye of different colors. Hydrogen-bubble wires were used to mark the flow away from the wall. Images of both the dye and bubble wire traces were stored on high-speed TV tapes and in color movies. The heat transfer wall contained 14 panels of liquid-crystal material, each 14 × 17 in. The liquid-crystal material was applied on the back surface of a Mylar sheet, 0.004 in. thick, made electrically conductive ($R = 3 \Omega$ per square) by a transparent plating of gold only a few angstroms thick. The whole assembly was encapsulated in a waterproof unit. When power is applied to such a panel, the local heat release is known from the power and area. The liquid-crystal paint changes color from black to yellow/green at 90°F ± 0.5°F, and if the free-stream temperature is known, the heat transfer coefficient is known along the yellow-green isochrome.

Work is continuing on this technique, and it has been successfully applied in air in other studies.

F. THE MIXED CONVECTION TUNNEL

An open-return, draw-through wind tunnel was built for these studies, shown schematically in Fig. 12. The entire tunnel is about 15 m long and 5 m high. The inlet aperture is 4.9 m wide and 4.3 m high, and equipped with five screens of 58% open area spaced 30 cm apart. A two-dimensional nozzle of 3 : 1 contraction feeds the test section, 1.2 × 4.3 m in cross section and 4.3 m in length. The working surface is 3 × 3 m and occupies one vertical wall. The opposite wall is black iron, water cooled from the outside by a spray-bar.

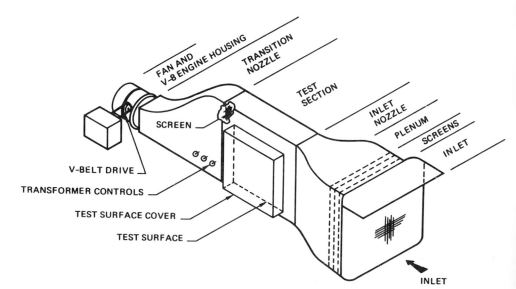

FIG. 12. The mixed convection wind tunnel.

The test plate is shown in Fig. 13, along with the traversing equipment used for making the boundary layer studies. The test surface consists of 21 strips of stainless steel, 0.0127 cm thick and 13.32 cm wide stretched tightly over a slightly crowned, well-insulated backing structure. The plate surface is heated to 600°C using ac power (up to 180 A).

Boundary layer suction is used to establish a stagnation line at the leading edge, and a small amount of suction is provided along the top edge

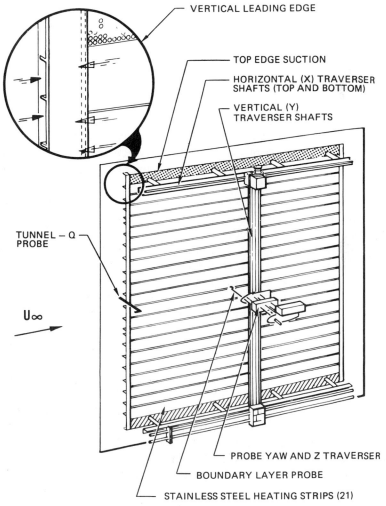

FIG. 13. Schematic diagram of the front of the test surface, including the traverse and the leading edge suction.

of the plate, to reduce temperature stratification in the tunnel at low velocities.

Extensive developmental testing was conducted to assure the accuracy of the surface temperature measurement and the corrections for radiation and conduction loss. These are discussed by Siebers *et al.* [43]. Base-line tests were conducted for free convection and forced convection heat transfer with small temperature differences. The results agreed within ±2% with the accepted correlations in both domains. Energy balance tests further confirmed the accuracy.

Boundary layer velocity and temperature traverses were taken with remotely controlled probe-positioning equipment using sensitive pressure transducers and boundary layer thermocouple probes.

G. DATA REDUCTION AND UNCERTAINTY ANALYSIS

Beginning with the first program on HMT-1, computerized data reduction has been the rule on all experimental work in this series. This has permitted the evolution of extended data reduction programs capable of acknowledging all of the corrections and refinements that could be identified.

Much of the scatter in reported heat transfer data arises because of uncontrolled and unobserved variations in parameters thought to have negligible impact on the result. The fact is that if one seeks data with less than 2% scatter, there are no "negligible" effects; if a mechanism capable of causing an error can be identified, it should either be corrected in the apparatus or acknowledged in the extended data reduction program. For example, consider the effect of conduction heat loss from a heat transfer plate to its support structure. If, from one day to the next, the support structure temperature varies (as it well might, being exposed to the room temperature), the conduction loss would be different on different days, giving rise to an unexplained "scatter" in the heat transfer data from that plate. This scatter can be removed either by modifying the apparatus or by incorporating a correction for heat loss into the program, using measurements of the support structure temperature and the plate temperature to calculate the heat loss as a function of the test conditions. Such an effort considerably complicates the data reduction program but greatly reduces the scatter in the results. In calibrating HMT-1, for example, each of the 24 segments was tested for conduction heat loss, and 24 different conduction-loss coefficients were stored. These were then used with the appropriate values of plate-to-base temperature difference to correct the data from each individual plate for its own heat loss. The result of such detail in the data reduction programs has been a significant reduction in the scatter in reported data.

The continuity of this program has had a significant effect. Factors such as the influence of ambient humidity on reported Stanton number, frequently neglected, have been built into these extended data reduction programs and handed down from project to project, thus accumulating the experience and insight of each succeeding researcher into each new data reduction program.

Another factor that has been important in guiding the experiments described here is uncertainty analysis, used as a planning and diagnostic tool. Uncertainty analysis has been built into the experiment planning and data reduction throughout the program. Following the early work of Kline and McClintock [19], the general principles of single-sample uncertainty analysis have been extended by several conceptual advances and used systematically to reduce the uncertainties in these experiments. The present data reduction programs calculate the uncertainty in each reported variable and display the uncertainty along with the value. This is done using the method of sequential perturbation [26], a method that uses the main data reduction program to calculate the uncertainty interval associated with each calculated result. The uncertainty interval associated with each input data bit is entered with the data and used by the program to calculate the uncertainty in the results. When the test program encounters a combination of test conditions for which the uncertainty in the calculated result is outside the allowable range, this fact is immediately apparent.

Uncertainty predictions are particularly important in the developmental phase, when the apparatus is being qualified for data taking. Without an estimate of the expected uncertainty, it is difficult (if not impossible) to assess the significance of a small change in rig behavior. Is the difference a random excursion or is it significant? This question cannot be answered unless the experimenter has an accurate estimate of the expected uncertainty interval for that result, at those test conditions. In the experiments reported here, the energy balances and momentum balances were expected to close within the first-order uncertainty (defined as an estimate of the expected scatter on repeated trials with the same apparatus and instrumentation). Failure to close within that interval was taken as evidence that there remained some unrecognized defect in the apparatus, or in the data reduction program, that had to be identified and corrected before the rig would be considered qualified.

Work continues on refining the techniques of experiment planning and uncertainty analysis, aimed at improving the future experiments in this program.

IV. Factors Affecting the Boundary Layer

A. Transpiration

The transpired boundary layer began to attract attention in the early 1950s, and work in that area developed rapidly. Before discussing the Stanford program, some general comments are in order, to set the stage.

In a typical transpiration cooling situation, the surface is constructed of porous material and cooling fluid is forced through the surface. This is a boundary layer problem for which the normal component of velocity at the solid–fluid interface is nonzero, but otherwise the same momentum and energy differential equations must be solved as for the nontranspired boundary layer. A variation on this problem occurs when the cooling fluid is a chemically different specie than the free-stream fluid. In this case, the mass-diffusion equation of the boundary layer must be solved, in addition to the momentum and energy equations. There are obvious similarities between these two types of problems, but also fundamental differences. Both are "mass transfer" problems in the sense that mass is transferred across the fluid–solid interface; but the latter is also a mass-diffusion problem, while the former is not. Evaporation or sublimation from an interface into a boundary layer or condensation onto the interface also constitute transpiration situations.

In any of the cases cited, the flow normal to the surface at the interface could be into the surface or out of it. The terms *blowing* and *suction* will be used to denote the direction of the flow at the interface, while *transpiration* will embrace both cases.

In the problem first considered at Stanford, the surface was assumed to be aerodynamically smooth (the "rough" surface with transpiration was studied later), with the holes in the surface small and close together relative to the boundary layer thickness. The velocity normal to the surface was treated as uniform, everywhere. The situation with large holes, at large spacing, results in a boundary layer that, though it still has some of the characteristics of a transpired layer, is different enough to warrant a separate heading: Section IV,E.

The transpired boundary layer may be laminar or turbulent, but the laminar boundary layer with transpiration has been extensively studied, and there are a large number of exact mathematical solutions for this situation. Experimental investigation of the laminar boundary layer did not appear very fruitful, except for very special cases. The turbulent boundary layer, on the other hand, was not nearly so well understood (even without transpiration), and at the time this program was started there were remarkably few experimental data available for a transpired turbulent boundary layer. This led to what became a very comprehensive program that extended over many years.

It is one thing to talk about transpiration, but quite another to build a uniformly porous surface that acts hydrodynamically as though it were smooth. A large amount of effort went into the development of the first porous surface, not only into its fabrication but also into the "proof testing" to show that the surface was uniform ($\pm 6\%$ in the center 6 in. span) and that it did behave like a smooth surface. Once the surface was established, the program went rapidly.

The first cases were flat-plate uniform-wall-temperature studies of the hydrodynamic and heat transfer effects of transpiration with constant free-stream velocity and uniform blowing. Equilibrium suction layers were studied, with blowing increased to the point of boundary layer blow-off. In essence, the data covered all the "boundary layer" regime with transpiration. For blowing higher than $F = 0.010$ ($F = v_0/U_\infty$), the boundary layer is blown free of the surface, while for suction greater than $F = -0.004$, the boundary quickly reverts to a laminar, asymptotic suction layer.

With the groundwork thus laid, the range of boundary conditions was broadened to include acceleration, deceleration, roughness, turbulence, steps in wall temperature, steps in blowing, and other perturbations that were felt likely to help elucidate the basic transport mechanisms in the boundary layer. The programs were laid out to study equilibrium conditions first, then to look at the transition between two equilibrium states, and lastly to examine a few cases of arbitrary variations of conditions. In all the work, the experimental results were used to deduce constants for the modified mixing-length model used in a finite-difference prediction program (presently STAN5, then STAN1).

The overall program from which the present data were taken covered accelerating and decelerating flows as well as variations in wall temperature and blowing. This variety of boundary conditions precluded the use of x-Reynolds number as a useful correlating parameter and led early in the program to the use of the local boundary layer thickness Reynolds number (either thickness).

It was found early in the program that the turbulent boundary layer came to "equilibrium" so rapidly that uniform blowing (constant F flows) produced virtually the same value of Stanton number as did constant B_h flows for a given enthalpy-thickness Reynolds number. This established the validity of "local equilibrium," and the rest of the experiments were done with the (experimentally) simpler boundary condition of constant F rather than the true equilibrium condition of constant B_h and B_m.

Accelerating flows were set up with a constant value of K [defined as

$(\nu/U_\infty^2)(dU_\infty/dx)]$. Such a flow is of fundamental value, as shown earlier, but it also has the merit of being easy to establish experimentally, since it naturally occurs in a channel between two converging planes. It was hypothesized at first, and then confirmed experimentally, that the asymptotic-accelerating boundary layer would have a constant value of $c_f/2$. This means that constant F also yields constant B_h, and the constant F accelerating flows were also constant B_h equilibrium boundary layers.

The decelerating flows were set up with nearly constant pressure gradients by fixing $U_\infty = Ux^m$ with m negative and x measured from the virtual origin. Blowing was set by fixing constant F, relying on the experience cited above.

Constant wall temperature was used throughout the baseline tests reported here, with an average temperature difference of about 25°F between the free stream and the wall. No correction was made for the effects of variable fluid properties, which were all evaluated at free-stream temperature.

1. *Correlation of Results*

The principal correlations deduced from the present data set are the differential ones: the correlations used to model the momentum mixing length and the turbulent Prandtl number in the finite-difference computing method to be described later. However, there is use for integral correlations such as Eqs. (1) and (2). The range of validity of such integral correlations is necessarily less than that of the differential correlations, since the latter have the differential equation to help cope with changing boundary conditions. Within their limited range and accuracy, however, integral correlations are extremely useful, particularly when confined to simple functional forms.

Therefore, the results of the present data set are presented in both ways: integral and differential correlations. Discussion of the differential correlations is collected in the later section on the differential models; the integral correlations are discussed with the presentation of the data, since they help to clarify the organization of the data.

2. *Heat Transfer*

The principal effects of transpiration on heat transfer through a constant-velocity boundary layer are shown in Figs. 14–17. Figure 14 shows Stanton number as a function of x-Reynolds number for uniform blowing and suction. The blowing fraction, $F = \dot{m}''/G_\infty$ (or v_0/U_∞ for constant density), is the ratio of the mass velocity through the surface ($\dot{m}''$) to the free-stream mass velocity (G_∞). It is apparent that blowing ($\dot{m}''/G_\infty > 0$) and suction ($\dot{m}''/G_\infty < 0$) both have large effects: blowing tends to drive

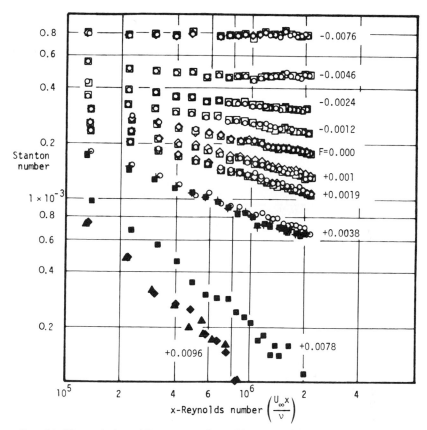

FIG. 14. The variation of Stanton number with x-Reynolds number at constant F.

the Stanton number toward zero as the blowing fraction increases whereas suction tends to force the Stanton number to an asymptotic value numerically equal to $(-F)$. This behavior is illustrated in Fig. 12, which shows St as a function of $\dot{m}''/G$ parametric in x-Reynolds number. The region of boundary layer behavior is thus seen to lie between $\dot{m}''/G$ of roughly $F = -0.006$ to $+0.01$ for this range of Reynolds numbers.

It is worth noting that all the features shown in Figs. 14 through 17 can be recovered, with good accuracy, from a finite-difference program using a damped mixing-length closure: the damping needs only to be made a function of the transpiration rate—no other change need be made. The evidence is that the main effects of blowing are confined to the inner portion of the momentum boundary layer: blowing changes the shear stress distribution and reduces the effect of the sublayer. The diminished influence of the sublayer can be simulated by making the damping factor, A^+, smaller as the blowing fraction increases, producing velocity profiles

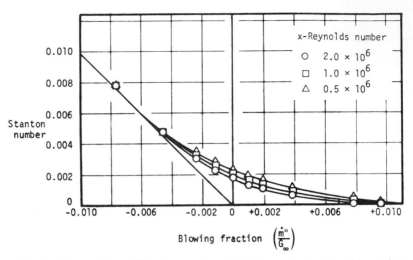

FIG. 15. The variation of Stanton number with F at constant x-Reynolds number.

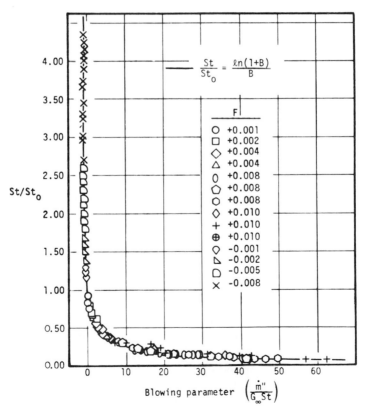

FIG. 16. The ratio St/St_0 versus the heat transfer blowing parameter B_h.

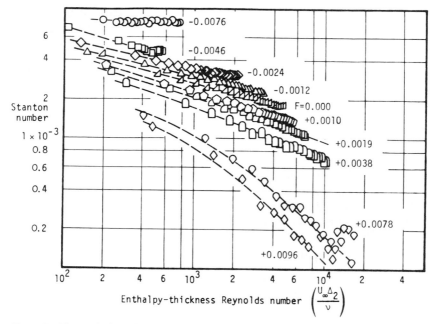

FIG. 17. The variation of Stanton number with enthalpy thickness Reynolds number at constant F.

that closely resemble those shown in Fig. 20. Experiments show that the turbulent Prandtl number is not much affected by blowing, and the A^+ variation alone results in satisfactory prediction of the principal features of the heat transfer.

The ratio of Stanton number with blowing, St, to Stanton number without blowing, St_0, is a unique function of the blowing parameter B_h. The comparison can be made at the same x-location (same Re_x) as given by Eqs. (20) and (21):

$$\left.\frac{St}{St_0}\right|_{Re_x} = \frac{\ln(1 + B_h)}{B_h} \tag{20}$$

or, equivalently,

$$\left.\frac{St}{St_0}\right|_{Re_x} = \frac{b_h}{e^{b_h} - 1} \tag{21}$$

where

$$B_h = \dot{m}''/G_\infty St \tag{22}$$

and

$$b_h = \dot{m}''/G_\infty St_0 \tag{23}$$

Equations (20) and (21) were presented during the early 1950s by several workers in the field as "stagnant film theory" or "Couette flow" models. The agreement between the data and Eq. (20) is shown in Fig. 13. In part, the good agreement results from the implicit nature of Eq. (20): diminishing St reduces both St/St_0 and $\ln(1 + B_h)/B_h$, hence preserving "good agreement."

The data of Fig. 14 have been recast to show Stanton number versus enthalpy-thickness Reynolds number and plotted as Fig. 17. Values of enthalpy thickness for this plot were calculated from the measured Stanton number data and the blowing fraction by integrating the two-dimensional energy integral equation, but values derived by this method agree well (5–6%) with values deduced by traversing the boundary layers. An empirical form could be deduced from this figure but a better guide is at hand. Whitten (1967) showed that Eq. (20) could be combined with the two-dimensional energy integral equation and with an equation describing the variation of St_0 with Re_x to yield the following form:

$$\left.\frac{St}{St_0}\right|_{Re_h} = \left[\frac{\ln(1 + B_h)}{B_h}\right]^{1.25}(1 + B_h)^{0.25} \tag{24}$$

Equation (24) was developed from Eq. (20); hence all the data for $U_\infty =$ constant fit Eq. (24). Implicit in Eq. (24) is the notion of local equilibrium; it is presumed that knowledge of Re_h and B_h will fix St/St_0, regardless of the upstream history. The validity of this hypothesis was investigated by experiments in the vicinity of a step change in blowing, as shown in Fig. 18. The boundary layer is seen to respond very rapidly to the step, with Stanton number dropping almost all the way from the unblown value to the uniformly blown value within one plate width (4 in. in the flow direction). The boundary layer thickness at the point of the step (99% velocity thickness) was approximately 1 in.; hence the local equilibrium was reestablished in about four boundary layer thicknesses. Figure 18 shows a strong "local equilibrium" tendency, favoring the use of Eq. (24) for cases of variable blowing as well as cases of uniform blowing, opening the door for substantially local predictions of the boundary layer behavior. An example of a more complex case is shown in Fig. 19, in which a linearly decreasing blowing, $F = 5 \times 10^{-5}(X)$, was combined with a sharply variable wall temperature.

3. Velocity and Temperature Profiles

Velocity and temperature profiles in inner coordinates are shown in Figs. 20 and 21 for the case of constant velocity (about 40 fps) and constant wall temperature (ΔT about 25°F) with injection of air into air. The

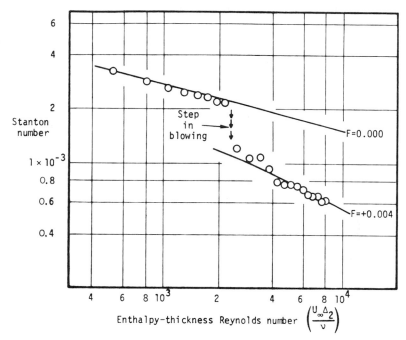

FIG. 18. The response of Stanton number to a step increase in blowing.

general features of the region from $y^+ = 10$ to $y^+ = 100$ can be deduced from a Couette flow analysis using a mixing-length model assuming no effect of blowing on the mixing-length distribution. The dramatic "uplifting" in the outer regions reflects mainly the effect of the transpiration flow on the shear stress distribution in the layer—not a drastic change in the mechanism of momentum transfer.

Figure 20 shows data from two projects (Simpson [47]; Andersen, [1]) that used different methods of evaluating the friction factor (used in both the u^+ and the y^+ coordinate definition). The difference shown is, again, due to the difference in reported values of the friction factor. Present opinion favors values near the high sides of the bands shown, rather than the low side.

Data inside of $y^+ = 10$ are suspect because of the possibility of probe errors due to wall displacement effects and shear effects. No definitive studies have been made concerning probe corrections in the presence of transpiration, hence no corrections were made to these data. To some extent, the situation is ameliorated by the fact that the finite difference program is required only to "bridge the gap" between $y^+ = 0$ ($u^+ = 0$)

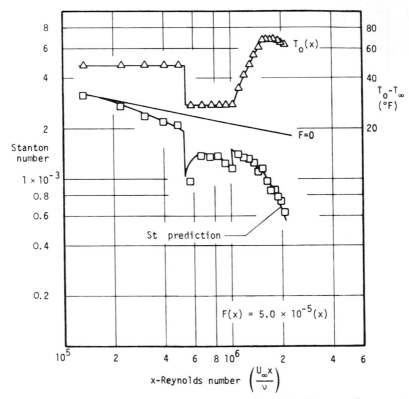

FIG. 19. A demonstration case: linearly varying blowing with arbitrary wall temperature.

and $y^+ = 10$ (u^+ known) by some reasonable means to get into a region of reasonably well-known behavior.

Relatively little had been done to this point in terms of measuring the temperature distributions in turbulent boundary layers with transpiration. Figure 21 shows some results of Moffat [25], Blackwell [4], Thielbahr [52], and Kearney [17] for some cases of blowing and suction. The parameter t^+ is defined as:

$$t^+ \triangleq \frac{(t - t_0)}{(t_\infty - t_0)} \frac{\sqrt{c_f/2}}{\mathrm{St}} = \bar{t} \frac{\sqrt{c_f/2}}{\mathrm{St}} \tag{25}$$

This differs from t_d^+, defined in Eq. (18).

B. ACCELERATION

In the early phases of the program [30], it was observed that the Stanton number was dramatically reduced by a strong acceleration. The relation-

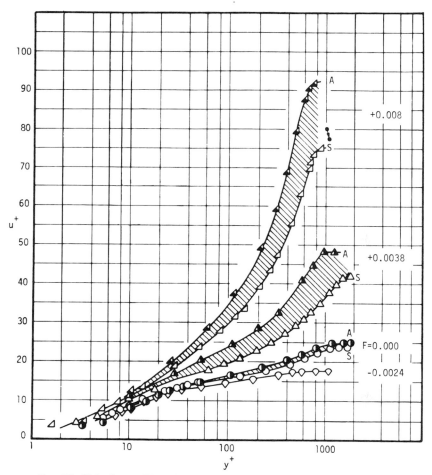

FIG. 20. Velocity profiles with blowing and suction, at constant F and U.

ship between Stanton number and enthalpy-thickness Reynolds number strongly resembled the behavior expected of a laminar boundary layer. As a result of this similarity the phenomenon was labeled *relaminarization* and occupied a number of workers throughout the late 1960s and early 1970s. There was general agreement that the acceleration parameter K could be used to characterize acceleration strength, though some felt that a better form would include $c_f/2$ to some power in the denominator. As has been shown earlier, constant K boundary layer flows offer a possibility for asymptotic or equilibrium boundary layers. Such a possibility is attractive, experimentally, on three counts: (1) it is relatively easy to accomplish (a constant K flow can be achieved using convergent planar

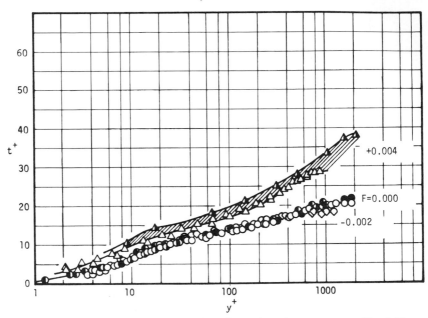

FIG. 21. Temperature profiles with blowing and suction, at constant F and U_∞.

walls); (2) it produces a possibly simpler family of responses by the boundary layer, with a better chance of revealing the fundamental effects; and (3) it helps resolve the dilemma of which possible cases, out of the infinite number of possibilities, to choose. Evidence of "local" behavior already mentioned suggests that slowly varying K conditions can be treated as quasi-equilibrium states.

1. Heat Transfer

Figures 22 through 25 show the effects of acceleration on heat transfer for values of K between 0.57×10^{-6} and 2.55×10^{-6} with transpiration controlled to yield constant F along the surface. The intention was to achieve and hold the asymptotic accelerated state for as long as possible; hence it was desired to start the acceleration at the particular value of momentum thickness Reynolds number appropriate for the values of K and F being used, that is, at the asymptotic Reynolds number. The momentum integral equation was used as a guide in choosing the starting value of momentum thickness Reynolds number. The validity of this assumption and the accuracy of the "set point" can be judged by looking ahead to Fig. 28, which shows that momentum thickness Reynolds num-

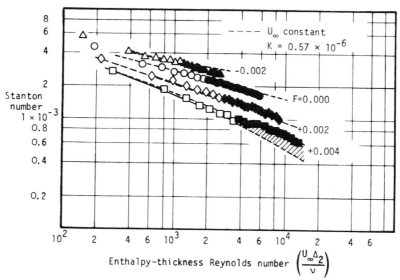

FIG. 22. The variation of Stanton number with enthalpy thickness Reynolds number at constant F for mild acceleration.

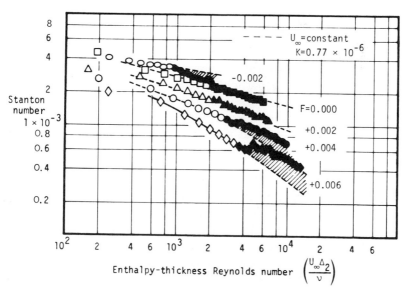

FIG. 23. The variation of Stanton number with enthalpy thickness Reynolds number at constant F for moderate acceleration.

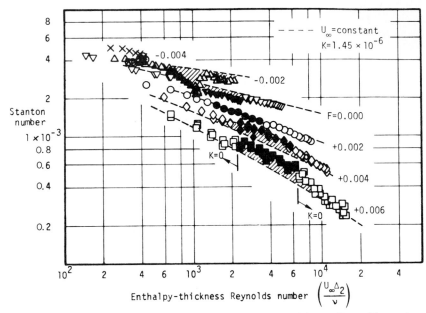

FIG. 24. The variation of Stanton number with enthalpy thickness Reynolds number at constant F for strong acceleration.

ber did, in fact, remain substantially constant throughout the test section in a typical run.

In every case shown in Figs. 22 through 25, a flat-plate, turbulent boundary layer with uniform transpiration was established in the test section and allowed to grow with length until the desired momentum-thickness Reynolds number was reached. At that location, the top wall of the test section was adjusted to set in the convergence required to yield the desired value of K. Total and static pressure measurements were made at 4 in. intervals along the test section to check that K was, in fact, constant throughout the test region.

Two effects were present in these tests: (1) acceleration and (2) high-velocity flow. To separate these effects, constant U_∞ tests were conducted at velocities up to 126 fps to ensure that Stanton number remained the same function of enthalpy-thickness Reynolds number at the high velocity end of the test section as at the low velocity end. At 126 fps the Stanton number correlation was indistinguishable from its values at 40 fps, though the friction factor was high by 5–7% (momentum-thickness method). This demonstrated that no contaminating effects were present; the changes in Stanton number were due to the acceleration itself, not the velocity level.

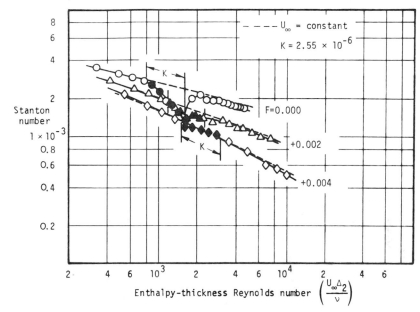

FIG. 25. The variation of Stanton number with enthalpy thickness Reynolds number at constant F for very strong acceleration.

Figures 22–25 show Stanton number versus enthalpy thickness Reynolds number for different values of K, compared with reference curves for constant U_∞ behavior. As a general comment, for a given enthalpy-thickness Reynolds number, acceleration combined with suction reduces Stanton number, while acceleration combined with blowing increases Stanton number with respect to the transpired flat-plate correlation. To illustrate this trend, note the progression of Stanton number behavior for $F = -0.002$ (moderate suction) shown in the four figures. At $K = 0.57 \times 10^{-6}$, the Stanton number slowly falls away from the constant U_∞ result, being low by about 10% at the end of the test section. At $K = 0.77 \times 10^{-6}$, the decline is more pronounced, with the terminal value low by almost 20%. At $K = 1.45 \times 10^{-6}$, the drop is nearly 40%, and the boundary layer seems to return only slowly to its constant U_∞ behavior. For suction, the stronger the acceleration the greater the depression of Stanton number.

With blowing at $F = +0.004$, a more complex behavior is noted. With moderate acceleration, the Stanton number rises above the constant U_∞ case by from 40 to 60%, while for strong acceleration, it returns to the constant U_∞ correlation. A general trade-off can be inferred between the stabilizing effect of acceleration and the destabilizing effect of blowing.

For positive values of F and K, there seems to be a neutral value relating K and F: if F exceeds 1.5×10^3 K, the Stanton number will be increased by the joint effect, and if F is less than that, the Stanton number will be reduced.

Figure 24 also shows data for $F = -0.004$ (strong suction) with $K = 1.45 \times 10^{-6}$ (moderate acceleration). The trajectory shows that an asymptotic suction layer was attained for these conditions, in the presence of strong acceleration. The first few data points show Stanton number diminishing very slightly in the approach region, in a typical turbulent suction layer fashion. Acceleration begins at an enthalpy-thickness Reynolds number of 400, and the Stanton number immediately begins to drop. With suction at $F = -0.004$, thermal equilibrium at the surface requires that the Stanton number be at least as large as $-F$ and the decline of Stanton number is stopped at that level. With Stanton number numerically equal to $-F$ and with a constant wall temperature, the energy content of the boundary layer ceases to change. The increasing values of U_∞ with distance then slowly drop the value of enthalpy-thickness Reynolds number, and the data points move sequentially to the left, at constant Stanton number.

No attempt has been made to devise an empirical formulation for predicting Stanton number in terms of enthalpy-thickness Reynolds number, K and F. Complex as it is seen to be in these figures, this is still only part of the story. All the data in these four figures are from asymptotic accelerated flows, where the flow entered the accelerating region at, or nearly at, the asymptotic value of momentum-thickness Reynolds number. As will shortly be seen, "overshot" or "undershot" layers, in which the entering values are either larger or smaller than the asymptotic values, behave much differently in the accelerating region.

Figures 26 and 27 illustrate the effects of inlet conditions on the response of the boundary layer to a strong acceleration. In Fig. 26, the accelerations began at momentum and enthalpy thicknesses between 800 and 1000, with the high-K runs beginning at the lower values. Looking ahead to Fig. 28 shows these to be nearly the asymptotic values. The solid symbols in Fig. 26 show the behavior of the Stanton number within the accelerated region and display a regular progression of slopes. With these curves as a baseline, a series of tests was run at a fixed value of $K = 2.55 \times 10^{-6}$, varying the initial momentum and enthalpy-thickness Reynolds numbers. The results, in Fig. 27, show that the slope of the Stanton number correlation is not a unique function of K but depends on the initial conditions. In Fig. 27, the square symbols represent a near-equilibrium combination, with momentum-thickness and enthalpy-thickness Reynolds numbers within 100 units of one another and of the same approximate values as shown in Fig. 26. It is worth noting that if the

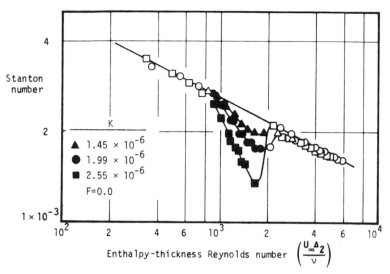

FIG. 26. The effects of acceleration on Stanton number with no transpiration.

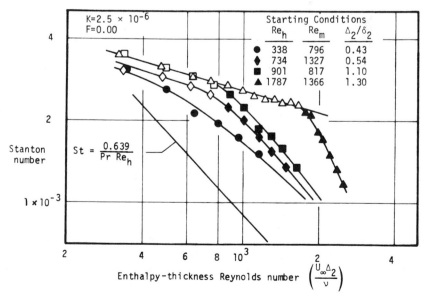

FIG. 27. The effects of initial boundary layer thickness on Stanton number during strong acceleration with no transpiration.

enthalpy thickness is kept small (in this case by delaying the heating), the response of the Stanton number to the acceleration is diminished. On the other hand, if the enthalpy thickness is held nearly constant and the momentum thickness increased, there is relatively less change in behavior from the reference case. When both the enthalpy and momentum thicknesses are increased to large values prior to the acceleration (an "overshot" case), the Stanton number comes down abruptly in the accelerating region. For the strongly "overshot" entrance conditions, it is not possible to obtain a long run at equilibrium conditions; hence most of the data shown are in the region where the boundary layer is still adjusting to the acceleration.

The solid line shown for comparison represents the similarity solution for a laminar accelerating flow with a very thick thermal boundary layer. It seems clear, from these data, that the same relative variation of Stanton number could be attained by different nonequilibrium combinations of enthalpy thickness Reynolds number and acceleration. The effects of a nonequilibrium combination of momentum thickness and acceleration cannot be uniquely identified by the value of K alone. Heat transfer in nonequilibrium accelerations is inherently responsive to all three variables: the acceleration parameter, the momentum thickness, and the enthalpy thickness.

The combination of acceleration and blowing has been shown, by these experiments, to be strongly nonlinear in its effect on the boundary layer. Empirical descriptions of this behavior are difficult to assemble in "output" notation. To express Stanton number as a function of Re_h, K, and F in such a way as to recover all the aspects shown in Figs. 22 through 27 would require a good deal of ingenuity. It is, however, relatively straightforward to predict this data set using a modified damping factor in a mixing-length formulation. If the damping factor, A^+, is expressed as a relatively simple function of v_0^+ and p^+ (transpiration level and pressure gradient, expressed in "inner" coordinates), the principal features of all the preceding are recoverable within about 5%.

2. Momentum Transfer

The decision to test equilibrium-accelerated flows places some constraints on the behavior of the momentum boundary layer, as illustrated in Fig. 28. Here, for accelerations at a value of $K = 0.75 \times 10^{-6}$, are trajectories of the boundary layer behavior for different values of F. The broken lines suggest the behavior of the skin-friction and momentum-thickness Reynolds number at various stations along the plate prior to the beginning of the acceleration. The vertical bar shown for each set is the last point in the unaccelerated flow. Considering the data for $F = 0.006$, after the

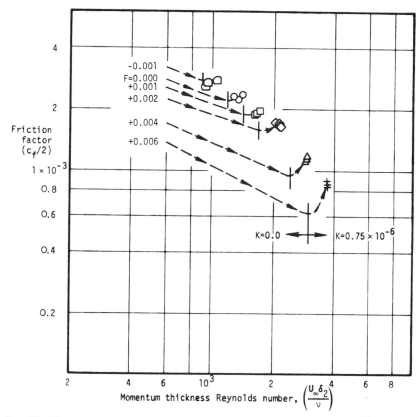

FIG. 28. The behavior of the friction factor and the momentum thickness Reynolds number after initiation of an asymptotic acceleration condition.

acceleration begins, the momentum-thickness Reynolds number grows only slightly, from 3000 to 3700, and does not change further with length along the plate. At the same time, the value of the skin friction coefficient rises quickly to a final value above the flat-plate value and then remains unchanged. The equilibrium point thus established is characteristic of this combination of acceleration and blowing. In each of these data sets, the acceleration was begun at or near the predicted value of the asymptotic momentum thickness Reynolds number, to reduce the transient effects as much as possible. The way in which the asymptotic values of momentum-thickness Reynolds number vary with K and F is shown in Fig. 29. Each symbol represents an experimentally achieved equilibrium state. Asymptotic values of $c_f/2$ established by these equilibrium flows are shown in Fig. 30. Again, each symbol represents an experimentally achieved equilibrium state. Some confusion existed in the data sets for $K = 2.5 \times 10^{-6}$

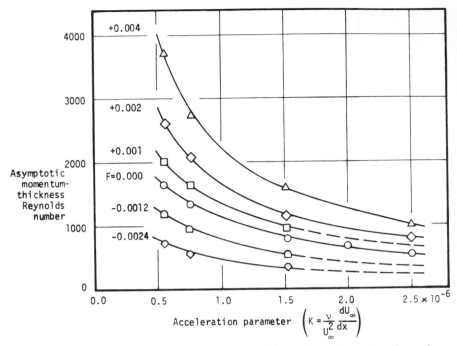

FIG. 29. The asymptotic values of momentum thickness Reynolds number for various values of K and F.

and $F = 0$, and four different terminal states were achieved. All are shown, but symmetry with the other data sets suggests that the highest value be used.

The momentum boundary layer for an asymptotic accelerated flow thus has a relatively simple description. Being uniquely specified by F and K, there is no need for a "size dependence" and, in essence, the complexity of description is reduced by one variable. The asymptotic value of friction factor with blowing can be predicted with reasonable accuracy from the unblown asymptotic value at the same K.

3. Mean Velocity and Temperature Profiles

In a constant-K, asymptotic boundary layer, the momentum-thickness Reynolds number seeks a characteristic level, as does the friction factor, while the velocity profile assumes a stationary shape in u^+, y^+ coordinates. This is illustrated in Figs. 31a, 31b, and 31c, which show the profiles as they developed in the streamwise direction. For the two lower values of K, the momentum-thickness Reynolds number was a constant

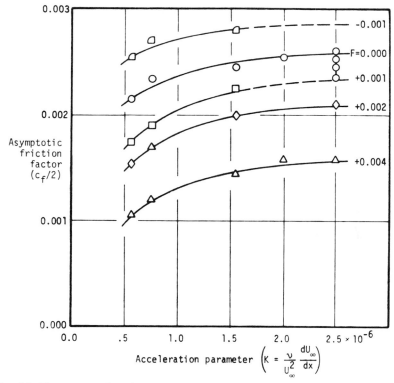

FIG. 30. The asymptotic values of friction factor $c_f/2$ for various values of K and F.

throughout the length of the test section to within ±10% and close to the asymptotic values shown in Fig. 29. The profiles are similar in both inner and outer regions. At $K = 2.6 \times 10^{-6}$, the boundary layer was "overshot," entering with a momentum-thickness Reynolds number of 750 compared to the asymptotic value of 480 (from Fig. 29). As can be seen, the boundary layer did not reach an asymptotic state, though the last profile (in the streamwise direction) could be taken as representative. The corresponding temperature profiles are shown in Figs. 32a, 32b, and 32c. For the two lower accelerations, the temperature profiles remain reasonably similar, showing small changes in the outer region (for $y^+ > 100$) that can be seen to increase in magnitude as K increases. Reviewing Figs. 22 and 24 shows that the Stanton number values for these conditions were only slightly affected by the acceleration. When the value of K reaches 2.6×10^{-6}, however, as shown in Fig. 32c, the temperature profile shows a drastic difference, with the profiles strongly nonsimilar in the streamwise direction. The effect is felt all the way in to y^+ of about 10.

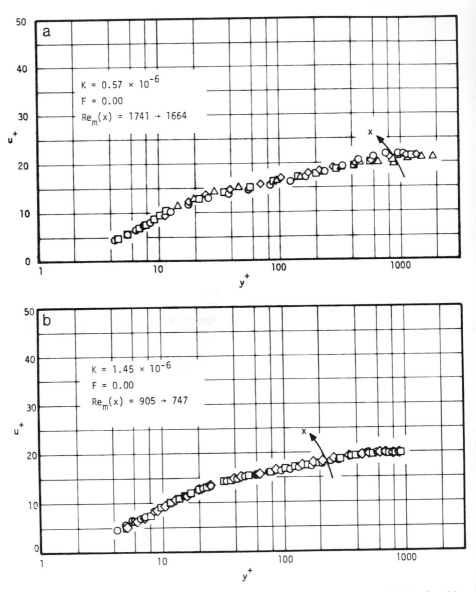

FIG. 31. Sequential velocity profiles within an equilibrium acceleration with $F = 0$ at (a) $K = 0.57 \times 10^{-6}$ and (b) $K = 1.45 \times 10^{-6}$; (c) profiles within a strong acceleration ($K = 2.6 \times 10^{-6}$) with significant changes in momentum thickness Reynolds number and a large reduction in Stanton number.

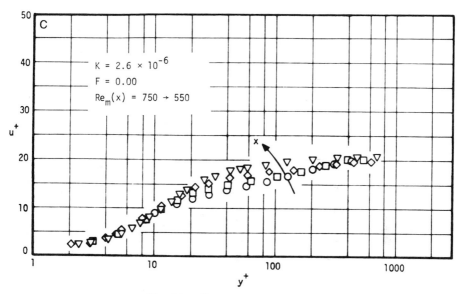

FIG. 31c. (See legend on p. 292.)

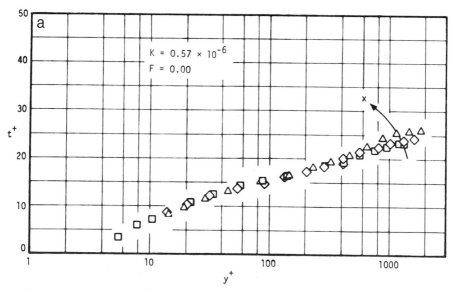

FIG. 32. Sequential temperature profiles within an equilibrium acceleration with $F = 0$ at (a) $K = 0.57 \times 10^{-6}$ and (b) $K = 1.45 \times 10^{-6}$; (c) profiles within a strong acceleration ($K = 2.6 \times 10^{-6}$) with significant changes in momentum thickness Reynolds number and a large reduction in Stanton number.

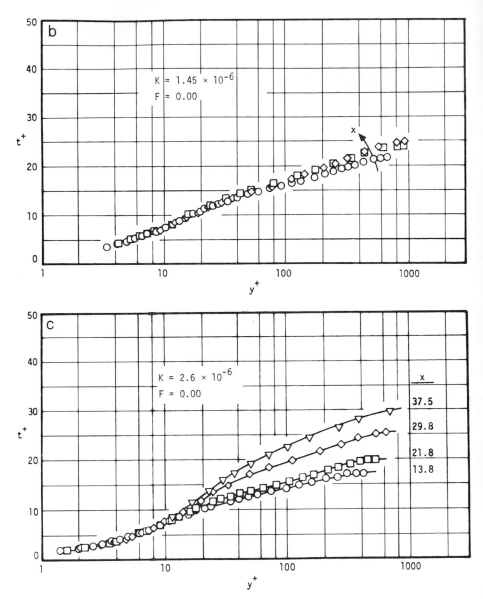

FIG. 32b and c. (See legend on p. 293.)

The Stanton number data in Fig. 26 show that this combination of conditions results in a drop in Stanton number that reaches 40% at the downstream end of the test section.

The "terminal states" of the velocity and temperature profiles are shown for high and low accelerations ($K = 0.57 \times 10^{-6}$ and 2.6×10^{-6}) in Figs. 33a and 33b and 34a and 34b for various values of blowing. The phrase "terminal states" is used because, while the profiles shown for velocity are representative asymptotic profiles, those shown for temperature are simply the last measured profiles; the energy boundary layer continues to grow, and a longer test section would have yielded a different "last" profile.

The temperature profiles shown for $K = 2.6 \times 10^{-6}$ display inner-region similarity out to about y^+ of 100; no such coherence is visible in the data for $K = 0.57 \times 10^{-6}$.

The velocity and temperature profiles shown in Figs. 31–34 illustrate the main structural features of the accelerated turbulent boundary layer. These data have been used as guides in refining the physical models used in the Stanford finite-difference computer program for boundary layer calculations, described in a later section.

C. DECELERATION

Decelerating flows differ from accelerating flows in that no asymptotic boundary layer state is approached, even though an equilibrium flow is established. The condition of equilibrium between the pressure-gradient force and the shear force is expressed by the parameter β, described earlier in Eq. (6). It was known, from earlier studies, that boundary layers for which β is constant with length display a constant value of G, the Clauser shape factor. An extension of this observation, based on the present data set, is that G remains constant with length whenever ($\beta + B_m$) remains constant, where B_m is the momentum blowing parameter. Thus decelerating flows can be studied under equilibrium conditions by holding constant these two parameters.

The experimental boundary conditions that produce flows with constant β are those for which the free-stream velocity varies with distance to some power, m. This introduces an experimental difficulty centering around the identification of the virtual origin of the boundary layer. It would not be appropriate simply to measure "x" from the leading edge of the test section unless the boundary layer were of zero thickness at that point and fully turbulent from the leading edge onward. Andersen [1] reported that, when "x" distance is measured from the virtual origin and

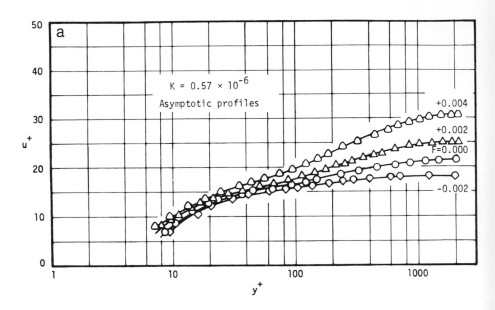

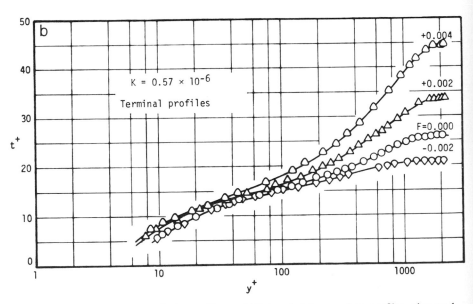

FIG. 33. (a) Asymptotic velocity profiles and (b) terminal temperature profiles, observed for mild acceleration with transpiration.

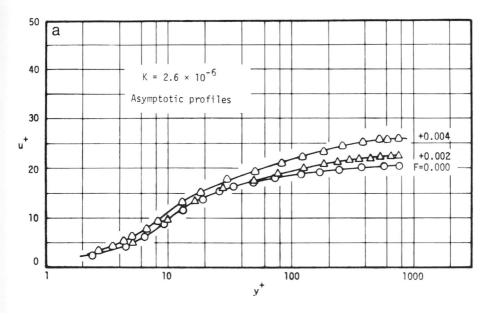

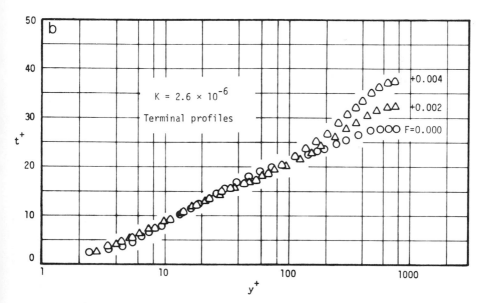

FIG. 34. (a) Asymptotic velocity profiles and (b) terminal temperature profiles, observed for strong acceleration with transpiration.

velocity varies with some power of *x*, then equilibrium boundary layers are achieved: both *B* and *G* remain substantially constant with length along the test section, after a brief accommodation. The decision was also made to avoid separation in these experiments. For this reason, only small negative values of "*m*" were used. High blowing would have encouraged separation; hence data were taken only for suction and small blowing.

Deceleration leads to thick boundary layers, and a thick boundary layer typically has a low heat transfer coefficient. Thus it was not surprising that low values of *h* were found in these tests. What was surprising, however, was to learn that decelerating flows offered no surprises at all; when the Stanton number data were examined as a function of enthalpy-thickness Reynolds number, the same correlation described both constant-velocity and decelerating flows, even including transpiration.

Figures 35 and 36 show the variation of Stanton number with enthalpy-thickness Reynolds number for moderate and strong decelerations with blowing and suction. The solid lines through the data represent constant U_∞ behavior. The same correlation that was recommended for the flat-plate case applies to these cases of decelerated flow. The effects of blow-

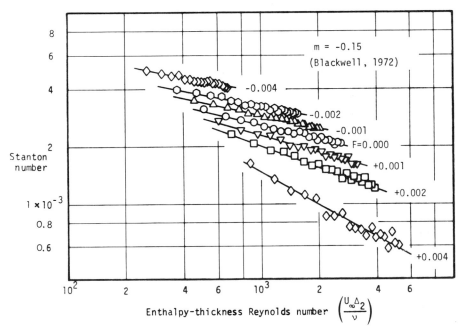

Fig. 35. Variation of Stanton number with enthalpy thickness Reynolds number for mild acceleration with transpiration.

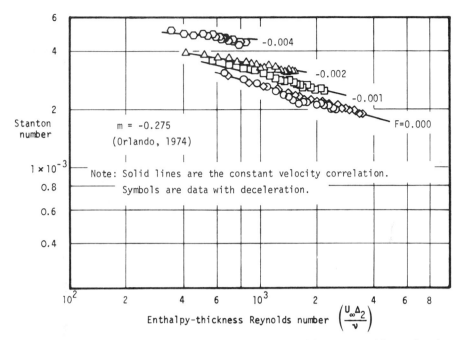

FIG. 36. Variation of Stanton number with enthalpy thickness Reynolds number for strong deceleration with transpiration.

ing are to reduce Stanton number but, again, exactly as was observed for the constant U_∞ case.

In terms of the surface heat transfer behavior of the boundary layer, adverse pressure gradients pose no new problems, within the range of conditions encountered in this study. Whatever effects the adverse pressure gradient may have on the structure of the boundary layer, in terms of changing the sublayer thickness or the turbulent transport mechanisms, the net effect is the same as for a constant U_∞ situation, in enthalpy-thickness Reynolds number coordinates. Whatever internal correlations are proposed to describe the effects of pressure gradient on the boundary layer must then produce this same behavior for decelerating flows.

Velocity and Temperature Profiles

Velocity and temperature profiles for moderate deceleration ($m = -0.15$) are shown in Figs. 37 and 38 in wall coordinates. The velocity profile is relatively unaffected by the deceleration inside y^+ of 200 either for suction or for no blowing. Blowing at 0.004 has a very pronounced effect, however, raising the values of u^+ at every y^+ greater than about 10.

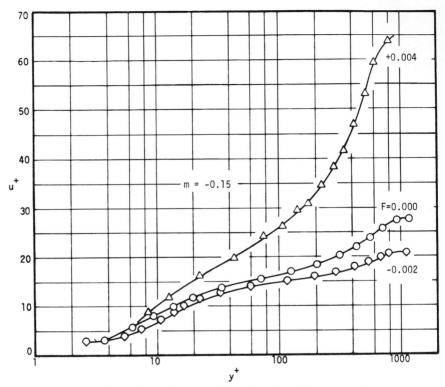

FIG. 37. Velocity profiles for mild deceleration with transpiration.

The profiles of t^+ are less affected by the deceleration than are the profiles of u^+, being slightly lower across the board.

D. ROUGHNESS

Roughness affects both the heat transfer and the hydrodynamics of the boundary layer; this has long been known. What has not been known is what happens when roughness and transpiration are present at the same time. This combination became significant in the late 1960s with the increasing use of ablative materials for thermal protection in rocket nozzles and on space vehicles. Ablative materials are designed to outgas at high temperatures, protecting their surface both by transpiration and by species diffusion. At the same time, however, the char layer that develops on the surface is rough.

The first problem was to construct a surface. These studies required a surface that would be uniformly porous and, at the same time, rough. The surface finally chosen was a brazed assembly of 8.5 million copper balls,

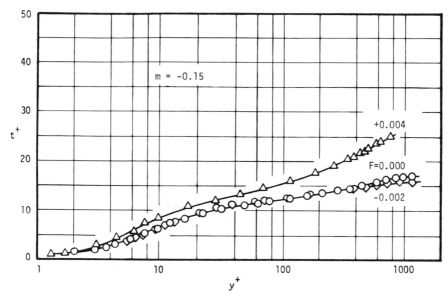

FIG. 38. Temperature profiles for mild deceleration with transpiration.

each 1.25 mm in diameter, packed carefully into mold cavities and fired in an atmosphere furnace. Twenty-four individual plates of this material were made, enough to make a test surface 8 ft long and 18 in. wide. Viewed from the boundary layer side, the surface exposed a regular array of hemispheres to the flow and was uniform in porosity, to within the resolution of modern hot-wire anemometry.

This surface can be regarded as an idealization of the "sand-grain" roughness concept, a rough surface with only one length scale. It was recognized that the regularity might affect the hydrodynamic behavior, but, at the same time, a reference roughness was thought to be valuable in establishing a fixed point in such studies.

The heat transfer results soon showed that the combination of roughness and transpiration had some surprising properties. The properties of the "fully rough" state were first determined, in which the heat transfer behavior was shown to be a function of boundary layer thickness only (i.e., independent of velocity); then transpiration was added. Blowing reduced the roughness Reynolds number (by reducing the friction factor) but did not take the boundary layer out of the "fully rough" state, as far as surface heat transfer was concerned; the results remained substantially independent of velocity, even at very low roughness Reynolds numbers.

These results led to more detailed studies of the structure of the rough-wall boundary layer by Pimenta [32], and finally to the development of techniques for artificially creating very thick boundary layers, described by Ligrani [21].

Roughness had long been known to reduce the effect of the laminar sublayer, and our results had agreed with the earlier work in that regard. On the other hand, the work of Julien *et al.* [16] with accelerating flows on a smooth wall had shown that acceleration increased the thickness of a laminar sublayer. It seemed only logical, then, for Coleman *et al.* [9] to test the effect of acceleration on a rough surface, to see which tendency would dominate.

1. *Heat Transfer*

Stanton number data are shown in Fig. 39 plotted against enthalpy thickness normalized on the sphere radius. Data are shown for velocities from 9.6 m/sec to 72.6 m/sec and for values of Δ/r up to 10 from naturally developed rough-wall boundary layers [27].

The natural coordinate for these fully rough boundary layer heat transfer data appears to be Δ_2/r, rather than x-Reynolds number or enthalpy-thickness Reynolds number. Use of either of these alternatives resulted in separation of the data directly related to the magnitude of velocity.

The boundary layer was laminar approaching the rough surface for the case of 9.6 m/sec, and the transition event is clearly visible. The momentum-thickness Reynolds number at the bottom of the saddle was 434. The Stanton number data for 9.6 m/sec lie somewhat above the data for the higher velocities. This is believed to be significant, based on later structural studies [21, 33].

One conclusion from Fig. 39 is that the Stanton number correlates well using enthalpy thickness alone, not enthalpy-thickness Reynolds number, for all velocities above 27 m/sec, and that, as far as heat transfer is concerned, even the data for 9.6 m/sec do not deviate much from this "fully rough" correlation. This conclusion is consistent with expectations based on the early work of Nikuradse [30].

Stanton number data with blowing are shown in Fig. 40 for the transition regime at 32 ft/sec. Uniform blowing from the origin of the rough plate caused the transition to move upstream to lower x-Reynolds numbers, but the momentum-thickness Reynolds number at the saddle point was relatively constant.

Figure 41 shows Stanton number data for the fully turbulent regimes for a range of velocities and blowing, in rough-wall coordinates. Roughness Reynolds numbers (based on measured Stanton numbers rather than friction factor) ranged from 28.8 to 9.6 for the 32 ft/sec data (including the effect of blowing), and from 88 to 24 for the 90 ft/sec data. For higher

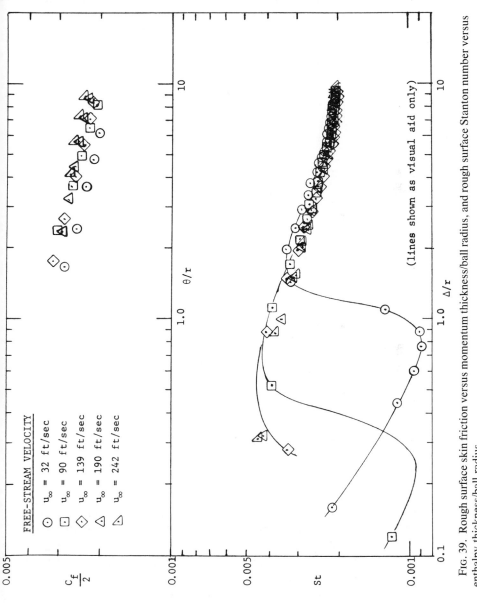

FIG. 39. Rough surface skin friction versus momentum thickness/ball radius, and rough surface Stanton number versus enthalpy thickness/ball radius.

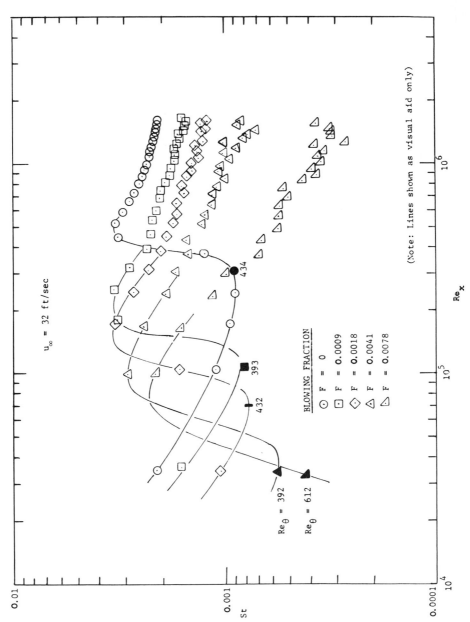

Fig. 40. Rough surface Stanton number versus x-Reynolds number at $U_\infty = 32$ fps.

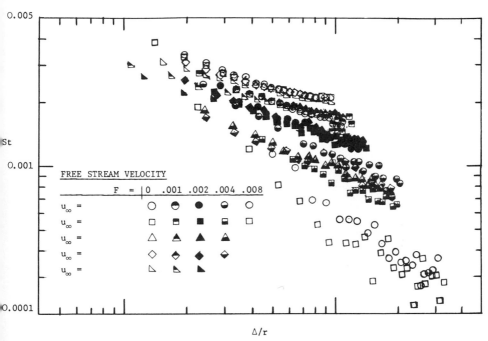

FIG. 41. Stanton number versus enthalpy thickness/ball radius for all boundary layers.

velocities, roughness Reynolds number was always greater than 57. The heat transfer data for velocities of 90 ft/sec and above are well correlated in terms of F and Δ/r only, with no velocity dependence. The 32 ft/sec data lie noticeably high at $F = 0.004$ but at all other blowing values seem indistinguishable from the fully rough data, even though the roughness Reynolds number gets as low as 9.6 with blowing at $F = 0.008$.

Stanton number data with no blowing fit the equation:

$$St = 0.00317(\Delta/r)^{-0.175} \qquad (26)$$

The data in Fig. 39 show a noticeable curvature in log coordinates, and the present recommendation of a slope of -0.175, following Pimenta [32], is weighted toward the high Δ/r data. In an earlier study, Healzer [14] recommended -0.25 as an average slope over the whole range.

Blowing reduces the Stanton number, and the effect can be predicted in terms of the blowing parameter B by the following equation:

$$St/St_0|_{\Delta/r} = \ln(1 + B_h)/(1 + B_h) \qquad (27)$$

The comparison is made at the same value of enthalpy thickness in this

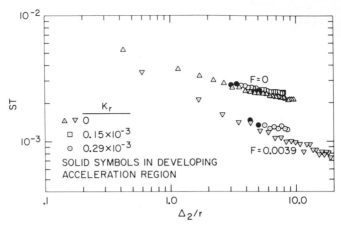

FIG. 42. Equilibrium acceleration Stanton number data versus enthalpy thickness/ball radius.

case, rather than at constant x-Reynolds number or constant enthalpy-thickness Reynolds number.

When acceleration is combined with roughness and transpiration, the response differs from that on a smooth wall; acceleration on a rough wall causes an increase in Stanton number, not a decrease. This response is shown in Fig. 42. In this figure the parameter held constant is K_r, the acceleration parameter for fully rough flows, defined by Coleman et al. [9] as

$$K_r = \frac{r}{U_\infty} \frac{dU_\infty}{dx} \tag{28}$$

where r is the radius of the spherical particles making up the surface.

2. Momentum Transfer

Friction factor shows the same behavior as Stanton number for fully rough behavior, as illustrated in Fig. 43.

The unblown data are well fit by the equation

$$c_f/2 = 0.00328(\delta_2/r)^{-0.175} \tag{29}$$

These $c_f/2$ values were deduced from hot-wire measurements of the shear stress within the boundary layer, extrapolated to the wall. Thus the observation that the global Reynolds analogy applies is based on two independently measured properties. For fully rough flow, the Stanton number is lower than $c_f/2$ by about 4%.

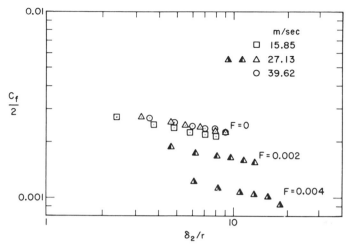

FIG. 43. Friction factor variation with blowing on a rough wall.

Blowing reduces the friction factor, and the effect is predicted by the same form as in Eq. (27) for the Stanton number, using momentum thickness to replace enthalpy thickness as the parameter held constant.

3. *Mean Velocity and Mean Temperature Profiles*

The literature abounds with descriptions of the velocity distribution in fully rough boundary layers, and there is no need to show more of them here. The Stanford data showed the same depression of the log law and the same agreement with the Law of the Wake that other works have shown.

Figures 44 and 45, however, show two aspects of the situation that have not been discussed before. In Fig. 44, the transition from laminar to turbulent state is seen to be essentially complete (as far as accommodation of the mean profile is concerned) within a very short distance: less than five boundary layer thicknesses. At $x = 1.0$ m the profile matched the Blasius relationship (using a y shift of 0.13 mm based on the work of Monin and Yaglom [28]), while at 1.1 m it matched the fully rough profile (the Law of the Wake). To some extent the abruptness of this transition may be due to the regularity of the surface used. The surface is composed of spheres of identical size, packed with perfect regularity. No element is different from any other; there is no particular place (such as "the largest element") to serve as a starting point for transition. Further evidence on this will be seen in Fig. 46.

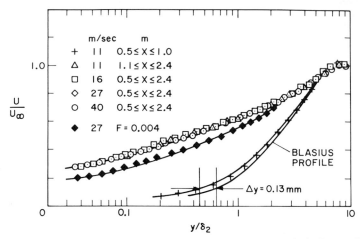

FIG. 44. Mean velocity distribution near transition on a deterministically rough wall.

Figure 45 illustrates an interesting relationship between the temperature and velocity distribution, in coordinates that are independent of the location of the "virtual" wall: T versus U. In these coordinates the rough-wall data form a straight line that intercepts the T axis at about $T = 0.1$ for all velocities and all blowing values tested. The "temperature step at the wall" thus indicated is related to the wall-layer resistance to heat transfer. Smooth-wall data, plotted in these coordinates, do not show this offset but pass through the origin, as shown. The fact that the line is straight reveals that the turbulent Prandtl number is uniform within the boundary

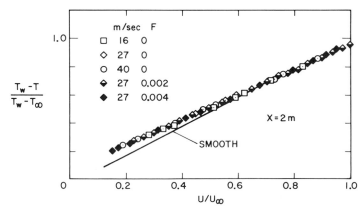

FIG. 45. Mean temperature versus mean velocity in a rough-wall turbulent boundary layer with and without blowing.

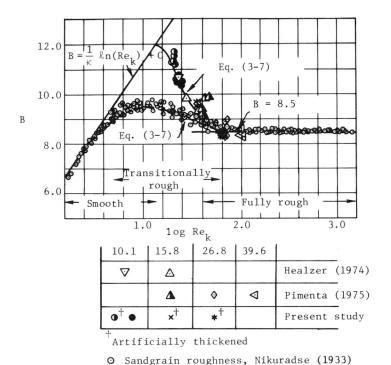

10.1	15.8	26.8	39.6	
▽	△			Healzer (1974)
	▲	◊	◁	Pimenta (1975)
o† ●	x†	*†		Present study

† Artificially thickened

⊙ Sandgrain roughness, Nikuradse (1933)

FIG. 46. Variation of B with roughness Reynolds number.

layer. This observation is confirmed by calculations of Pr_t based on separately measured profiles of temperature and velocity.

The mean velocity profile for rough walls is usually described in terms of a modified Law of the Wall, using an offset parameter B:

$$u^+ = \frac{1}{\kappa} \ln y/k_s + B \tag{30}$$

The B parameter for the present surface behaves differently from that of sand-grain roughness as a function of roughness Reynolds number. This is shown in Fig. 46, where results from the present surface are compared with those from sand-grain data presented by Schlichting [40]. On the deterministic surface, B rises much more abruptly in the transition region, although sharing the same asymptotic value. This abrupt response is believed to be due to the uniformity of the roughness elements. Whereas sand grains have a range of sizes and a variety of shapes, the present elements are all spheres of the same diameter, within 0.005 mm. The abruptness with which B goes from the fully rough to the smooth value may be responsible for the fact that the boundary layer displayed the "fully rough" profile immediately downstream of transition, discussed in connection with Fig. 44.

4. *Turbulence Properties*

Measurements inside the fully rough boundary layer by Pimenta [32] and Ligrani *et al.* [22] showed the turbulence properties to be the same as in the outer region of a normal turbulent boundary layer. The measured turbulent Prandtl number was nearly constant, changing only from about 0.96 near the wall ($y/d = 0.05$) to 0.82 at the outer edge. The Reynolds shear stress, normalized on q^2 (i.e., $u'v'/q^2$) was 0.14 throughout the layer, for all velocities, with and without blowing (up to the highest value tested, $F = 0.004$). This is the same value used in smooth-wall work. The correlation coefficient $u'v'/(u'^2v'^2)$ was uniform at 0.045, and for $v't'/(v'^2t'^2)$ rose from 0.58 at the wall to 0.68 at the outer edge.

The mixing length approaches the wall along the line

$$l = 0.41(y + \delta y) \tag{31}$$

where δy was taken to be 0.15 mm, for all velocities and all values of blowing.

Figure 47 shows the distribution of u'^2/U_τ^2 through the boundary layer, as a function of the free-stream velocity. At low velocities, u'^2 rises abruptly near the wall, a behavior characteristic of a smooth wall. At 40 m/sec, the maximum occurs far from the wall, near $y/d = 0.10$, with u'^2 relatively lower near the wall. This is the fully rough characteristic. As velocity decreases, the value of u'^2 near the wall rises steadily toward the smooth-wall profile. There is no abrupt change from a "rough" to a "smooth" profile, even though the mean velocity profiles, the heat transfer, and the friction factor all show significant (and relatively abrupt) changes.

E. Full-Coverage Film Cooling

Film cooling using slots and rows of holes is a practical alternative to transpiration cooling, and it has been extensively studied. Some programs have concentrated on the effects of one or two rows of holes. The Stanford project has focused on full-coverage film cooling. This term refers to the use of an array of slots or holes that extends far enough in the streamwise direction that a fully developed state develops in the boundary layer.

Full-coverage film cooling can be regarded as an extension of the transpiration problem, with one extra complication: the injected fluid need not

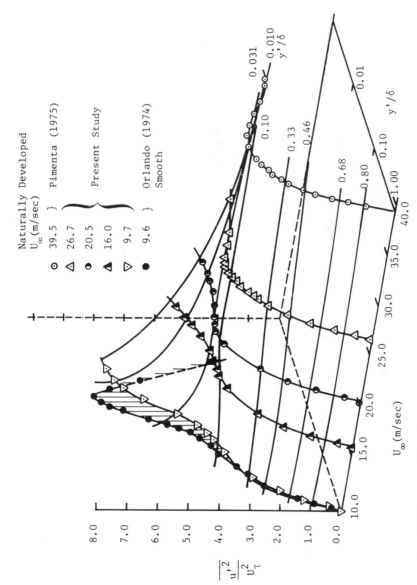

Fig. 47. Summary of profiles of the longitudinal component of turbulence intensity, normalized by using the friction velocity, for transitionally rough and fully rough turbulent boundary layers.

be at wall temperature. Whereas the transpiration situation was fully described by specifying the geometry, the mainstream flow, and the injection velocity, in the full-coverage film-cooling situation there is an additional independent variable: the injectant temperature. The method of superposition is used to deal with this complication in the present work. For each hydrodynamic situation (i.e., a fixed geometry and flow condition), the heat transfer coefficient must be measured for two temperature conditions: one with the injectant at wall temperature and one with the injectant at free-stream temperature. From the values of "St(1)" and "St(0)" thus determined, the effect of any other injection temperature can be determined exactly, as long as the problem remains linear. The heat transfer coefficient is based on the local heat transfer rate and the difference between the actual wall temperature and the actual gas temperature; the concept of "effectiveness" is not used.

As of the present writing, four situations have been covered: perpendicular injection [7], 30° slant-angle injection [11], 30° × 45° compound-angle injection [18], and 30° slant-angle injection on a convexly curved wall [12]. Heat transfer and hydrodynamic data have been taken both within the full-coverage region and in the recovery region downstream of it.

Crawford et al. [11] have reported some success modeling the full-coverage film-cooled boundary layer data reported here, but the work has not reached a definitive stage. Two effects have been introduced: an increased mixing induced by the injection process and the effect of the mass addition distributed through the mixing region of the boundary layer. A mixing-length approach has been used, augmenting the usual mixing length to account for the increased turbulence caused by the injection. The injected fluid is distributed into the boundary layer in such a way as to satisfy conservation of mass, momentum, and energy.

The goal of obtaining a simple predictive model led to definition of "spanwise-averaged" properties of the boundary layer and the surface heat transfer. Surface heat transfer is "area-averaged," while the boundary-layer-conserved properties are "mass-flow-averaged" to ensure global satisfaction of mass, momentum, and energy balances.

The data set is extensive, and only representative samples are included. The data are presented in coordinates of Stanton number versus x-Reynolds number, to facilitate comparison with future predictions.

Heat Transfer

Figures 48, 49, and 50 present Stanton number data for full-coverage film cooling with a pitch to diameter ratio of 5 and a wide range of values of the injection parameter M for three different hole configurations: per-

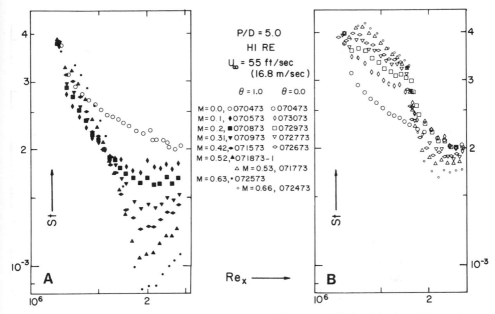

FIG. 48. Stanton number versus x-Reynolds number; perpendicular injection. (A) $\theta = 1$; (B) $\theta = 0$.

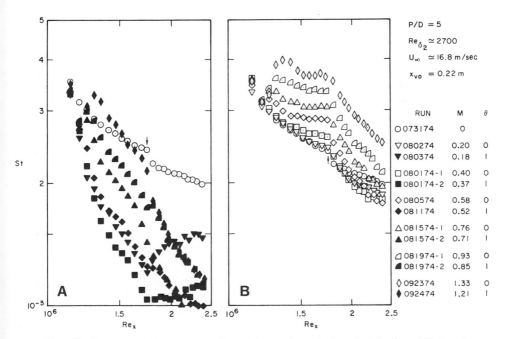

FIG. 49. Stanton number versus x-Reynolds number; slant angle injection. (A) $\theta = 1$; (B) $\theta = 0$.

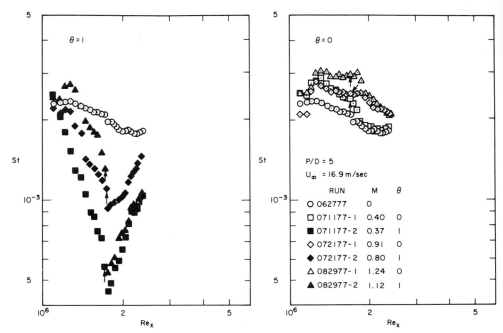

FIG. 50. Stanton number versus x-Reynolds number for $\theta = 0$ and $\theta = 1$ with $P/D = 5$, heated foreplates; compound angle injection ($30° \times 45°$).

pendicular injection, $30°$ slant injection, and $30° \times 45°$ compound-angle injection. Figure 51 shows the comparable data for $P/D = 10$ for the compound-angle case.

Each figure shows Stanton number data for two values (1.0 and 0.0) of the temperature parameter θ, thus providing the necessary data for calculation of the Stanton number for any other injection temperature, by superposition. The superposition equation, discussed by Choe *et al.* [7], is

$$St(\theta) = St(0) - \theta[St(0) - St(1)] \tag{32}$$

There are five descriptors for the full-coverage situation: mainstream flow condition, initial state of the boundary layer, injection geometry, injection mass flow parameter M, and injection temperature parameter θ.

All the data reported here were taken with the free stream at normal room ambient conditions of temperature and pressure.

The ratio of boundary layer thickness to injection hole diameter ranged from 1 to 5, depending on the initial conditions being tested. Changing the initial boundary layer thickness had only a small effect on the values of Stanton number. Thin inlet boundary layers showed higher Stanton numbers by about 10% or less near the upstream edge of the array, with the

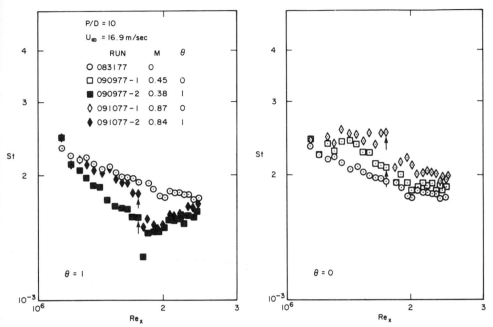

FIG. 51. Stanton number versus x-Reynolds number for $\theta = 0$ and $\theta = 1$ with $P/D = 10$, heated foreplates; compound angle injection ($30° \times 45°$).

effect diminishing after a few rows of injection holes. The trends of the data were the same for thin and for thick initial layers.

The injection geometry is described by the array pattern, the P/D ratio, the angularity, and the number of rows. The injection holes were 1.03 cm in diameter for all test plates, spaced 5.15 cm apart. The holes in alternate rows were staggered, and the rows were 5.15 cm apart. The data in Figs. 48 through 50 are for $P/D = 5$, while those for Fig. 51 are for $P/D = 10$, achieved by covering alternate holes with thin tape.

The primary variable in this study was the injection parameter M, the ratio of injection mass velocity to mainstream mass velocity. For each value of M, two runs were made: one with the injection air near plate temperature ($\theta = 1$) and one near free-stream temperature ($\theta = 0$). The actual values used were between 0.0 and 0.2 for the "cold" run and 0.9 to 1.1 for the "hot" run. Based on the measured Stanton numbers and the measured values of the two θs, values for $\theta = 0.0$ and 1.0 were calculated using superposition. It is the calculated values that are shown in all the figures.

Figure 48 shows data for injection perpendicular to the surface. Open circles represent the baseline case: no injection. The high initial values and sharp curvature in the low Reynolds number portion of the baseline data are effects of an unheated starting length in the test section.

Consider first the data for injection temperature equal to plate temperature ($\theta = 1.0$). Injection at $M = 0.1$ immediately drops the Stanton number about 20%. As injection continues, the Stanton number falls further and further below the unblown data until, after 11 rows of blowing, it lies nearly 30% below the unblown data for the same x-Reynolds number. Higher values of M do not simply drop the Stanton number further; during the first seven rows, increasing the blowing above $M = 0.1$ raises the Stanton number. There is a "crossover point" at the eighth row, through which all the St(1) data pass. Beyond the eighth row, increasing M decreases St(1); before the eighth row, increasing M increases St(1). The effect of M is simpler in the recovery region, after the last blown row. Here the effect is unambiguous; the higher M's value has been, the lower is St(1) in the recovery region. For every case, St(1) begins to recover toward the unblown line as soon as the blowing stops. The faster the approach, the larger the initial difference between the blown and unblown values.

The data for $\theta = 0$ lie above the unblown data throughout the blown region, for all values of M, and they are ordered directly with increasing M. There is an abrupt rise within the first two rows of injection holes, sufficient to double the Stanton number for $M = 0.66$ and proportionately less for the lower values. After the rise, the Stanton number falls again at about the same slope as for the unblown data, remaining high throughout the blown region. Once again there is a "crossover point" in the data, but this time it is after the end of blowing. In the recovery region, the high values of M produce low values of St(0).

The trends in data shown in Fig. 48 can be rationalized in terms of two competing tendencies: increased turbulence mixing, caused by the interaction of the jets with the boundary layer, and the thermal effect of the injected air. A low injection rate of fluid at wall temperature ($M = 0.1$, with $\theta = 0.1$, for example) provides cooling without much increase in turbulence. As a consequence, St(1) is low. As the injection velocity rises, the interaction between the jets and the boundary layer becomes more intense, the turbulent mixing goes up, and the Stanton number goes up. For St(0), the injected air is at free-stream temperature. The hydrodynamic effects are the same, but now the injected air tends to increase Stanton number by distorting the temperature distributions in the inner regions of the boundary layer.

Figures 49 through 52 show data for different hole geometries in format similar to that of Fig. 48.

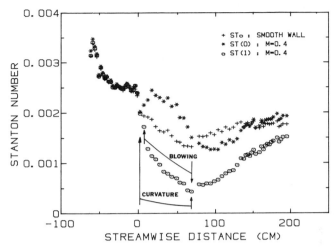

FIG. 52. Stanton number for full-coverage cooling with $m = 0.4$ on a convex curved wall with $St/R = 0.10$.

The general trends are the same in all the data sets. Slanting the injection holes at 30° to the surface (Fig. 49) decreases the turbulence generated for a given value of M. This shifts the "breakeven" point to higher values of M and allows more thermal protection—hence lower values of St(1). Compound-angle injection (30° downstream and 45° cross-stream) still further reduces the turbulence associated with a given injection rate and results in the lowest Stanton number values of the three geometries, as shown in Fig. 50.

Increasing the pitch–diameter ratio from 5 to 10 by closing every other hole altered the level of the results, but did not introduce any new features to the data. Figure 51 shows some of the data from tests of compound-angle injection at $P/D = 10.0$. Injection at $M = 0.4$ produced about the same reduction in St(1) as might have been expected for $M = 0.1$ or 0.2 for $P/D = 5.0$. The reduction in St(1) is not simply proportional to the injected mass flow; injection at $M = 0.8$ for $P/D = 10$ corresponds to injection at $M = 0.2$ for $P/D = 5$, in terms of the total mass injected. The data in Fig. 51 show that $M = 0.8$ was above the optimum injection rate [since St(1) is larger for $M = 0.8$ than for $M = 0.4$]. Thus it can be inferred that injection at $P/D = 10$ generates more turbulence for a given amount of thermal protection than $P/D = 5$.

The combination of slant-angle injection and convex curvature is presented in Fig. 52. The same planform geometry was used in these studies, but the wall was convexly curved at a constant radius of 45 cm. The curvature results without injection are described in Section IV,F, Surface Curvature. Figure 52 shows that injection on a convexly curved surface is more effective in protecting the wall than injection on a flat surface. This conclusion may be subject to modification in systems in which the injected air might be considerably colder than the surface. The behavior in the recovery region, after simultaneous cessation of blowing and curvature, is complex but can be explained in terms of the three effects present—turbulence augmentation, curvature, and thermal protection.

Modeling of the spanwise-averaged heat transfer characteristics of full-coverage film cooling has been pursued using a mixing-length model that augments the Van Driest mixing length in accordance with the hole geometry and injection ratio. The injection process is modeled as an abrupt increase in the mass flow in the inner stream tubes of the boundary layer, using a heuristic model for "peeling off" mass from the injection jet until all the injected mass has been deposited into the boundary layer. The location of the peak value in the augmented mixing length is related to the maximum penetration distance of the jet into the boundary layer. The variation of mixing length is indicated qualitatively in Fig. 53. This modeling work is not considered well enough developed to be reported here in detail.

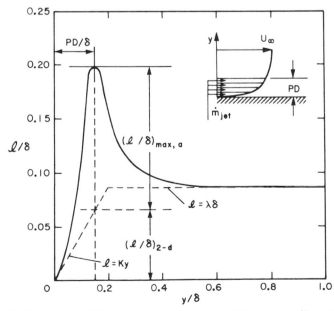

FIG. 53. Illustration of the mixing length model for full-coverage film cooling.

To date, it has been established that a two-dimensional finite-difference code can make usefully accurate predictions of the heat transfer coefficient distribution in a full-coverage situation, using only three additional modeling constants beyond those required for a normal flat-plate boundary layer.

The recovery region on a flat plate has been successfully modeled using the concept of an "inner" and an "outer" boundary layer. The inner layer develops as a normal turbulent boundary layer under a highly turbulent mainstream (actually, the residue of the outer region of the blown boundary layer). The outer region behaves as a convective flow of decaying isotropic turbulence, setting the outer boundary conditions for the inner layer. For this situation, it was possible to predict the variations in the mean velocity profiles and the turbulence kinetic energy distribution, using constants evaluated in independent tests of jets and wakes.

The modeling of full-coverage film cooling and its recovery region is still in a primitive state, and it will require a great deal of work before any definitive proposals can be put forth.

F. SURFACE CURVATURE

1. *Convex Curvature*

Even though there were some warnings in the literature in the early 1930s, it was not until the work of Thomann [53] that the boundary layer heat transfer community sat up and took notice of surface curvature as a significant variable. Studies at various industrial laboratories soon showed that convex curvature significantly depressed heat transfer, while concave curvature augmented it.

The Stanford program began in the mid-1970s with the objective not of finding out whether or not curvature affected heat transfer, because that was already known, but of determining the mechanisms by which it acted and establishing a workable model for prediction.

The first program concerned the effects of convex curvature on heat transfer from a turbulent boundary layer. In a joint program with J. P. Johnston, the effects of boundary layer thickness, free-stream velocity, and transition on the heat transfer, momentum transfer, and turbulence structure were investigated.

It might seem that the concave problem should have been taken up first, since that seems to be the more important problem (anything that increases the heat transfer coefficient to a turbine blade is cause for alarm, whereas a decrease usually brings a sigh of relief). The convex wall problem was done first, based on the following arguments. First, it was well established that convex curvature reduced heat transfer, and it seemed certain that this effect was due to some modification of the turbulence structure in the boundary layer. With this in mind, it was reasonable to assume that there would be a corresponding effect in a concave flow, causing an increase in heat transfer. However, in the concave situation, suggestions were already being heard that the increase was due to streamwise vortices like those that had been observed in curved laminar flows by Taylor and Görtler. It would be difficult to determine how much of the increase in heat transfer was due to each of these two mechanisms unless there was prior knowledge of one of them. Studying the convex case first should reveal the interaction between curvature and turbulence, without interference from any other flow structures, and it should then be possible to carry this understanding into the concave study.

In view of the importance of discrete-hole injection as a cooling technique for aircraft gas turbines, we soon incorporated a study of discrete-hole injection effects into the curvature program with the work of Furuhama and Moffat [12].

The Stanford project concerned with convex curvature effects on turbulent boundary layer heat transfer began with the parallel works of Gillis *et al.* [13] and Simon *et al.* [44]. Working on a large-scale apparatus (45 cm radius of curvature with boundary layers about 5 cm thick), they investigated the hydrodynamic and heat transfer behavior of turbulent boundary layers under a wide variety of conditions.

2. Heat Transfer

Figure 54 shows the variation of Stanton number with streamwise distance for a case in which the ratio of boundary layer thickness to radius of curvature, $\delta_{0.99}/R$, was 0.10 and the velocity was 14.8 m/sec. This is in the range of $\delta_{0.99}/R$ regarded as "strong curvature." Throughout the runs shown here, the pressure coefficient was kept at 0.00 within ±2% of the velocity head throughout the curved region. This restriction effectively isolates the curvature effect from the streamwise pressure gradient that frequently accompanies it. The momentum-thickness Reynolds number of the boundary layer was 4173 at the beginning of the curved region, and the shape factor was 1.41, indicating a mature turbulent boundary layer. Also shown is a line drawn from an accepted correlation for isothermal, turbulent heat transfer on a smooth, flat plate. The effect of curvature is dramatic. The Stanton number drops about 15% within the first 2.5 cm after curvature is introduced; it continues to decrease through the curve until, at the end of the curved region, it is 35–40% below the flat-plate correlation. At the end of the curved region, there is an abrupt rise of about 15%, followed by a very slow recovery toward the flat-plate values.

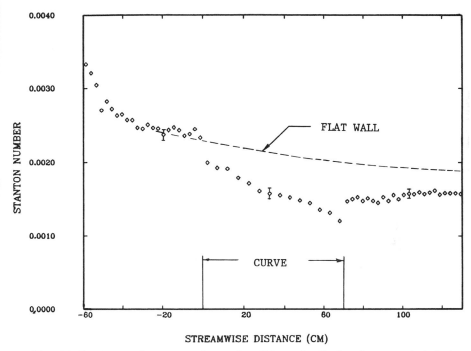

FIG. 54. Stanton number versus streamwise distance for the baseline case: turbulent, δ_{99}/R $(0 = 0) = 0.10$, $U_{pw} = 14.8$ m/s, $K \simeq 0.0$.

After 60 cm of recovery, the Stanton number is still some 15% below the accepted correlation for that location. The Stanton number recovery follows the same trend as the friction factor, for which a slow recovery was also noted by Gillis.

The data can be recast into enthalpy-thickness Reynolds number coordinates, as shown in Fig. 55, employing the two-dimensional energy integral equation to calculate the enthalpy thickness from the Stanton number data. This formulation is less dependent on prior history of the boundary layer than is the x-Reynolds number presentation. Within the curved region, the data follow a line of -1 slope. This characteristic has also been observed in highly accelerated turbulent boundary layer flows, in which it is called relaminarization. The boundary layer on the curved wall is not laminar, as Gillis's data clearly showed, but the turbulence structure has been significantly altered. Turbulence production is limited to a thin layer next to the surface of approximately constant thickness throughout the curved region. The outer region is also turbulent, but the turbulence is uncorrelated, and the Reynolds stresses are zero, or nearly zero, throughout the outer region. The evidence in Fig. 55 is that, within a convex

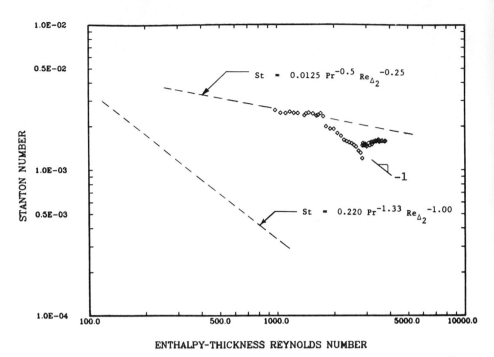

Fig. 55. Stanton number versus enthalpy thickness Reynolds number for the baseline case: turbulent, δ_{99}/R $(0 = 0) = 0.10$, $U_{pw} = 14.8$ m/s, $K \simeq 0.0$.

curve, Stanton number is significantly lower for a given enthalpy thickness than it would be on a flat plate. The depression persists for a relatively long distance downstream after the cessation of curvature. This is further evidence of the low Reynolds stresses in the outer region of the boundary layer.

The data in Figs. 54 and 55 have introduced the characteristic response of convex curvature; qualitatively, at least, this behavior has been seen in every reported study of convex curvature effects on heat transfer.

Tests were conducted at three values of $\delta_{0.99}/R$ (0.02, 0.05, and 0.10). The results for all three cases were within $\pm 6\%$ of the mean for the three cases. Thus if $\delta_{0.99}/R$ is a parameter of this problem, it has a very weak effect at these levels. In enthalpy-thickness Reynolds number coordinates, the data for $\delta_{0.99}/R$ of 0.05 and 0.10 assume the characteristic -1 slope almost immediately on entering the curved region, whereas the data for $\delta_{0.99}/R$ of 0.02 reach that slope only near the end of the curved region. Gillis [13] reported that the region of active turbulence was approximately $0.03R$ whenever the boundary layer was thicker than that before entry into the curve. A thin boundary layer may not have time to grow to the required thickness within the curve, hence may not display the characteristic slope.

Tests conducted at three different free-stream velocities (7.0, 14.8, and 26.4 m/sec) showed a difference of only about 20% between the Stanton numbers from the lowest to the highest velocity, in the latter portion of the curved region, as shown in Fig. 56. Since the velocity changed by almost a factor of 4, the heat flux must have varied almost linearly with velocity to yield such a small net change in Stanton number. The recovery region data were also very similar. It appears that free-stream velocity is not a strong factor in setting the response of the boundary layer to convex curvature.

In enthalpy-thickness Reynolds number coordinates, data from all three velocities showed the -1 slope at the end of curvature, though their approaches to it were different. The low-velocity data approached the -1 slope from above, the high-velocity data approached it from below, while the mid-velocity data followed it throughout the curve.

Three cases of streamwise acceleration were studied, each holding a constant value of the acceleration parameter K. For each test, the momentum thickness of the boundary layer at the start of curvature was set

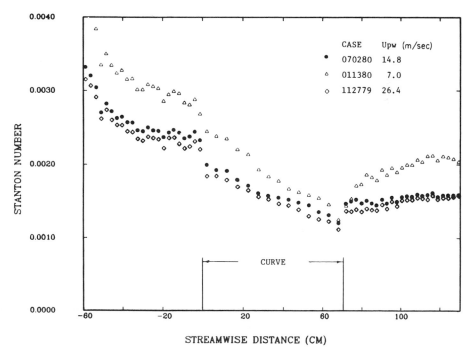

FIG. 56. The effect of U_{pw} Stanton number versus streamwise distance.

at the appropriate value for an equilibrium acceleration. The results are shown in Figs. 57 and 58. Before examining these figures, the effect of acceleration on a flat surface should be recalled. For K values of 2.5×10^{-6} or more, the flat-plate Stanton number displayed a characteristic -1 slope in enthalpy-thickness Reynolds number coordinates. Values of K less than about 1.0×10^{-6} have little effect on Stanton number.

In regions of convex curvature the effects of acceleration are augmented considerably. For $K = 0.57 \times 10^{-6}$, there is already a noticeable change in the Stanton number behavior, whereas for $K = 1.25 \times 10^{-6}$, which would have caused about a 20% reduction on a flat plate, the Stanton number falls rapidly in the curved region. Surprisingly, the data for $K = 1.25 \times 10^{-6}$ assume a slope of -2 in enthalpy-thickness Reynolds number coordinates, as though the effects of curvature and acceleration are simply additive.

Acceleration on a flat surface is known to act mainly on the inner layer, while curvature seems to act mainly in the outer region.

Convex curvature suppresses the tendency for transition from the laminar to the turbulent state. Figures 59 and 60 show data for three

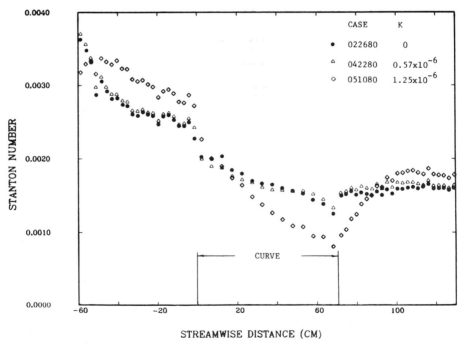

FIG. 57. The effect of free-stream acceleration on the curved boundary layer Stanton number versus streamwise distance.

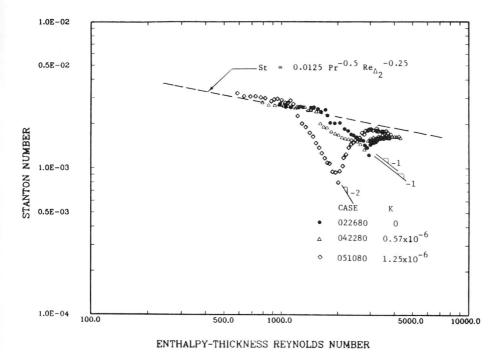

ENTHALPY-THICKNESS REYNOLDS NUMBER

FIG. 58. The effect of free-stream acceleration on the curved boundary layer Stanton number versus enthalpy thickness Reynolds number.

cases, spanning momentum-thickness Reynolds numbers of 222 to 724 at the beginning of curvature. In each case the boundary layer has begun the transition event just before curvature. The immediate effect of curvature is to depress the Stanton number. There then follows a slow rise until near the end of curvature, when the Stanton number begins to fall again along the line of -1 slope.

Figure 61 shows an isometric view of the shear-stress distributions measured by Gillis. Of particular importance are the rapid drop at the beginning of the curve and the fact that the surface shear stress stays low throughout the curved region. There are noticeable regions of negative Reynolds stress in the outer regions of the boundary layer within the curve. Only slowly does the shear stress return to a normal distribution after curvature ends.

The effect of curvature seems to be to compress the active turbulence into a thin layer near the wall, approximately $0.03R$ in thickness. This effect is visible in the distribution of mixing lengths calculated from measured profiles of shear stress and mean velocity, as shown in Fig. 62. Based on these data and other data that support the same conclusions, it

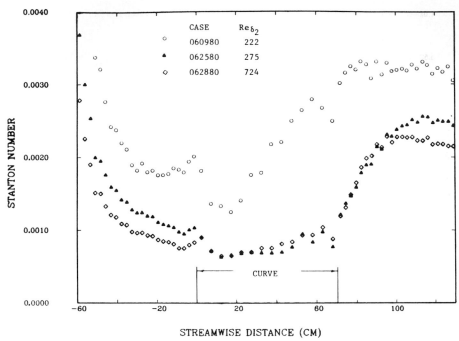

FIG. 59. The effect of the maturity of the momentum boundary layer Stanton number versus streamwise distance for the laminar and early transitional cases.

appears that the boundary layer on a convex wall should be modeled as a thin boundary layer buffered from the free stream by a thick layer of turbulent, but uncorrelated, fluid through which all heat transfer is by conduction alone.

3. Concave Curvature

When the boundary layer heat transfer community began to consider concave curvature, one of the earliest theories advanced was that the acknowledged increase in heat transfer was directly caused by streamwise vortices similar to the Taylor–Görtler vortices seen in laminar flow. Indeed, several studies reported evidence of spanwise regular disturbances in the mean velocity profiles, which seemed to be evidence that an array of vortices existed. To accept the notion that the vortices were responsible for the increased heat transfer would have committed future experiments and future modeling efforts to some very difficult problems indeed—three-dimensional boundary layers with significant large-scale structural elements. There was still no irrefutable evidence, however, that

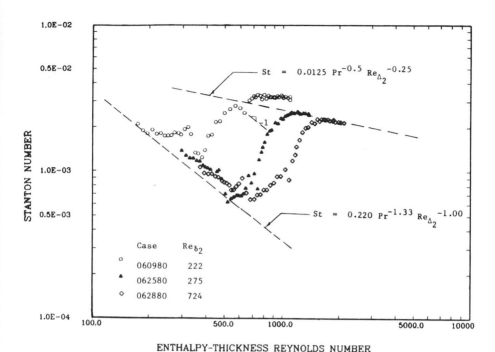

FIG. 60. The effect of the maturity of the momentum boundary layer Stanton number versus enthalpy thickness Reynolds number for the laminar and early transitional cases.

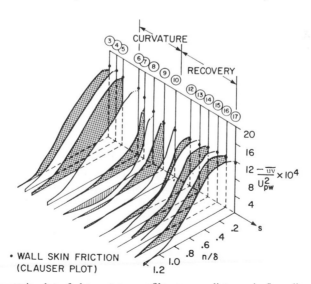

FIG. 61. Isometric plot of shear-stress profiles versus distance in flow direction for the second experiment.

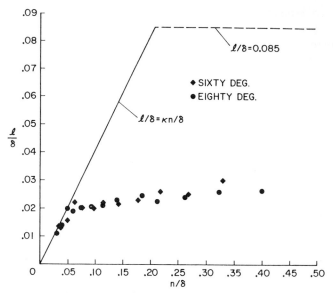

Fig. 62. Mixing-length profiles at stations near the end of curvature.

the vortices were responsible for the increase or, for that matter, that the vortices were even there. In view of the importance to future research plans and the lack of direct evidence in favor of the vortex theory, the first study focused on the question of whether or not the vortices were present and, if present, whether they were responsible for the increase in heat transfer.

The first research problem was to develop a field-mapping technique that would allow measurement of the heat transfer effects by streamwise vortices, if they were there. The problem was complicated by the fact that the vortices might not be stationary; they might meander across the surface, and the measurement technique had to prepare for that possibility.

The experiments were planned for a water channel, operating at 15 cm/sec with a test section that developed a uniform turbulent boundary layer with a momentum-thickness Reynolds number of about 2000 at the entry to the concave curved section. At this velocity, the dominant events showed frequencies of about 1 Hz.

A visualization technique was developed using a liquid-crystal layer on top of a constant heat flux surface that made visible (and photographable) regions of uniform heat transfer coefficient as regions of uniform color on the surface [45].

It was important that the surface be able to signal the presence of a meandering vortex, and a good deal of effort was put into testing the dynamic response of the visualization surface. The criterion for "fast enough" was that the surface would show the presence of a change in heat transfer coefficient as small as 15% occurring at 2 Hz or lower.

The steady-state and dynamic characteristics of the visualization system were sufficient to assure that, if streamwise vortices were responsible for the increase in heat transfer, their traces on the wall would be visible, whether they were stationary or not.

The visualization program reported by Simonich and Moffat [45] showed no evidence of streamwise vortices developing inside the curved region. It appears that two-dimensional computations will be successful in the concave region.

A companion program by Jeans and Johnston [15] investigated the hydrodynamics in detail and supported the finding that streamwise vortices were not present.

Subsequent studies by Barlow and Johnston [2] have shown that, although a uniform boundary layer may not generate new vortices, even small amounts of residual streamwise vorticity in the flow approaching the curve will cause the emergence of detectable patterns within the curved region.

Typical results are shown in Fig. 63 for a turbulent boundary layer on a flat wall. Flow was from left to right at 15 cm/sec, and the wall was of uniform temperature. The average Stanton number in this view was about 0.0010. The light-colored areas show where the Stanton number was 10% low, and the dark regions mark 10% high. The spanwise width of the streaks was 0.3–0.6 cm (z^+ from 25 to 50), while the spanwise spacing was about 1.1 cm (z^+ of about 90). The spanwise spacing agrees reasonably well with the classical spacing of $z^+ = 100$ reported for the spacing of hydrodynamic streaks in the near-wall region.

Experiments were then conducted with the liquid-crystal visualization surface applied along the concave wall. A fully turbulent boundary layer of normal properties and good spanwise uniformity was generated upstream of the curved wall.

The average heat transfer coefficient was higher in the curved region than flat-plate values for the same x-Reynolds number, as had been expected, but there was no evidence of streamwise roll cells. The same pattern of undulating streaks was observed on the curved wall as had been seen on the flat wall, but superimposed on that background pattern were occasional large, isolated, divergent patterns unlike anything seen in the flat-wall boundary layer studies.

Had the increase in average Stanton number been due to locally high heat transfer associated with roll cells, there would have been streamwise traces in the wall patterns. None was ever seen, even though the liquid-

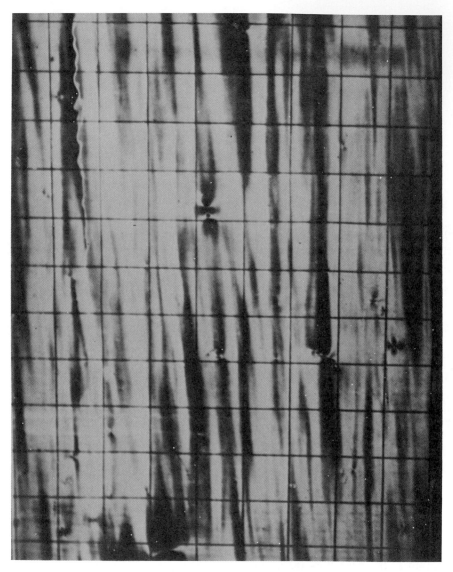

Fig. 63. Photograph of streamwise streak patterns obtained in a turbulent boundary layer by using the liquid-crystal technique.

crystal wall was judged to have adequate time response to make them visible had any been present. The small-scale spanwise variations in Stanton number were no larger in the curved region than in the flat.

Instead of streamwise streaks, evidencing roll cells, isolated large-scale patterns were seen on the curved wall having the forms shown in Fig. 64 (p. 332). The way in which these images appeared and disappeared suggested that they were caused by relatively large elements of fluid from the outer region of the boundary layer impacting on the wall and extruding wall region fluid into the "starburst" pattern illustrated. Flow visualization by Jeans and Johnston supports this idea. They suggest that a high level of turbulence is generated, and at large scale, when the low-momentum fluid ejected from the wall region is mixed into the outer region. This increase in turbulence is believed to be responsible for the increase in heat transfer.

These results are encouraging from the standpoint of modeling heat transfer on concave walls; it appears that a two-dimensional code should work as well in the concave region as in the convex. The isolated events observed are simply lower frequency and larger events than typically encountered. If they are random in space and time, their effects can be absorbed into conventional mixing length or turbulence kinetic energy models with relative assurance; they extend the low-frequency spectrum of turbulence and increase the mixing but do not change its nature.

G. FREE CONVECTION

The prospect of central solar power plants with external receivers brought to light new areas needing research. One particular problem which attracted attention was that of mixed convection with the buoyant force orthogonal to the shear force. This led to the research discussed in the following section on mixed convection. As part of that same program, as baseline data, heat transfer data were taken at high Grashof number, using both low- and high-temperature differences in pure free convection.

The end use of the mixed-convection research was to be in the design of large-scale central receivers: 10–15 m tall and 10–15 m in diameter. To minimize the extrapolation and because of the ambiguities that could arise in scaling up results from small models, a large-scale test was planned using a vertical plate 3 m tall and 3 m wide and operating at about 600°C.

When operated in pure free convection, this system yielded a Grashof number of 2×10^{12}, about one order of magnitude higher than any data in the literature. In addition, since the temperature of the plate was controllable, the same high Grashof numbers could be attained at different combinations of plate temperature and elevation on the plate. This provided

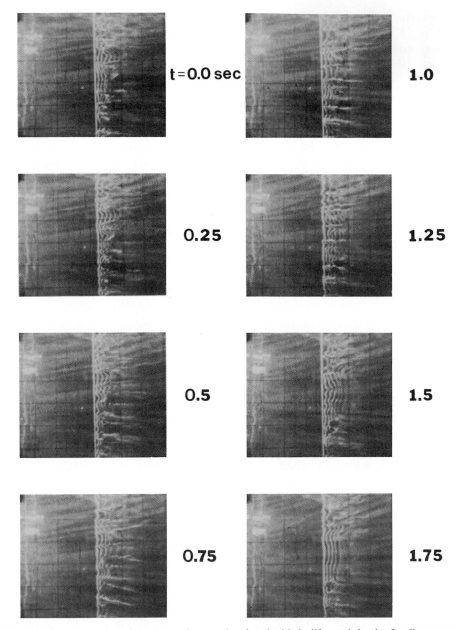

FIG. 64. Sequence of motion picture frames showing the birth, life, and death of a divergence.

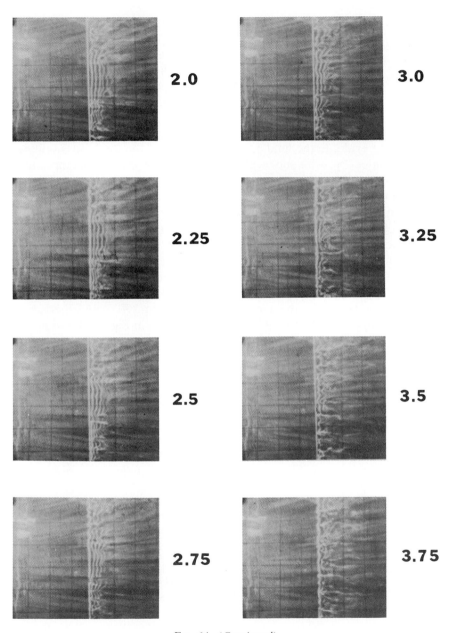

FIG. 64. (*Continued*)

an unparalleled opportunity to investigate the effects of variable fluid properties on free-convection heat transfer.

The results for low-temperature difference confirmed existing correlations, as far as they went, and showed that they extended accurately to higher Grashof numbers than had been established before. Of more importance, however, were the results relating to the variable properties effects.

In every calculation of heat transfer, one must choose the temperature at which to evaluate the fluid properties. Each of the three possibilities is recommended by someone: wall temperature, free-stream temperature, or an intermediate reference temperature. In every case, an additional correction is applied—the ratio of the chosen temperature to the free stream or the wall temperature, raised to some exponent. The better the choice of reference temperature, the lower will be the exponent needed on the temperature-ratio correction.

The present results clearly favor the free-stream temperature as the reference temperature for high-temperature free convection. As shown by Siebers *et al.* [43], an exponent of 0.14 will collect all the high- and low-temperature data onto the same line within 1 or 2%. Any other choice requires a larger exponent. The results are shown in Fig. 65. Fluid properties were evaluated at free-stream temperature in calculating both Grashof number and Nusselt number. With this choice of reference temperature, the effects of high wall temperature can be accounted for using a

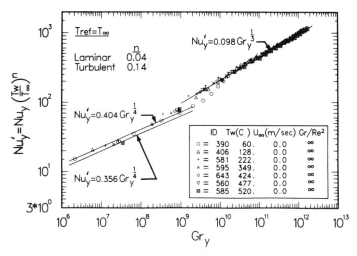

FIG. 65. Correlation of variable properties effects on free convection from a vertical surface in air ($T_\infty = 20°C$).

temperature-ratio correction (T_w/T_∞) with very small exponents: 0.05 in the laminar region and 0.14 in the turbulent region. Any other choice of reference temperature requires larger exponents to account for high wall temperatures.

The results, including the wall-temperature effect, are correlated by the following equation, within 6%:

$$\text{Nu}_y = 0.098 \text{Gr}_y^{1/3} \left(\frac{T_w}{T_\infty}\right)^{-0.14} \tag{33}$$

H. MIXED CONVECTION

Solar-powered central power plants received considerable attention during the late 1970s, some using internal or cavity receivers and some using external receivers. A typical external receiver is a cylinder 15 m in diameter and 15 m high, mounted on top of a tower about 100 m tall. The external surface is the boiler tube sheet, running at anywhere up to 600°C.

There are several interesting possibilities for research in that general situation: short cylinders in cross flow, transition on a high-temperature surface, roughness, curvature, and mixed convection. The mixed-convection problem was chosen because it was the one part of that picture for which there was no experimental or analytical support, because it fit well into the overall program of boundary layer heat transfer at Stanford, and because the results would add a whole new area to the convection literature.

Reviewing the work on three-dimensional boundary layers and boundary layers with more than one force mechanism soon revealed that scaling up results from small-sized tests to "solar receiver size" would introduce more uncertainty than could be tolerated. Consequently, a large-scale experiment was chosen: 3 m high and 3 m wide, in a 6 m/sec wind, and operated at 600°C. These conditions are close to the full-scale ones and, furthermore, offer the advantage that the boundary layers are thick enough for detailing traversing of the velocity and temperature fields in support of future modeling efforts.

A special wind tunnel was constructed for these tests, built around the requirements of accurate measurement of free-convection and mixed-convection heat transfer from the 3 m × 3 m plate. Boundary layer traversing was limited to temperature and mean velocity; no turbulence data were taken.

The free-convection baseline results in the preceding section as well as the mixed-convection results here were reported by Siebers et al. [43].

1. *Heat Transfer*

Experiments were conducted on a 3.0 m × 3.0 m vertical plate parallel to a horizontal wind to investigate the mixed convection problem with the buoyant force orthogonal to the shear force. Grashof number based on plate height ranged from 2×10^6 to 2×10^{12}, while the Reynolds numbers based on plate length in the flow direction ranged from 0 to 2×10^6. In calculating these values, fluid properties were evaluated at free-stream temperature.

Heat transfer coefficients were measured at 105 points on the surface, for each flow combination. The distribution of h was affected by both the Reynolds number and the Grashof number, as shown in Figs. 66, 67, 68. These figures show surfaces generated from bicubic splines that fit the 105 data points for each condition within 2%, on average. The largest differences, about 15%, occur in the transition regions, where streamwise gradients are largest.

The figures represent three different "cuts" through the operating domain of the experiment: one at constant velocity, one at constant wall temperature, and one at constant Gr/Re^2.

In Fig. 66 the lowest surface represents a situation dominated by forced convection. The ratio Gr/Re^2 was 0.6, with a velocity of 6.1 m/sec and a wall temperature of 234°C. The average h was 15.6 W/(m² °C) over the entire surface. Flow was from the right to the left ($X/L = 0.0$ is the convective leading edge), and a free-convection flow would move toward the front of the image ($Y/H = 0.0$ is the free-convection leading edge). There is a well-defined transition saddle near the convective leading edge and little evidence of free-convection effects.

Decreasing the velocity to 2.5 m/sec has three effects: the transition saddle is deepened, the turbulent forced-convection region flattens out (i.e., displays a lower slope), and a significant disturbance is seen near the free-convection leading edge, at the downwind end of the plate. At 1.5 m/sec, h is uniform over nearly all the plate, except in the forced-convection transition region, where a positive ridge appears.

At zero velocity a typical free-convection distribution is observed, with h uniform over the entire turbulent region.

Figure 67 shows a set of surfaces for constant velocity, with the wall temperature increasing from 54°C on the lowest sheet to 588°C on the top. The first effect of increasing the wall temperature was to compress the transition region dramatically.

Figure 68 shows four surfaces at constant Gr/Re^2. The distributions of h over the surface are similar, even though the average value of h increases from 8.5 to 11.0 as velocity and temperature rise.

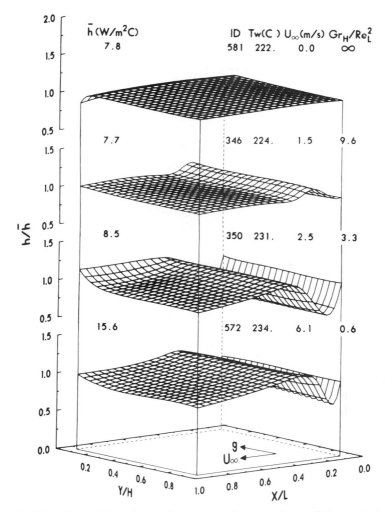

FIG. 66. The effect of U_∞ on the local convection heat transfer coefficient variation at a nominally constant T_w of 225°C.

The effects of mixed convection can be summarized by looking at the behavior of the average Nusselt number over the entire surface (laminar and turbulent regions) as a function of Reynolds number and Grashof number. Such a view is illustrated in Fig. 69.

The ratio Gr/Re^2 is significant. When Gr/Re^2 is less than 0.7, the average Nusselt number can be calculated within 5%, assuming that free convection has no effect. When Gr/Re^2 exceeds 10, the forced convection

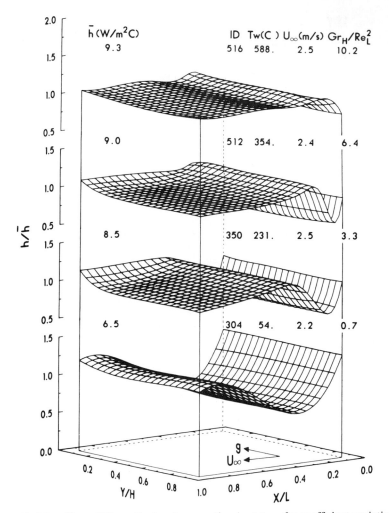

FIG. 67. The effect of T_w on the local convection heat transfer coefficient variation at a nominally constant U_∞ of 2.4 m/s.

can be ignored. Between 0.7 and 10.0, the flow must be regarded as mixed, and both effects considered.

Fluid properties were evaluated at free-stream velocity in all calculations. The reference length for Reynolds number was the plate length in the flow direction, whereas plate height was used in the Grashof number and a composite length used for Nusselt number. Since the plate was 2.95 m × 3.02 m, no significant error in understanding will result if the same length (3.0 m) is considered to have been used for all three length scales.

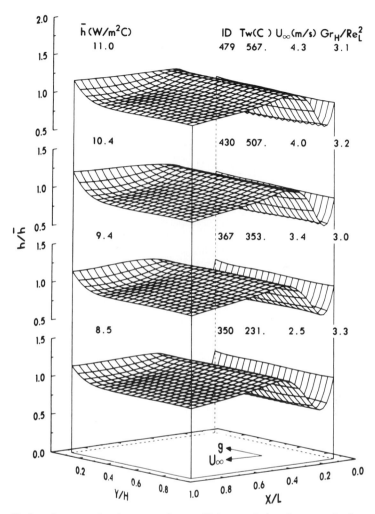

FIG. 68. Local convection heat transfer coefficient variation for a nominally constant Gr_H/Re_L^2 of 3.2.

To apply these results to cases in which the surface is not square, one should go to the original document [43]. Within the mixed-convection regime, the average h can be calculated from the pure forced and pure free values for the appropriate Reynolds numbers and Grashof numbers by a simple technique, like root-sum-square, but using an exponent of 3.2 instead of 2.0. The exponent is not critical; changing from 3.0 to 3.4 changes the prediction only 3%.

$$h_{mixed} = (h_{forced}^{3.2} + h_{free}^{3.2})^{1/3.2} \qquad (34)$$

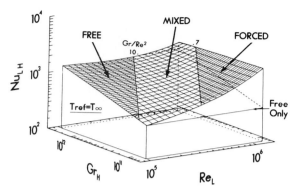

FIG. 69. The average Nusselt number based on LH versus the test surface Grashof and Reynolds number.

2. Velocity and Temperature Profiles

This same study investigated the distribution of velocity and temperature in the boundary layer, and the flow angle at the wall as functions of the Reynolds number and Grashof number.

Figure 70 shows measured and predicted wall-flow angles for a case of $Gr/Re^2 = 3.5$, about in the middle of the mixed-convection regime. The arrows represent measured wall-flow angles, while the lines were calculated from an algebraic model based on the assumption that the x-component of particle velocity was equal to the free-stream velocity and the y

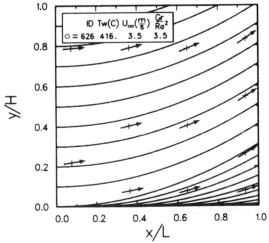

FIG. 70. Measured flow angles and predicted streamlines of the constant-angle region for $T_w = 416°C$, $u_\infty = 3.5$ m/s, and $Gr_H/Re_L^2 = 3.5$.

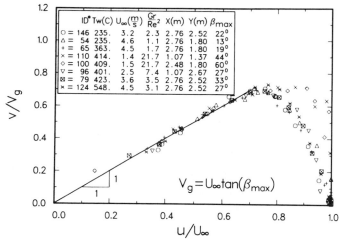

ID	Tw(C)	$U_\infty(\frac{m}{s})$	$\frac{Gr}{Re^2}$	X(m)	Y(m)	β_{max}
O = 146	235.	3.2	2.3	2.76	2.52	22°
△ = 54	235.	4.6	1.1	2.76	1.80	13°
+ = 65	363.	4.5	1.7	2.76	1.80	19°
× = 110	414.	1.4	21.7	1.07	1.37	44°
◇ = 100	409.	1.5	21.7	2.48	1.80	60°
▽ = 96	401.	2.5	7.4	1.07	2.67	27°
⊠ = 79	423.	3.6	3.5	2.76	2.52	33°
⋈ = 124	548.	4.5	3.1	2.76	2.52	27°

$$V_g = U_\infty \tan(\beta_{max})$$

FIG. 71. The horizontal velocity component versus the vertical velocity component for turbulent flow for several test conditions and various x and y locations.

component reflected the buoyant acceleration hence increased linearly with time.

The ratio of the horizontal component to the vertical component of velocity within the boundary layer is shown in Fig. 71 for eight different situations covering a range of Gr/Re^2 from 1.1 to 21.7. Remarkably, the profiles are identical in the inner region and clearly show the flow to be colinear. From the wall out to the location where $U/U_\infty = 0.7$, the ratio of the horizontal component to the vertical component is constant. This means that a large region of uniform flow angle exists near the wall within the turbulent mixed-convection boundary layer.

V. A Model for the Solution of the Momentum and Energy Equations

During the past two decades, enormous strides have been made in our ability to solve the partial differential equations of the boundary layer, using finite-difference techniques and the power of the digital computer. To all intents and purposes, mathematically exact solutions to the boundary layer equations can be obtained for virtually any kind of boundary conditions, provided that the turbulent transport processes are adequately modeled.

Although the primary emphasis of the Stanford turbulent boundary layer heat transfer program has been on obtaining definitive experimental data, a parallel effort since 1967 has been concerned with the develop-

ment of mathematical models that can be used, together with finite-difference computation procedures, to predict those data, and by inference to
predict boundary layer behavior in other applications, that is, to develop
an engineering design tool. A basic philosophy of the program that has
been consistently followed, however, is that the experimental data should
come first and the analytic work later, and not the other way around.

Various types of models have been investigated, but it has been found
that a relatively simple mixing-length model will predict virtually all the
experimental cases as well as or better than any of the higher-order
models investigated. This is partially because the experimental emphasis
has been heavily on equilibrium boundary layers, but the mixing-length
model has worked remarkably well for those nonequilibrium boundary
layers for which experimental data have been obtained. The fact is that
the turbulent boundary layer generally comes to equilibrium very rapidly,
at least in the inner region, and rather simple schemes can be used to
model the transition.

The mixing-length model described here provides a single continuous
scheme for calculating most of the boundary layers that have been investigated experimentally at Stanford. This includes the original constant-
velocity, constant-surface-temperature work in the 1950s, accelerated
flows, the transpired boundary layer with and without acceleration, adverse-pressure-gradient flows with and without transpiration, and the
rough-surface boundary layer with and without transpiration. The model
has not yet been perfected for a rough surface with strongly accelerated
flows, or for flow over curved surfaces. An extension of the model has
been developed for discrete-hole blowing, but this extension is fairly complex and will not be discussed here.

First the model will be described, and then several examples will be
given demonstrating the ability of the model to reproduce the previously
discussed experimental data for a variety of cases.

The time-averaged momentum equation of the boundary layer, particularized to constant fluid properties and neglecting normal turbulence
stresses, may be written as follows:

$$\bar{u}\frac{\partial \bar{u}}{\partial x} + \bar{v}\frac{\partial \bar{u}}{\partial y} - \frac{\partial}{\partial y}\left[\nu\frac{\partial \bar{u}}{\partial y} - \overline{u'v'}\right] + \frac{1}{\rho}\frac{d\bar{P}}{dx} = 0 \qquad (35)$$

If the turbulent shear stress $\overline{u'v'}$ is known at all points in the boundary
layer, the momentum problem becomes simply one of solution of Eq. (35)
for any desired boundary condition.

The concept of eddy diffusivity for momentum, ε_m, is used as a convenient way of expressing the turbulent shear stress.

$$\overline{u'v'} = \varepsilon_m \, \partial\bar{u}/\partial y \qquad (36)$$

Next, the Prandtl mixing length is defined:

$$\varepsilon_m = l^2 |\partial \bar{u}/\partial y| \tag{37}$$

It is convenient to visualize the turbulent boundary layer as consisting of an inner wall-dominated region and an outer region that actually occupies most of the thickness of the boundary layer.

The inner region may be subdivided further into a region immediately adjacent to the wall in which viscous forces predominate (ε_m approaches zero), and a region farther out in which momentum transfer is almost entirely by turbulent transport processes, but in which the scale and intensity of the turbulence are still strongly dependent on the proximity of the wall.

Outside the viscous-dominated region immediately adjacent to the wall, the mixing length in the inner part of the boundary layer is found to be proportional to distance from the wall, with a proportionality factor k, that is independent of transpiration, pressure gradient, and wall roughness. Figure 72 shows some measurements of the mixing length for a number of cases of transpiration, both blowing and suction, with no pressure gradient, and with an adverse pressure gradient. Results for favorable pressure gradients and wall roughness are similar. Note that all the data in the region near the wall converge on a single linear relation with

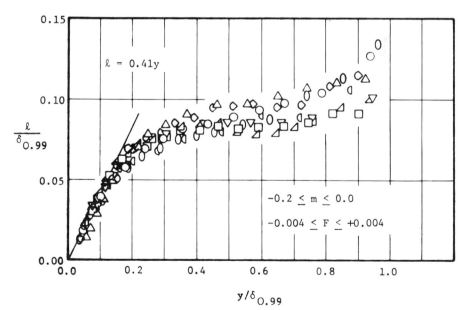

FIG. 72. The behavior of the mixing length for flat plate and decelerating flows with transpiration.

$k = 0.41$. The region outside the viscous near-wall region (called the viscous sublayer), but inside the outer, or "wake," region is modeled by

$$l = ky \tag{38}$$

where $k = 0.41$.

The viscous sublayer immediately adjacent to the wall can be adequately modeled by introducing a damping function that forces the mixing length to naught at the wall. Designating the damping function as D, the mixing length over the entire inner region may then be expressed as:

$$l = kyD \tag{39}$$

The damping function D can be satisfactorily expressed in a number of different ways. A popular scheme, which was first suggested by Van Driest (1956), is an exponential function that leads to mean velocity profiles that correspond quite well to those observed experimentally.

$$D = 1.0 - \exp(-y^+/A^+) \tag{40}$$

where y^+ is the nondimensional distance from the wall, expressed in "wall" coordinates, $y^+ = (y/\nu)\sqrt{\tau_0/\rho}$, and A^+ is the effective thickness of the viscous sublayer expressed the same way.

The effective thickness of the viscous sublayer is probably the single most important parameter in a mixing-length model. Failure to model it correctly is the main reason that for a long time mixing-length models were in bad repute. If this region is modeled accurately, only a very approximate scheme is needed throughout the rest of the boundary layer.

The thickness of the sublayer (the magnitude of A^+) is evidently determined by viscous stability considerations. The experimental evidence is that a favorable pressure gradient (dP/dx negative) results in increased thickness, while an adverse pressure gradient has the opposite effect. Transpiration into the boundary layer (blowing) decreases the thickness if it is expressed in nondimensional wall coordinates, while suction has the opposite effect. Surface roughness decreases the sublayer thickness, ultimately to zero for a "fully rough" surface.

The effects of pressure gradient and transpiration on A^+ are conveniently expressed in terms of a nondimensional pressure-gradient parameter p^+ and a nondimensional blowing parameter v_0^+, both of which can be either positive or negative. The effects of a rough surface can be expressed in terms of the roughness Reynolds number, $\mathrm{Re}_k = u_\tau k_s/\nu$, where k_s is the equivalent "sand-grain" roughness size. Note that all three of these parameters are expressed in "wall" coordinates.

The functional dependence of A^+ upon these parameters has been deduced experimentally by examination of a very large number of velocity profiles obtained as part of the Stanford program. Some reasonably adequate theories have been developed, based on stability considerations, to relate A^+ to p^+ and v_0^+, but it has proved simpler and computationally more satisfactory to use an empirically based algebraic equation:

$$A^+ = \frac{25.0R}{a[v_0^+ + bP^+/(1 + cv_0^+)] + 1.0}$$
(41)

where

$a = 7.1,$ $b = 5.24,$ $c = 10.0$

If $p^+ > 0.0,$ $b = 2.9,$ $c = 0.0$

If $v_0^+ < 0.0,$ $a = 9.0$

If $Re_k \leqq 7,$ $R = 1.0;$ if $7 < Re_k < 55,$ $R = (4.007 - \ln Re_k)/2.061$

If $Re_k > 55,$ $R = 0.0.$

    Equation (41) is plotted on Fig. 73, where the effects of pressure gradient and transpiration can be clearly seen (this plot is for a smooth wall). Note that a strong favorable pressure gradient forces A^+ to very high values, and that blowing lessens this effect, while suction increases it. If A^+ becomes very large, the viscous sublayer simply overwhelms the entire boundary layer, and this is the "laminarization" discussed earlier. In

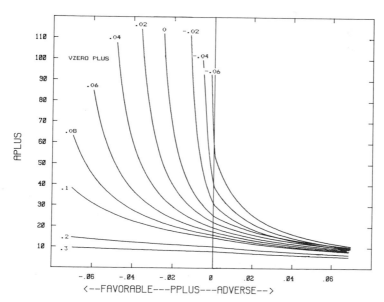

FIG. 73. The variation of A^+ with P^+ and v_0^+.

fact, most of the trends noted in connection with the experimental data on Stanton number are recoverable by varying the values of A^+. The thickening of the sublayer caused by a favorable pressure gradient (accelerating flow) results in a decreased Stanton number, simply because the major resistance to heat transfer is in the viscous sublayer.

Note that an adverse pressure gradient causes a decrease in sublayer thickness, in y^+ coordinates. Interestingly, where these results are used to compute velocity profiles for adverse pressure gradients with no transpiration, and when the velocity profiles are plotted on u^+, y^+ coordinates, they tend to fall on approximately the same line obtained for no pressure gradient in the near-wall region, but outside the sublayer: the "law of the wall." This has long been noted to be universal for no-pressure-gradient and adverse-pressure-gradient flows. The conventional "law of the wall" does not apply, however, for strong favorable pressure gradients. The applicability of the "law of the wall" for adverse pressure gradients results from compensating effects of the decreased sublayer thickness and a shear stress that rises with distance from the wall.

Turning now to the effect of roughness, recall that three regions are identifiable. For $\mathrm{Re}_k < 7$, the surface is "smooth" and there is no effect. For $7 < \mathrm{Re}_k < 55$, there is a transition region where A^+ tends to zero. For $\mathrm{Re}_k > 55$, $A^+ = 0.0$ and there is effectively no sublayer. A logarithmic function is used in the transition, based on the Stanford experiments in which the roughness was obtained by closely packed spheres in a regular rectangular pattern. As shown earlier, in Fig. 46, there is evidence that other types of roughness will lead to a somewhat different function in the transition region, but the two limits of the transition region seem to be fairly universal. It should be noted also that transpiration seems to have an influence on the effective value of roughness Reynolds number. The following modified definition of roughness Reynolds number models this influence, which really becomes of more importance in the fully rough region:

$$\mathrm{Re}_k = k_s(u_\tau \rho + 14\dot{m}''/\mu \tag{42}$$

In effect, this modification suggests that, as blowing becomes stronger, the actual transpiration velocity v_0 replaces the shear velocity u_τ as the significant velocity for normalizing the roughness size.

A^+ as represented by Eq. (41) has been evaluated under essentially equilibrium conditions, that is, conditions under which v_0, p^+, and Re_k are invariant or, at worst, are varying only slowly along the surface. This is a case of inner-region equilibrium. It is probable that, when a sudden change of external conditions is imposed, the inner region comes to equilibrium more rapidly than the outer region, although this has not been proved. In any case, under nonequilibrium conditions where p^+ or v_0^+ are changing rapidly, it has been observed that the sublayer does not change instantaneously to its new equilibrium thickness, that is, A^+ does not immediately assume its new equilibrium value. A reasonably satisfactory expedient is to use a rate equation of a type suggested by Launder and Jones [21]:

$$dA_{\text{eff}}^+/dx^+ = (A_{\text{eff}}^+ - A_{\text{eq}}^+)/C \tag{43}$$

where $x^+ = xu_\tau/\nu$.

A_{eff}^+ is the locally effective value of A^+, while A_{eq}^+ is the equilibrium value obtained from Eq. (41). A value of C of about 3000 has been found to be reasonable. (An alternative procedure is to use the same rate equation to evaluate effective values of p^+ and v^+ separately before evaluating A^+. This removes any lag in the effect of Re_k, which is probably realistic.)

For the case of the rough surface, there remains another problem. In the fully rough regime, the flow is observed experimentally to be independent of viscosity. Thus viscous shear cannot be an important mechanism transmitting shear stress to the wall. The shear is transmitted to the wall in this case by pressure forces acting on the roughness elements. This can be easily modeled by providing for a nonzero value of mixing length at the wall. A simple way to do this is to modify Eq. (39) as follows:

$$l = k(y + \delta y_0)D \tag{44}$$

At $y = 0$ in the fully rough region, $D = 1.0$ and the mixing length becomes δy_0, which can be logically expressed in nondimensional form using "wall" coordinates:

$$\delta y_0^+ = \delta y_0 u_\tau/\nu \tag{45}$$

It seems plausible that δy_0 would scale on k_s. Thus one might expect δy_0^+ to be proportional to Re_k. This is indeed found to be the case, at least at sufficiently high values of Re_k that all effects of viscosity can be neglected. The following equation models the experimental data very well for $\text{Re}_k > 43$. Below 43, δy_0^+ may be assumed to be zero.

$$\delta y_0^+ = 0.031(\text{Re}_k - 43) \tag{46}$$

All the discussion up to now has been concerned with the inner region of the boundary layer. The outer region, comprising the greater part of the boundary layer thickness, is of considerably less importance in predicting performance and thus can be handled successfully using more gross approximations. This statement may not be valid for strongly nonequilibrium boundary layers, especially under adverse-pressure-gradient conditions. Its validity for accelerating flows, with and without transpiration and for rough-surface boundary layers, will be demonstrated later. In any case, for equilibrium or near equilibrium boundary layers, either the assumption of a constant value of eddy diffusivity over the entire outer region or the assumption of a constant value of mixing length over the entire outer region yields approximately the same result and is quite adequate for most applications. If a constant eddy diffusivity is used, an empirical correlation of eddy diffusivity as a function of either displacement or momentum-thickness Reynolds number can be obtained. However, if mixing length is used in the inner regions, it is computationally simpler to use the mixing-length concept for the entire boundary layer.

Figure 72 shows measured mixing-length data for the outer region for a number of cases of transpiration with and without pressure gradient. The adequacy of an assumption that the mixing length is constant in the outer region may be judged from these data. A further simplification is also illustrated in this figure. The outer-region mixing-length scales approximately on the total boundary layer thickness. A satisfactory computation scheme is to express the outer-region mixing length as a fixed fraction λ of the 99% boundary layer thickness.

$$l = \lambda \delta_{0.99} \tag{47}$$

A value of $\lambda = 0.085$ works reasonably well over the entire range of experimental data discussed here, including favorable and adverse pressure gradients, blowing and suction, and rough walls. One then simply evaluates l from Eq. (44) until the value obtained equals that given by Eq. (47), and then uses the latter value for the remainder of the boundary layer.

There is some evidence that the effective value of λ is larger than 0.085 for boundary layers for which the momentum-thickness Reynolds number is less than about 5500. This may be the result of the fact that at low Reynolds numbers the sublayer is a larger fraction of the boundary layer and the approximation of a constant mixing length over the remainder of the boundary is less valid. For strong blowing, even at low Reynolds numbers, λ again appears to be close to 0.085, and this is consistent with the preceding explanation because the sublayer is then thinner. The following equation describes the observed low Reynolds number behavior quite well:

$$\lambda = 0.025 \, \mathrm{Re}_m^{-1/8}(1 - 67.5\dot{m}''/u_\tau \rho) \tag{48}$$

if $\lambda < 0.085$; $\lambda = 0.085$.

The time-averaged energy equation of the boundary layer, particularized to constant fluid properties and negligible viscous dissipation and neglecting turbulent conduction in the streamwise direction, may be written:

$$\bar{u}\frac{\partial \bar{t}}{\partial x} + \bar{v}\frac{\partial \bar{t}}{\partial y} - \frac{\partial}{\partial y}\left[\alpha\frac{\partial \bar{t}}{\partial y} - \overline{t'v'}\right] = 0 \tag{49}$$

This equation can be solved for any desired boundary conditions, provided the velocity field has been established first by solution of the momentum equation and provided that we have information on the turbulent heat transfer rate, $\overline{t'v'}$.

The concept of eddy diffusivity for heat, ε_h, is used, analogous to the method of solution of the momentum equation:

$$\overline{t'v'} = -\varepsilon_h\, \partial\bar{t}\, \partial y \tag{50}$$

Although it might be fruitful to attempt to evaluate either $\overline{t'v'}$ or ε_h on the basis of assumptions that are independent of the turbulent shear stress, it does seem plausible that there is some kind of correlation between $\overline{t'v'}$ and $\overline{u'v'}$, or ε_h and ε_m. Therefore, most analysts have found it convenient to introduce the concept of a turbulent Prandtl number, Pr_t, defined as follows:

$$Pr_t = \varepsilon_m/\varepsilon_h \tag{51}$$

Introducing Eqs. (50) and (51) into Eq. (49), we obtain:

$$\bar{u}\frac{\partial \bar{t}}{\partial x} + \bar{v}\frac{\partial \bar{t}}{\partial y} - \frac{\partial}{\partial y}\left[\left(\alpha + \frac{\varepsilon_m}{Pr_t}\right)\frac{\partial \bar{t}}{\partial y}\right] \tag{52}$$

If Pr_t were known, Eq. (52) could be solved for any desired boundary conditions, so long as the momentum equation had been previously solved.

A very simple physical model of the turbulent momentum and energy-transfer processes leads to the conclusion that $\varepsilon_h = \varepsilon_m$, that is, $Pr_t = 1.00$ (the Reynolds analogy). Slightly more sophisticated models suggest that $Pr_t > 1.00$ when molecular Prandtl number, Pr, is less than unity. Still other models suggest that Pr_t equals 0.7 or 0.5 in turbulent wakes.

The experimental data are not abundant, but Figs. 74–79 show the measurements, respectively, of Simpson et al. ([48]), Kearney ([17]), Blackwell ([4]) (two graphs), Orlando et al. ([31]), and Pimenta ([32]), all with air as the working fluid. These were all evaluated from measurements of the slopes of mean velocity and temperature profiles, together with estimates of shear stress and heat flux, and the experimental uncertainty is high, especially near the wall ($y^+ < 20$) and near the outer edge of the boundary layer. The data on Fig. 74 are all for constant free-stream

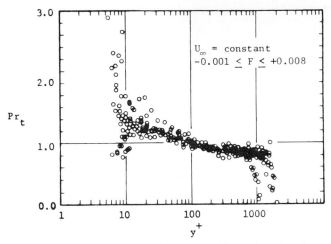

FIG. 74. The variation of turbulent Prandtl number within the boundary layer for flat plate flow with transpiration.

velocity, but cover a wide range of blowing and suction conditions. The data in Fig. 75 are for accelerated flows with a considerable range of blowing. Figure 76 shows three separate test runs with no transpiration, the first with no pressure gradient, and then two successively stronger cases of equilibrium adverse pressure gradients. Figure 77 shows three test runs for an adverse pressure gradient with three cases of successively stronger blowing. Figure 78 shows the results of a series of test runs, starting with no pressure gradient and no transpiration, and then proceed-

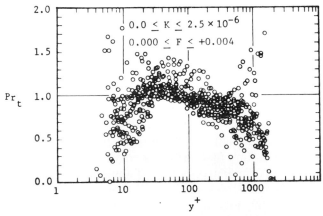

FIG. 75. The variation of turbulent Prandtl number within the boundary layer for accelerating flows with transpiration.

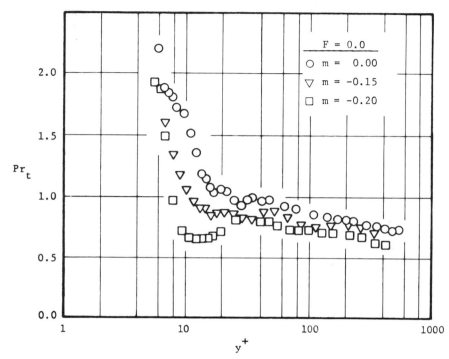

FIG. 76. The variation of turbulent Prandtl number within the boundary layer for decelerating flows with transpiration.

ing with four runs at a rather strong adverse pressure gradient and successively greater suction. Finally, Fig. 79 shows the results of one test run for a fully rough surface with no pressure gradient and no transpiration. In general, it can be assumed that the more recent data have a smaller experimental uncertainty because of improvements in experimental technique.

Despite the very considerable scatter of data, a few conclusions seem definitely warranted. First, in the region around y^+ of 50–500, the turbulent Prandtl number, at least for air, seems to be about unity: the Reynolds analogy ($Pr_t = 1.0$) is not a bad approximation.

The second conclusion is that Pr_t seems to go to a value higher than unity very near the wall, and less than unity in the outer region. The situation very close to the wall is especially vexing, because it is extremely difficult to make accurate measurements in this region, and yet it seems evident that something interesting and important is happening in the range of y^+ from 10.0 to 15.0. The behavior of Pr_t at values of y^+ less than about 10.0 is highly uncertain but fortunately not very important,

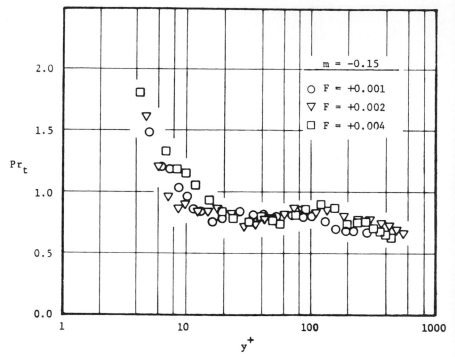

FIG. 77. The variation of turbulent Prandtl number within the boundary layer in a mild deceleration with transpiration.

because molecular conduction is the predominant transfer mechanism in this region. At the other extreme, in the wake region Pr_t does not need to be known precisely, because the heat flux tends to be small there.

Another conclusion, for which the evidence is not very strong, is that there is some small effect of axial pressure gradient. Figures 76 and 77 suggest that an adverse pressure gradient tends to decrease Pr_t in the middle region.

The results in Fig. 74 suggest that blowing does not influence Pr_t. This conclusion is also implied by the results in Fig. 75. On the other hand, the data of Orlando in Fig. 78 indicate that suction tends to increase Pr_t while at the same time confirming the pressure gradient effect seen in the Black-well data. The results of Pimenta in Fig. 79 for a fully rough surface suggest that roughness has no effect. The Pimenta data are all for y^+ considerably in excess of 100.

Most analysts have been content to assume that turbulent Prandtl number is a constant throughout the boundary layer; indeed, the assumption that $Pr_t = 0.9$, for air, will generally yield satisfactory predictions of

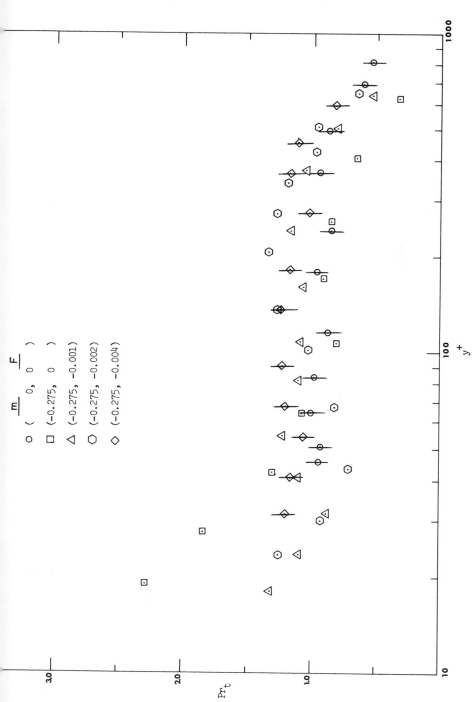

FIG. 78. Distribution of turbulent Prandtl number in a turbulent boundary layer in an adverse pressure gradient with suction.

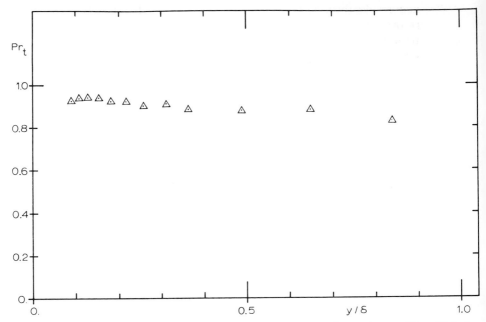

FIG. 79. Distribution of the turbulent Prandtl number for a fully rough state ($U_\infty = 89$ fps).

overall heat transfer rates. However, the assumption of constant Pr_t yields temperature profiles that do not correspond well with experiment except in the regions very close to the wall and near the outer edge of the boundary layer. Temperature profiles can be much more accurately predicted if some attempt is made to introduce a variation of Pr_t with y^+ that at least approximates the variation seen in the experimental data. The following equation has been developed on the basis of a "thermal" model of the turbulent heat transfer process, that is, a model that takes into consideration the thermal conductivity of the fluid as the mechanism for transferring heat to and from a turbulent eddy. It is an empirical equation in the sense that two unknown constants are evaluated from the experimental data.

$$Pr_t = \frac{1}{\dfrac{1}{2C_2} + C_1(\varepsilon_m/\nu)\,Pr\,\sqrt{\dfrac{1}{C_2}} - (C_1(\varepsilon_m/\nu)\,Pr)^2}\left[1 - \exp\!\left(-\frac{1}{C_1(\varepsilon_m/\nu)\,Pr\,\sqrt{C_2}}\right)\right] \qquad (53)$$

$C_1 = 0.2, C_2 = 0.86$

Although it may be for the wrong reasons, Eq. 53 does in fact reproduce rather accurately the rise in Pr_t that is observed in the sublayer region. Furthermore, it does seem to correspond fairly well with most of the experimental data for low Prandtl number fluids (liquid metals) where Pr_t appears to be generally well above 1.00. On the other hand, this model does not include the apparent effect of pressure gradient discussed above nor the possibility that strong suction may increase Pr_t. In the section to follow, one effect of the omission of a pressure gradient effect will be illustrated.

VI. Some Examples of Boundary Layer Predictions

The quality of boundary layer predictions that can be made using the mixing-length model described in the previous section will now be demonstrated. Six examples have been chosen for illustration. The first is the case of the simple, smooth, impermeable wall with no pressure gradient. This is both an equilibrium momentum boundary layer and an equilibrium thermal boundary layer. The second is an adverse-pressure-gradient equilibrium boundary layer. The third is an adverse pressure boundary layer with strong blowing that is not precisely an equilibrium boundary layer but shows near-equilibrium characteristics. The fourth is a strongly accelerated boundary layer with strong blowing, in which both blowing and acceleration are abruptly stopped at different points along the surface to yield nonequilibrium conditions. The final two are fully rough surfaces, the first with constant free-stream velocity and no transpiration, and the second with strong blowing.

A modification of the Spalding and Patankar [49] finite-difference program was used for all predictions, although any good finite-difference procedure should yield similar results.

Figure 80 shows $c_f/2$ as a function of momentum-thickness Reynolds number for the simple, smooth, impermeable wall with no pressure gradient. Shown for comparison is the recommendation of Coles [10], based on an extensive examination of the available data, and also two sets of data from the Stanford program—the earlier results of Simpson [47], and the more recent results of Andersen [1]. The predicted friction coefficients coincide closely with Cole's, and, indeed, the auxiliary functions in the model were chosen to force this coincidence.

The corresponding heat transfer results are shown in Fig. 81, where comparison is made with two sets of data from the Stanford program—the results of Whitten [56] and of Blackwell [4]. The Blackwell data are a little lower than would be expected for a corresponding equilibrium thermal

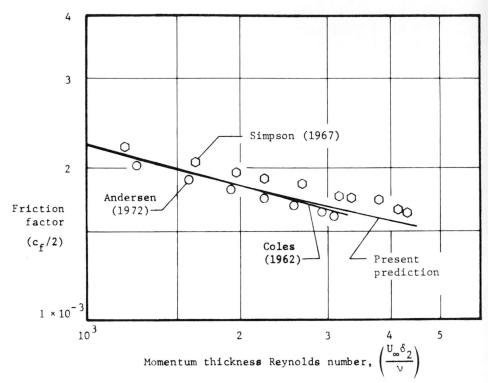

FIG. 80. Comparison of measured and predicted values of friction factor for a flat plate flow without transpiration.

boundary layer, because in his tests the thermal boundary started out at the beginning of the test section much thinner than the momentum boundary layer; that is, Re_h was consistently lower than Re_m. The prediction also corresponds almost precisely with the equation in Fig. 1.

It should also be added that all the experimental results shown were obtained using low-velocity air with temperature differences from 25 to 35°F. Although the influence of temperature-dependent fluid properties have not been systematically investigated as a part of this program (indeed, small temperature differences were deliberately used to avoid this problem), calculations with the computer program using real properties of air suggest that the temperature-difference effect would be to reduce the experimental Stanton number by about 1 or 2% for the conditions used. This effect has not been considered in any of these results; the predictions have been made using constant properties, and the experimental data have not been corrected for any variable properties effects, with the sole exception of the original data of Reynolds et al. [34–37].

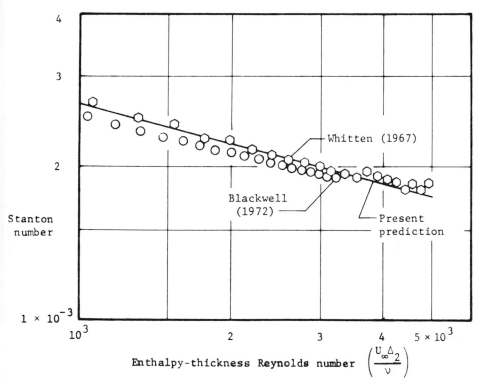

FIG. 81. Comparison of measured and predicted values of Stanton number for a flat plate flow without transpiration.

Note that if the friction coefficient prediction in Fig. 80 is acceptable, the heat transfer prediction in Fig. 81 is entirely dependent on the turbulent Prandtl number function employed, because everything else in the model is identical.

In the upper part of Fig. 82, both friction and heat transfer results are shown for an adverse pressure gradient test run with no transpiration. For this case, U_∞ was varied:

$$U_\infty = U_1 x^m \tag{54}$$

where $m = -0.15$. Both the Clauser shape factor G and β were found to be essentially constant for the experimental data over most of the test section, so this is believed to be an equilibrium boundary layer.

The prediction program also produced essentially constant values of G and β. The friction prediction is excellent, but the heat transfer prediction is about 5% low. Experimental uncertainty may account for this difference, but it is also quite possible that we see here evidence of a pressure

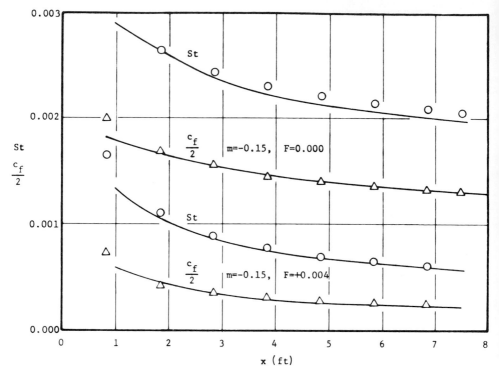

FIG. 82. Comparison of measured and predicted values of Stanton number and friction factor in a mild deceleration with transpiration.

gradient influence on turbulent Prandtl number. A 10% difference in Pr_t throughout the boundary layer would account for the difference.

The scheme described not only predicts $c_f/2$ and St quite adequately, but it does equally well for velocity and temperature profiles. Figure 83 shows a pair of profiles for the adverse-pressure-gradient, strong blowing case discussed earlier. These are presented in dimensional coordinates so that normalization will not tend to mask anything, and are presented for a point 70 in. downstream, so that a small percent drift of the predictions would show as a large effect. The results shown on this figure would be hard to improve upon.

The next illustration, Fig. 84, shows an example of prediction of a very difficult case. In this run the flow starts at constant free-stream velocity but with relatively strong blowing, $F = 0.004$. This flow is then subjected to a very strong acceleration starting at $x = 2$ ft. In approximately the middle of the accelerated region, the blowing is removed entirely. Then at

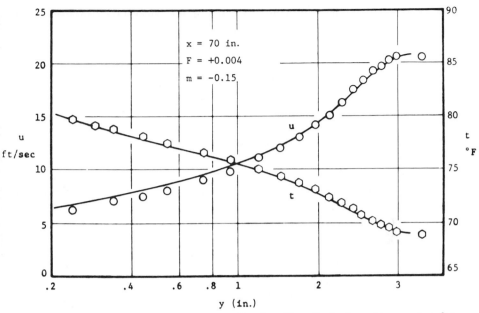

FIG. 83. Comparison of measured and predicted profiles of velocity and temperature for mild deceleration with transpiration.

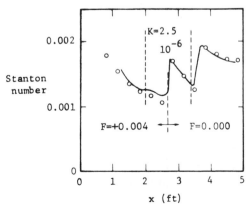

FIG. 84. Comparison of measured and predicted Stanton numbers for a step decrease in blowing within a strong acceleration.

about $x = 3.4$ ft the acceleration is removed, and for the remainder of the test section there is no blowing and no change in free-stream velocity.

The important thing to note here is that the model responds remarkably to the abrupt changes in boundary conditions and predicts the resulting nonequilibrium boundary layer very well indeed. Of particular significance is the abrupt rise in Stanton number following the removal of blowing. The ability of the prediction to follow the data at this point is heavily dependent on the use of the rate equation and lag constant, Eq. (43). This shows very graphically the importance of the sublayer and the fact that the sublayer does not instantaneously assume its new equilibrium thickness after an abrupt change of boundary conditions. There may well be significant nonequilibrium effects in the outer part of the boundary layer, but these have a relatively minor influence on overall heat transfer rates.

All the preceding predictions have been for a smooth surface. In Fig. 85 are shown the results of two test runs with a fully rough surface (spherical, regularly spaced roughness elements). Both are for constant free-stream velocity, but the second includes fairly strong blowing. In each case, Stanton number is plotted as a function of distance x along the surface. The prediction in both cases is close to perfect. Finally, in Fig. 86 is shown a prediction of both temperature and velocity profiles for the

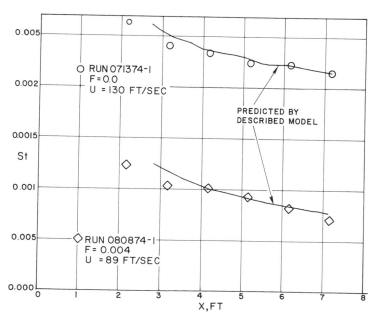

FIG. 85. Comparison of measured and predicted Stanton number distribution along a rough surface with and without blowing.

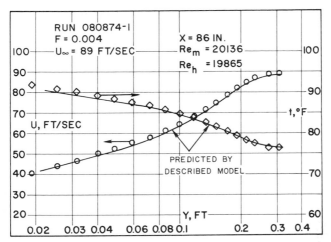

FIG. 86. Comparison of predicted and measured distributions of velocity and temperature in a turbulent boundary layer on a rough wall.

strongly blown case. These are plotted in dimensional coordinates to eliminate the usual advantages of normalization and are for a point near the trailing edge of the test surface. The excellence of the prediction is apparent.

NOMENCLATURE

A^+ dimensionless length scale for the damping function, Eq. (41)

B offset parameter in velocity distribution, Eq. (30)

B_h blowing parameter of the heat transfer problem, Eq. (22)

b_h blowing parameter of the heat transfer problem, Eq. (23)

B_m blowing parameter of the momentum problem, Eq. (6)

c specific heat at constant pressure, Btu/(lb_m °F)

c_f coefficient of skin friction

c_{f_0} coefficient of skin friction with no transpiration, other factors remaining constant

D damping function for mixing length

F blowing fraction, $\rho_0 v_0/\rho_\infty U_\infty$, or v_0/U_∞ if ρ = constant

g_c gravitational constant, ft/sec² lb_f/lb_m

G Clauser shape factor, Eq. (14)

G_∞ mass velocity of the free stream (e.g., lb_m/ft² sec)

Gr_y Grashof number based on y dimension $g\beta \Delta Ty^3/\nu^2$; also Gr

h heat transfer coefficient, Eq. (34)

h_{forced} heat transfer coefficient by forced convection

h_{free} heat transfer coefficient for pure convection

H shape factor, δ_1/δ_2, or total height of a free convection surface

k mixing length constant; also κ

k_s equivalent sand-grain roughness size, Eq. (30)

K acceleration parameter, $(\nu/U_\infty^2)(dU_\infty/dx)$, Eq. (3)

l mixing length

l_T length scale of turbulence

M injection parameter for discrete-hole injection, $\rho_0 v_0/\rho_\infty U_\infty$ in the jets

m	distance, in meters	St_0	Stanton number with no transpiration, other factors remaining constant
m	exponent describing free-stream velocity variation in decelerating flows, Eq. (4)		
		s	time interval, seconds
$\dot{m}''$	transpiration rate [e.g., lb_m/(sec ft^2)]	t	temperature
		t_0	surface temperature; also T_w
Nu_y	Nusselt number based on y dimension, Eq. (33)	$\bar{t}$	dimensionless temperature, Eq. (28)
P	pressure	t^+	dimensionless temperature, Eq. (25)
p^+	dimensionless pressure, $(\nu/\rho U_\tau^3)\dfrac{dp}{dx}$	t_d^+	dimensionless temperature defect, Eq. (18)
Pr	Prandtl number	t_∞	temperature of free stream; also T_∞
Pr_t	turbulent Prandtl number, Eq. (51)	U	velocity (e.g., ft/sec); also $\bar{u}$
$\dot{q}_0''$	surface heat transfer rate [e.g., Btu/(sec ft^2)]	u'	fluctuating component of u, ft/sec
R	radius of curvature of curved surface	$\bar{u}$	dimensionless velocity, u/U_∞, or time-averaged velocity; also U
r	radius of spherical elements on the Stanford rough surface	u^+	dimensionless velocity, u/U_τ
		$\overline{u'v'}$	turbulent shear stress, Eq. (35)
Re_h	enthalpy-thickness Reynolds number, $G_\infty \Delta_2/\nu$	U_∞	velocity of the free stream, ft/sec
Re_k	roughness Reynolds number, Eq. (42)	U_τ	shear velocity, $\sqrt{g_c \tau_0/\rho}$
		v_0	velocity of the transpired fluid, at the wall
Re_m	momentum-thickness Reynolds number, $G_\infty \delta_2/\nu$	v_0^+	velocity of the transpired fluid, at the wall, dimensionless, v_0/U_τ
Re_t	turbulence Reynolds number, Eq. (50)	v'	fluctuating component of v
Re_x	x-Reynolds number, $G_\infty x/\nu$	$\bar{v}$	dimensionless velocity, u/U_∞, or time-averaged velocity
St	Stanton number	x	distance in the streamwise direction
$St(0)$	Stanton number with injected fluid at wall temperature, Eq. (32)	y	distance normal to the wall
$St(1)$	Stanton number with injected fluid at free-stream temperature, Eq. (32)	y^+	dimensionless distance from the wall, yU_τ/ν
$St(\theta)$	Stanton number with injected fluid at temperature θ, Eq. (32)	z	distance normal to the flow, parallel to the surface, ft

Greek Symbols

α	thermal diffusivity, $k/\rho c$	ε_m	eddy diffusivity for momentum, Eq. (36)
β	pressure gradient parameter, $(\delta_2/\tau_0)(dP/dx)$, Eq. (9)	ξ	unheated length, Eq. (1)
$\delta_{0.99}$	velocity thickness (to 0.99 U_∞)	κ	mixing length constant, Eq. (30); also k
δ_2	momentum thickness		
δ_3	velocity defect thickness, Eq. (13)	θ	temperature ratio for discrete hole injection ($t - t_\infty/t_0 - t_\infty$)
Δ_2	enthalpy thickness (abbreviated as Δ in Eq. (26) et seq.)	ρ_0	density of fluid at the wall
Δ_3	temperature defect thickness, Eq. (19)	ρ_∞	density of fluid in the free stream
		τ_0	shear stress at the wall
ε_h	eddy diffusivity for heat, Eq. (51)	ν	kinematic viscosity

ACKNOWLEDGMENTS

The results presented in this report come from the dissertations of the heat transfer group from 1958 through 1983. Each of those researchers made a unique contribution to our present understanding.

The authors would like to express their thanks to the agencies that supported the work reported here: the National Science Foundation, the National Aeronautics and Space Administration, the Office of Naval Research, and the Air Force Office of Scientific Research.

REFERENCES

1. Andersen, P. S. (1972). "The Turbulent Boundary Layer on a Porous Plate: An Experimental Study of the Fluid Mechanics for Adverse Free Stream Pressure Gradients," Ph.D. Thesis, Stanford Univ., Stanford, Calif.
2. Barlow, R., and Johnston, J. P. (1983). Work in progress.
3. Black, T. J., and Sarnecki, A. J. (1965). "The Turbulent Boundary Layer with Suction or Injection," A.R.C. R & M 3387.
4. Blackwell, B. F. (1972). "The Turbulent Boundary Layer on a Porous Plate: An Experimental Study of the Heat Transfer Behavior with Adverse Pressure Gradients," Ph.D. Thesis. Stanford Univ., Stanford, Calif.
5. Bradshaw, P. (1967). The turbulence structure of equilibrium boundary layers. *J. Fluid Mech.* **29,** Part 4, 625–645.
6. Bradshaw, P. (1969). A note on reverse transition. *J. Fluid Mech.* **35,** Part 2, 387–390.
7. Choe, H., Kays, W. M., and Moffat, R. J. (1976). The turbulent boundary layer on a full-coverage, film-cooled surface: An experimental heat transfer study with normal injection. *NASA [Contract. Rep.] CR* **NASA—CR-2642.**
8. Clauser, F. H. (1954). Turbulent boundary layers in adverse pressure gradients. *J. Aero. Sci.* **21,** 91.
9. Coleman, H. W., Pimenta, M. M., and Moffat, R. J. (1978). Rough-wall turbulent heat transfer with variable velocity, wall temperature, and blowing. *AIAA J.* **16,** No. 1.
10. Coles, D. E. (1962). The turbulent boundary layer in a compressible fluid. *Rand Corp. [Rep. R]* **R-403-PR.**
11. Crawford, M. E., Kays, W. M., and Moffat, R. J. (1980). Full-coverage film cooling on a flat, isothermal surface: A summary report on data and predictions. *NASA [Contract. Rep.] CR* **NASA-CR-3219.**
12. Furuhama, K., and Moffat, R. J. (1982). Turbulent boundary layer heat transfer on a convex wall with discrete hole injection. *ASME–JSME Jt. Conf. Therm. Eng..* Honolulu, Hawaii, April.
13. Gillis, J. C., Johnston, J. P., Kays, W. M., and Moffat, R. J. (1980). "The Turbulent Boundary Layer on a Convex Curved Surface," Rep. No. HMT-31. Thermosci. Div., Stanford Univ., Stanford, Calif.
14. Healzer, J. M. (1974). "The Turbulent Boundary Layer on a Rough Porous Plate: Experimental Heat Transfer with Uniform Blowing," Ph.D. Thesis. Stanford Univ., Stanford, Calif.
15. Jeans, A. H., and Johnston, J. P. (1982). "The Effects of Streamwise Concave Curvature on Turbulent Boundary Layer Structure," Rep. No. MD-40. Thermosci. Div., Stanford Univ., Stanford, Calif.
16. Julien, H. L., Kays, W. H., and Moffat, R. J. (1971). Experimental hydrodynamics of the accelerated turbulent boundary layer with and without mass additive. *J. Heat Transfer* **93,** No. 4.

17. Kearney, D. W. (1970). "The Turbulent Boundary Layer: Experimental Heat Transfer with Strong Favorable Pressure Gradients and Blowing," Ph.D. Thesis. Stanford Univ., Stanford, Calif.

18. Kim, J. K., Moffat, R. J., and Kays, W. M. (1978). "Heat Transfer to a Full-Coverage Film-Cooled Surface with Compound Angle (30° and 445°) Hole Injection," Rep. No. HMT-28. Thermosci. Div., Stanford Univ., Stanford, Calif.

19. Kline, S. J., and McClintock, F. A. (1953). Describing uncertainty in single-sample experiments. *Mech. Eng.*

20. Launder, B. E., and Jones, W. P. (1968). On the prediction of laminarization. *ARC Heat Mass Transfer Subcomm. Meet., April.*

21. Ligrani, P. M. (1979). "The Thermal and Hydrodynamic Behavior of Thick, Rough-Wall, Turbulent Boundary Layers." Ph.D. thesis, Stanford Univ., 1979.

22. Ligrani, P. M., Moffat, R. J., and Kays, W. M. (1983). Artificially thickened turbulent boundary layers for studying heat transfer and skin friction on rough surfaces. (Trans. ASME). *J. Fluid Eng.* **105**(2), 146–153.

23. Mickley, H. S., and Davis, R. S. (1957). Momentum transfer for flow over a flat plate with blowing. *Natl. Advis. Comm. Aeronaut., Tech. Notes* **NACA-TN-4017.**

24. Mickley, H. S., Ross, R. C., Squyers, A. L., and Stewart, W. E. (1954). Heat, mass and momentum transfer for flow over a flat plate with blowing or suction." *Natl. Advis. Comm. Aeronaut., Tech. Notes* **NACA-TN-3208.**

25. Moffat, R. J. (1967). "The Turbulent Boundary Layer on a Porous Plate: Experimental Heat Transfer with Uniform Blowing and Suction," Ph.D. Thesis. Stanford Univ., Stanford, Calif.

26. Moffat, R. J. (1982). Contributions to the theory of single-sample uncertainty analysis. *J. Fluid Eng.* **104**, 250–260.

27. Moffat, R. J., Healzer, J. M., and Kays, W. M. (1978). Experimental heat transfer behavior of a turbulent boundary layer on a rough surface with blowing. *Trans. ASME* **100**, 134–142.

28. Monin, A. S., and Yaglom, A. M. (1971). "Statistical Fluid Mechanics," Vol. 1. MIT Press, Cambridge, Massachusetts.

29. Moretti, P. M., and Kays, W. M. (1965). Heat transfer to a turbulent boundary layer with varying free-stream velocity and varying surface temperature—An experimental study. *Int. J. Heat Mass Transfer* **8**, 1187–1202.

30. Nikuradse, J. (1950). Strömungsgestze in rauhen Rohren. *VDI-Forschungsh.* No. 361; Engl. transl., *NACA Tech. Memo.* **NACA-TM-1292.**

31. Orlando, A. F., Moffat, R. J., and Kays, W. M. (1974). Heat transfer in turbulent flows under mild and strong adverse pressure gradient conditions for an arbitrary variation of the wall temperature. *Proc. Heat Transfer Fluid Mech. Inst., 24th,* 91–104.

32. Pimenta, M. M. (1975). "The Turbulent Boundary Layer: An Experimental Study of the Transport of Momentum and Heat." Ph.D. thesis, Stanford Univ., Stanford, Calif.

33. Pimenta, M. M., Moffat, R. J., and Kays, W. M. (1979). The structure of a boundary layer on a rough wall with blowing and heat transfer. *J. Heat Transfer* **101**, 193–198.

34. Reynolds, W. C., Kays, W. M., and Kline, S. J. (1958a). Heat transfer in the turbulent incompressible boundary layer. I. Constant wall temperature. *NACA Memo.* **12-1-58W.**

35. Reynolds, W. C., Kays, W. M., and Kline, S. J. (1958b). Heat transfer in the turbulent incompressible boundary layer. II. Step-wall temperature distribution. *NACA Memo.* **12-2-58W.**

36. Reynolds, W. C., Kays, W. M., and Kline, S. J. (1958c). Heat transfer in the turbulent incompressible boundary layer. III. Arbitrary wall temperature and heat flux. *NACA Memo.* **12-3-58W.**

37. Reynolds, W. C., Kays, W. M., and Kline, S. J. (1958d). Heat transfer in the turbulent

incompressible boundary layer. IV. Effect of location of transition and prediction of heat transfer in a known transition region. *NACA Memo.* **12-4-58W.**

38. Rubesin, M. W. (1954). An analytical estimation of the effect of transpiration cooling on the heat-transfer and skin friction characteristics of a compressible, turbulent boundary layer. *Natl. Advis. Comm. Aeronaut., Tech. Notes* **NACA-TN-3341.**

39. Schlichting, H. (1968a). "Boundary Layer Theory," 6th Ed., p. 587. McGraw-Hill, New York.

40. Schlichting, H. (1968b). "Boundary Layer Theory," 6th Ed., p. 583. McGraw-Hill, New York.

41. Schraub, R. A., and Kline, S. J. (1965). "A Study of the Structure of the Turbulent Boundary Layer with and without Longitudinal Pressure Gradients," Rep. No. MD-12. Thermosci. Div., Stanford Univ., Stanford, Calif.

42. Siebers, D. L., Schwind, R. G., and Moffat, R. J. (1983a). Experimental mixed convection heat transfer from a large vertical surface in a horizontal flow. *Proc. Int. Heat Trans. Conf., 7th, Munich, 1982* **3,** 477–482.

43. Siebers, D. L., Schwind, R. G., and Moffat, R. J. (1983b). Experimental mixed convection heat transfer from a large vertical surface in a horizontal flow. *Sandia Lab. [Tech. Rep.] SAND* **SAND 83-8225.**

44. Simon, T., Johnston, J. P., Kays, W. M., and Moffat, R. J. (1980). "Turbulent Boundary Layer Heat Transfer Experiments: Convex Curvature Effects Including Introduction and Recovery," Rep. No. HMT-32. Thermosci. Div., Stanford Univ., Stanford, Calif.

45. Simonich, J. C., and Moffat, R. J. (1982). A new technique for mapping heat transfer coefficient contours. *Rev. Sci. Instrum.* **53,** 678–683.

46. Simonich, J. C., and Moffat, R. J. (1983). Liquid crystal visualization of surface heat transfer on a concavely curved turbulent boundary layer. *ASME–JSME J. Conf. Therm. Eng., Tokyo.*

47. Simpson, R. L. (1967). "The Turbulent Boundary Layer on a Porous Plate: An Experimental Study of the Fluid Dynamics with Injection and Suction," Ph.D. Thesis. Stanford Univ., Stanford, Calif.

48. Simpson, R. L., Whitten, D. G., and Moffat, R. J. (1970). An experimental study of the turbulent Prandtl number of air with injection and suction. *Int. J. Heat Mass Transfer* **13,** 125–143.

49. Spalding, D. B., and Patankar, S. V. (1967). "Heat and Mass Transfer in Boundary Layers." Morgan-Grampian, London.

50. Squire, L. C. (1970). The constant property turbulent boundary layer with injection; a reanalysis of some experimental results. *Int. J. Heat Mass Transfer* **13,** 939.

51. Stevenson, T. N. (1963). "A Law of the Wall for Turbulent Boundary Layers with Suction or Injection," Rep. No. 166. Cranfield College of Aeronautics, Cranfield, Engl.

52. Thielbahr, W. H. (1969). "The Turbulent Boundary Layer: Experimental Heat Transfer with Blowing, Suction, and Favorable Pressure Gradient," Ph.D. Thesis, Stanford Univ., Stanford, Calif.

53. Thomann, H. (1968). Effect of Streamwise Wall Curvature on Heat Transfer in a Turbulent Boundary Layer. *J. Fluid Mech.* **33**(2), 283–292.

54. Thomann, H. (1967). "Heat Transfer in a Turbulent Boundary Layer with a Pressure Gradient Normal to the Flow," Rep. No. 13. Aero. Res. Inst. Sweden.

55. Van Driest, E. R. (1956). On turbulent flow near a wall. *Heat Transfer Fluid Mech. Inst., Prepr. Pap.*

56. Whitten, D. G. (1967). "The Turbulent Boundary Layer on a Porous Plate: Experimental Heat Transfer with Variable Suction, Blowing, and Surface Temperature," Ph.D. Thesis. Stanford Univ., Stanford, Calif.

Author Index

Numbers in parentheses are reference numbers and indicate that an author's work is referred to although his name is not cited in the text. Numbers in italics show the pages on which the complete references are listed.

A

Abdeksalam, M.A., 95, *154*
Abdelsalam, M., 181(17), *238*
Abraham, G., 9, 11, 23, 24, 26, *56*
Abramov, A. I., 110(68), 111(68), 114(68), 128(68), 131(68), *155*
Ackermann, H., 114(78), *155*
Ades, M., 12, 24, 39, *57*
Afgan, N. H., 114(75), 117(75), 125, 128, 131, *155*
Albertson, M. L., 5, 9, 11, 31, *56*
Ametistov, Y. V., 216(27), 233, 234, *239*
Andersen, P. S., 279, 297, 355, 356, *363, 365*
Arshad, J., 150(132), *156*
Astruc, J. M., 236

B

Bailey, C. A., 233
Bajura, R. A., 10(12), 11(12), 14(12), 16(12), 19(2), 27(12), 29(12), 32(12), *57*
Bankoff, S. G., 64(7), 87, *153, 154*
Barlow, R., 329, *363*
Barstow, D., 46, *57*
Basarow, I. P., 68(18), *153*
Beaton, W., 70, *153*
Beaubouef, R. T., 232
Bell, K. J., 148, 149, *156*
Benjamin, J. E., 90, 91, *154*
Bennett, D. L., 144, *156*
Berenson, P. J., 139(109), 141(109), *156*
Berenson, P. P., 215, *238*
Bewilogua, L., 114(78), *155, 233*
Bier, K., 115, 117(80), *155*, 171, 172, 173, 193(5), 215(5), 220, 221(5), 234, *238*

Bird, R. B., 147(126), *156*
Black, T. J., *363*
Blackwell, B. F., 280, 298, 349, 352, 355, 357, *363*
Bland, M. E., 233
Bobrovich, G. I., 127, 128, 129, 141(114), *155, 156*, 235
Bonilla, C. F., 114(69, 70), 125, *155*, 233
Boom, R. W., 233
Borishanskii, V. M., 163, 181(14), 186, 187, 211, 226, 235, *237, 238*
Bowley, W. W., 15, 16, *57*
Bowman, H. F., 234
Bradshaw, P., *363*
Bromley, L. A., 138(106), 141, *156*
Brooks, N. H., 9, 12, 16, 19, 27, 29, *57*
Brown, L. E., 134, *155*
Bruijn, P. J., 83, *154*
Butterworth, D., 142, 143(117), 147(128), 148(128), *156*

C

Calus, W. F., 82(30), 117(30,82), 119, 120, 122, 125, *154, 155*, 237
Cess, R. D., 136, *156*
Chan, G. K. C., 53, *57*
Chen, C. J., 53, *58*
Chen, C. T., 55, *58*
Chen, J. C., 143, 144, *156*, 180, 181, *238*
Choe, H., 312(7), 314, *363*
Chojnowski, B., 232, 233
Chu, J. C., 61(3), *153*
Chu, V. H., 53, *57*
Cichelli, M. T., 114(69), 125, *155*, 233

Subject Index

A

Ablative material, 300
Acceleration
  combined with roughness and
    transpiration, effects on boundary
    layer, 306
  effect, on turbulent boundary layer,
    280–297
Acetone-butanol, A_0, 119
Acetone-cyclohexanol, film boiling, 133,
  138
Acetone-ethanol, A_0, 119
Acetone-ethanol-water, boiling, heat
  transfer, 116–117
Acetone-methanol, slope of saturation
  curve, 67
Acetone-methanol-water, boiling, 116
Acetone-water
  A_0, 119
  heat transfer coefficient, 107–108
    prediction, 127
  slope of saturation curve, 67
Ambient media
  characteristics, 2–3
  density stratification, determination,
    17–19
  flow conditions, 3
  flowing
    differential modeling of jets in, 10–13
    entrainment functions for jets in, 12
  properties, 17–20
  quiescent
    differential modeling of jets in, 9–10
    entrainment functions for discharges
      into, 11
  stratification modeling, 17–20
  stratifications, effects on jets, 42–51

  stratified quiescent
    jet behavior in, 43–46
    nondimensional differential form of
      governing equations, 23
  unstratified flowing
    jets in, comparative calculations for,
      36–42
    nondimensional differential form of
      governing equations, 22
  unstratified quiescent
    modeling of jets in, 25–36
    nondimensional differential form of
      governing equations for, 22
Arctic ocean, temperature, salinity, and
  density, 48–51
Asymptotic suction layer, 253
Atlantic ocean, tropical, temperature,
  salinity, and density, 48–50

B

Benzene, heat flow rates, effect of heater
  shape and material, 191, 192
Benzene-diphenyl, boiling, effect of
  subcooling, 110–111
Benzene-toluene, A_0, 119
Blowing. *See also* Transpiration
  definition, 272
Boiling, 60. *See also* Nucleate pool boiling
  bubbles in, 231–232
  convective, 142–150
    heat transfer coefficient, 144
    heat transfer region, 145
    research recommendations, 150
  film, 133–142
    on flat plates and tubes, 133–138
    at Leidenfrost point, 138–142

CONTENTS OF PREVIOUS VOLUMES

385